ASCE Manuals and Reports on Engineering Practice No. 90

ASCE American Society of Civil Engineers
345 East 47th Street
New York, New York, 10017-2398

Abstract:

Recent advances in computer methods for analysis and design have been of substantial aid to the designer in the creative process of designing the best (optimum) system. Optimization methods are seen to be at the heart of computer methods for designing engineering systems. With these methods, the designer can evaluate more alternatives, thus resulting in a better and more cost-effective design. The main purpose of this *Guide* is to describe the use of modern optimization methods with simple yet meaningful structural design samples. It emphasizes the formulation of design problems as optimization problems and includes discussion of the features and pitfalls of the statements of these problems. Optimum solutions are obtained and, where possible, compared with solutions obtained using traditional design procedures. Sample applications demonstrate and clarify the basic concepts and serve to show how and what can be achieved by optimum design methods.

Library of Congress Cataloging-in-Publication Data

Guide to structural optimization / by the Technical Committee on Optimal Structural Design of the Technical Administrative Committee on Analysis and Computation of the Technical Activities Division of The Structural Engineering Institute of the American Society of Civil Engineers ; edited by Jasbir S. Arora.
p. cm. -- (ASCE manuals and reports on engineering practice ; no. 90)
Includes bibliographical references and index.
ISBN 0-7844-0220-5
1. Structural optimization. I. Arora, Jasbir S. II. Structural Engineering Institute. Technical Committee on Optimal Structural Design. III. Series.
TA658.8.G85 1997 96-30071
624.1'7713--dc21 CIP

Library of Congress Catalog Card No: 96-30071
ISBN 0-7844-0220-5
Manufactured in the United States of America.

MANUALS AND REPORTS ON ENGINEERING PRACTICE

(As developed by the ASCE Technical Procedures Committee, July 1930, and revised March 1935, February 1962, April 1982)

A manual or report in this series consists of an orderly presentation of facts on a particular subject, supplemented by an analysis of limitations and applications of these facts. It contains information useful to the average engineer in his everyday work, rather than the findings that may be useful only occasionally or rarely. It is not in any sense a “standard,” however; nor is it so elementary or so conclusive as to provide a “rule of thumb” for nonengineers.

Furthermore, material in this series, in distinction from a paper (which expresses only one person’s observations or opinions), is the work of a committee or group selected to assemble and express information on a specific topic. As often as practicable the committee is under the direction of one or more of the Technical Divisions and Councils, and the product evolved has been subjected to review by the Executive Committee of the Division or Council. As a step in the process of this review, proposed manuscripts are often brought before the members of the Technical Divisions and Councils for comment, which may serve as the basis for improvement. When published, each work shows the names of the committees by which it was compiled and indicates clearly the several processes through which it has passed in review, in order that its merit may be definitely understood.

In February 1962 (and revised in April, 1982) the Board of Direction voted to establish:

> A series entitled ‘Manuals and Reports on Engineering Practice,’ to include the Manuals published and authorized to date, future Manuals of Professional Practice, and Reports on Engineering Practice. All such Manual or Report material of the Society would have been refereed in a manner approved by the Board Committee on Publications and would be bound, with applicable discussion, in books similar to past Manuals. Numbering would be consecutive and would be a continuation of present Manual numbers. In some cases of reports of joint committees, bypassing of Journal publications may be authorized.

#	MANUALS AND REPORTS OF ENGINEERING PRACTICE
10	Technical Procedures for City Surveys
13	Filtering Materials for Sewage Treatment Plants
14	Accommodation of Utility Plant Within the Rights-of-Way of Urban Streets and Highways
31	Design of Cylindrical Concrete Shell Roofs
33	Cost Control and Accounting for Civil Engineers
34	Definitions of Surveying and Associated Terms
35	A List of Translations of Foreign Literature on Hydraulics
36	Wastewater Treatment Plant Design
37	Design and Construction of Sanitary and Storm Sewers
40	Ground Water Management
41	Plastic Design in Steel-A Guide and Commentary
42	Design of Structures to Resist Nuclear Weapons Effects
45	Consulting Engineering-A Guide for the Engagement of Engineering Services
46	Report on Pipeline Location
47	Selected Abstracts on Structural Applications of Plastics
49	Urban Planning Guide
50	Planning and Design Guidelines for Small Craft Harbors
51	Survey of Current Structural Research
52	Guide for the Design of Steel Transmission Towers
53	Criteria for Maintenance of Multilane Highways
54	Sedimentation Engineering
55	Guide to Employment Conditions for Civil Engineers
57	Management, Operation and Maintenance of Irrigation and Drainage Systems
58	Structural Analysis and Design of Nuclear Plant Facilities
59	Computer Pricing Practices
60	Gravity Sanitary Sewer Design and Construction
62	Existing Sewer Evaluation and Rehabilitation
63	Structural Plastics Design Manual
64	Manual on Engineering Surveying
65	Construction Cost Control
66	Structural Plastics Selection Manual
67	Wind Tunnel Model Studies of Buildings and Structures
68	Aeration-A Wastewater Treatment Process
69	Sulfide in Wastewater Collection and Treatment Systems
70	Evapotranspiration and Irrigation Water Requirements
71	Agricultural Salinity Assessment and Management
72	Design of Steel Transmission Structures
73	Quality in the Constructed Project-a Guide for Owners, Designers, and Constructors
74	Guidelines for Electrical Transmission Line Structural Loading
75	Right-of-Way Surveying
76	Design of Municipal Wastewater Treatment Plants
77	Design and Construction of Urban Stormwater Management Systems
78	Structural Fire Protection
79	Steel Penstocks
80	Ship Channel Design
81	Guidelines for Cloud Seeding to Augment Precipitation
82	Odor Control in Wastewater Treatment Plants
83	Environmental Site Investigation
84	Mechanical Connections in Wood Structures
85	Quality of Ground Water
86	Operation and Maintenance of Ground Water Facilities
87	Urban Runoff Quality Manual
88	Management of Water Treatment Plant Residuals
89	Pipeline Crossings
90	Guide to Structural Optimization

TABLE OF CONTENTS

FOREWORD

The Technical Administrative Committee on Analysis and Computation (A&C–TAC) was established in 1992 by combining two former technical administrative committees: Electronic Computation and Analysis and Design of Structures. A&C–TAC now consists of five technical committees: Computer-Aided Analysis and Design, Emerging Computing Technology, Methods of Analysis, Optimal Structural Design, and Structural Control.

Among other A&C–TAC publications, *Guide to Structural Optimization* is the harvest of several years of diligent work by the Technical Committee on Optimal Structural Design (OSD–TC). This work was carried out with devotion and enthusiasm. Manuscript preparation involved former and current OSD members and officers.

Formerly the committee consisted of Richard J. Balling, Gordon S. Bjorkman, Scott A. Burns (Chair), Franklin Y. Cheng, Dan M. Frangopol, Ramon E. Gilsanz, Donald E. Grierson, Shahram Pezeshk, Hasan Kamil, and Kevin Z. Truman (Secretary).

Currently the committee consists of Jasbir S. Arora, Richard J. Balling, Scott A. Burns, Chien-Chi Chang, Franklin Y. Cheng, Nayan K. Das, Dan M. Frangopol (Chair), Donald E. Grierson, Mircea Iordanescu, C. W. Martin, Shahram Pezeshk (Secretary), Subramaniam D. Rajan, and Pramod B. Thanader.

OSD–TC's purpose is to further the development of the state-of-the art of optimal design of structures. The committee encourages and disseminates new developments in practical applications for structural design, and new computation techniques, methodology, and algorithms. This publication reflects a significant part of OSD–TC's objectives.

A&C–TAC acknowledges the encouragement and support of ASCE STD EXCOM contact members Barry J. Goodno (current) and Vernon B. Watwood, Jr. (former) as well as OSD–TC chairs Scott A. Burns and Dan M. Frangopol and especially the persistent effort and sensitivity of editor Jasbir S. Arora in finalizing this volume.

Franklin Y. Cheng, Chair
Technical Administrative Committee
on Analysis and Computation

PREFACE

A goal of every designer is to design the best (optimum) systems. Traditionally, this has been achieved through intuition, experience, and repeated trials, a process that has worked well as evidenced by the existence of many fine buildings and other structures. However, in the latter half of this century, considerable advances have taken place in computer methods for analysis and design that can be of substantial aid to the designer in the creative process of designing the best system.

Optimization methods are perceived to be at the heart of computer methods for designing engineering systems. With these methods, the designer can evaluate more alternatives, thus resulting in a better and more cost-effective design. A purpose of this Guide is to describe the use of modern optimization methods with simple yet meaningful structural design examples. The background knowledge required to use the Guide is covered in current undergraduate engineering curricula, such as in courses on calculus, matrix algebra, structural analysis, structural design, and probability and statistics. The approach followed in the Guide is to emphasize the formulation of the design problem as an optimization problem. The features and pitfalls of the statement of the optimization problem are discussed. Optimum solutions are obtained and, where possible, compared with the solutions obtained using traditional design procedures. Sample applications demonstrate and clarify the basic concepts, and serve to show how and what can be achieved by the optimum design methods. The numerical methods are not presented in detail. Instead, the basic ideas and concepts used in the methods are described so that the numerical algorithms can be used properly and effectively. References are cited for those readers interested in specific details of the methods.

In Chapter 1, the question of "what is optimization" is posed and addressed. Basic concepts of optimization are described using the design of a plate girder as an example. The process of formulating a design optimization problem is described and illustrated. A graphical optimization process is explained and some basic optimization concepts and terminology are introduced. Several optimum designs for different conditions are obtained and discussed to show the flexibility and power of the optimization techniques and the optimum design process.

Chapter 2 describes a problem of maximizing the bending strength of a cold-formed beam while maintaining a fixed volume of steel and section depth. The optimum design problem is formulated incorporating the AISI (1989) specifications for cold-formed steel members. Several optimum designs for the beam are presented and discussed.

The design of hybrid composite plate girders is formulated as an optimization problem in Chapter 3. The formulation is based on the load factor design method of the American Association of State Highway and Transportation Officials (AASHTO 1992) specifications for highway bridges. An optimum design for the plate girder is obtained and compared with the design obtained using the traditional trial-and-error approach.

Chapter 4 describes the problem of optimizing reinforced concrete beams and columns. The design problem is formulated and is shown to belong to a class of so-called discrete variable optimization problems. For such problems, some of the design variables must take on values that are integers and/or commercially available, such as the number of bars and bar sizes. Optimum designs for many sample problems are obtained and discussed.

The optimum design of prestressed concrete beams is considered in Chapter 5. It is shown that the problem can be decomposed into three simple suboptimization problems, and then each problem can be solved quite efficiently. An example problem is solved to demonstrate the solution procedure.

Chapter 6 describes the use of sensitivity analysis for structural design. This material is a precursor to the full scale optimization of structural systems presented in later chapters. Calculation of sensitivity coefficients is illustrated using simple truss and frame design problems. The concept of using design sensitivity coefficients to guide the design process is described and illustrated. The basic idea presented here is that the design sensitivity coefficients can be used in the conventional design process to reduce the number of iterations required to obtain an acceptable design.

The optimum design of reinforced concrete frames is described in Chapter 7. A two-phase approach is illustrated. In the first phase, concrete dimensions of the members are optimized using a continuous optimization procedure. The steel reinforcement is then optimized in each member in the second phase using the combinatorial procedure presented in Chapter 4.

The optimum design of structural steel frameworks is presented in Chapter 8. The design optimization problem is formulated and the solution procedure is described. The method determines the best member sizes from commercially available sections while satisfying all the design code constraints. Examples are used to illustrate the design optimization process.

The optimum design of tall steel building frameworks is described in Chapter 9. A minimum weight design problem is formulated and a computer-based optimization procedure is described. A procedure to systematically specify members from the database of available sections is presented. Design of a 50-story three-dimensional frame is presented to illustrate the optimization process.

Chapter 10 describes the concept of plastic design of frames. The basic concepts and assumptions underlying the optimum plastic design of structures are presented and discussed. The problem of optimum plastic design is formulated and the solution process is illustrated using a portal frame. The optimum plastic design of tall buildings is briefly discussed.

Chapter 11 addresses the issue of incorporating structural reliability into the optimum design process. The factors used in the Load and Resistance Factor Design (LRFD) approach have been developed based on modern structural reliability theory. As such, structural optimization and structural reliability are coupled to achieve practical optimum solutions. This chapter describes a framework for controlling structural reliability during the optimization process. The process is demonstrated using an example problem.

Chapter 12 concentrates on the problem of optimum design of framed structures to resist earthquake loads. Inclusion of earthquake loads makes the analysis and design problems more difficult to solve because of the transient nature of the loads and the constraints. The chapter describes a formulation for the design problem and numerical procedures to solve it. Numerical examples are used to illustrate the optimum design process.

Chapter 13 addresses the issue of shape optimum design of structures. Both discrete and continuum structures are treated. Changes in the shape of a structure can lead to substantial gains in its performance. The shape optimization problem is formulated using the finite element method of analysis. The optimization procedure is discussed and demonstrated for several shape design problems.

There are six appendices to the Guide. Appendix A describes basic concepts of optimization and presents the mathematical conditions that need to be satisfied by the optimum design. Appendix B describes fundamental concepts of computational methods for optimum design. The concept of iterative algorithms for continuous variable optimization is explained and the basic requirements of a robust algorithm are delineated. Sequential linear programming and quadratic programming methods are explained. Appendix C discusses numerical optimality criterion methods for weight minimization of structural systems. An iterative algorithm is described and illustrated using a simple truss design problem. Appendix D describes a Dual method for discrete variable structural optimization. The weight of the structure is taken as the minimization criterion and an iterative algorithm is described and illustrated using a simple design problem. Appendix E describes deterministic combinatorial search methods for discrete variable optimization problems that cannot be solved by gradient-based optimization methods. Appendix F presents stochastic search methods that are either purely random or based on principles of simulated annealing or genetic evolution.

This Guide has been prepared under the auspices of the Optimal Structural Design (OSD) Committee which is a Technical Committee (TC) of the Analysis and Computation, a Technical Administrative Committee (TAC) of the Structural Division (STD), ASCE. The initial idea of a *Guide to Structural Optimization* was conceived by Dr. Shlomo Ginsburg in 1985 when he was teaching at the University of Kansas. After considerable debate about the contents, in 1992 an outline for the Guide was prepared and the contributors to the Guide were selected and assigned the task of writing various chapters. Each chapter of the first draft of the Guide was reviewed by two members of the OSD Committee. The chapters were then revised by the authors based on the comments and suggestions of the internal reviewers. The second draft of the Guide was completed in March 1995 and submitted to the Executive Committee (EXCOM) of the Structural Division for review. Six independent external reviewers were selected by EXCOM and their input was solicited. This final draft of the Guide includes most of the suggestions of these reviewers.

I would foremost like to thank the contributors to the Guide for their diligence and hard work in preparing the final drafts of their chapters. I would also like to thank the external reviewers who devoted considerable time to reading the Guide and providing us with extremely thoughtful and useful suggestions. The external reviewers were Dr. Mark E. Botkin of GM Research and Development Center, Professor Theodore V. Galambos of the University of Minnesota, Professor Narbey Khachaturian of the University of Illinois, Dr. Narendra S. Khot of Wright-Patterson Air Force Base, Professor P. Benson Shing of the University of Colorado, and Dr. Mehdi S. Zarghamee of Simpson, Gumpertz & Heger, Inc.

8 March 1996 Jasbir S. Arora

Chapter 1

WHAT IS OPTIMIZATION

Basic concepts of optimization are described using the design of a simple plate girder as an example. First, a way to formulate the design optimization problem is presented. Then basic terminologies and concepts related to design optimization are explained by solving the plate girder problem using the graphical solution procedure. Several additional optimum designs for the plate girder are obtained using the iterative numerical optimization methods. The solutions for different design conditions are discussed and some important concepts related to the formulation of an optimum design problem and the solution process are explained. It is noted that once the optimization problem is formulated, numerical methods and software are available to solve the problem and study the effects of parameter variations. In this way, optimization methodology is viewed as a basic framework for computerization of the design process.

INTRODUCTION

Methods of optimization have been studied and developed over the last 35 years. These methods have matured to the point where they are beginning to be utilized in the design of realistic engineering systems. Although the methods are currently being used primarily by large organizations, they will undoubtedly be developed further so that even small design offices will have access to this new technology. The main purpose of this chapter is to address the question,

---

Jasbir S. Arora, M. ASCE, Professor, Civil and Environmental Engineering, The University of Iowa, Iowa City, IA 52242-1527, email: ⟨Jasbir-Arora@uiowa.edu⟩ (contact for correspondence); Scott A. Burns, M. ASCE, Associate Professor, General Engineering, 104 S. Mathews Avenue, #117, University of Illinois, Urbana, IL 61801-2996; and Min-Wei Huang, Post Doctoral Associate, Civil and Environmental Engineering, The University of Iowa, Iowa City, IA 52242-1527.

"What is optimization?" This is accomplished by formulating and solving the simple structural engineering problem of designing a welded plate girder. This particular example is chosen to explain the basic concepts related to design optimization of engineering systems with a simple, yet realistic, design problem. It is shown that once the optimum design problem has been formulated, it is relatively easy to obtain solutions for different conditions and requirements.

In the next section, the process of formulating an optimization problem is illustrated by using the design of a plate girder as an example. Then conventional and optimum design processes are compared and contrasted. Following that, a simplified version of the problem is solved using a graphical optimization procedure that serves to explain various terms and concepts used in structural optimization. Basic ideas of numerical optimization methods to solve design problems using the computer are described. These methods are very general and can be used to solve a wide variety of design optimization problems. To illustrate their power and flexibility, many optimum solutions for the plate girder are obtained and discussed.

OPTIMIZATION PROBLEM FORMULATION

Design of a Plate Girder

Correct formulation of a design optimization problem is an important first step. To illustrate this step, we consider the problem of designing a welded plate girder for a highway bridge. The objective is to design the girder with a minimum cross-sectional area (or, equivalently, minimum mass), and at the same time satisfy AASHTO Specifications (American Association of State Highway and Transportation Officials). The dead load for the girder consists of the weight of the pavement and self-weight of the girder. The live load consists of equivalent uniform load and concentrated loads based on HS-20(MS18) truck loading conditions. The cross-section of the girder is shown in Fig. 1-1.

Formulation of an optimum design problem uses the following steps (Arora 1989).

Step 1. Identification and definition of independent design variables.

As a first step in the problem formulation phase, the variables that describe the design of the system must be identified and defined very precisely. These variables are such that once their values are specified, a design of the system is known. Usually, the design variables are independent of each other. If some of the dependent variables are also designated as design variables, then there must be equality constraints between the variables to have a meaningful design of the system.

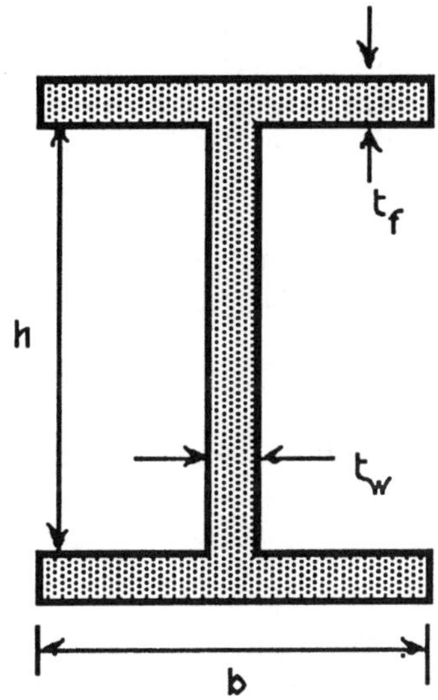

FIG. 1-1. Cross-Section of Plate Girder.

At this stage of the problem formulation phase, it is also necessary to collect all the design parameters and data related to the problem and define them precisely. These include material properties, loading conditions, and the like. Any additional problem variables that depend explicitly or implicitly on the design variables should be identified and defined as well. Such variables are called *dependent variables* (such as the cross-sectional properties of members, the bending stress, the shear stress, the deflection, etc.).

Step 2. Identification and definition of an objective function.

As a second step in the problem formulation phase, an objective function (called the *cost function* for the minimization problem) that measures the relative merit of alternate designs needs to be identified and defined. This is a scalar function that must depend explicitly or implicitly on the design variables. Once a design is specified, it should be possible to evaluate this function. Various objective functions are dollar cost of the system, mass of the system, maximum deflection, moment capacity of a section, natural frequency, and so on.

Step 3. Identification and definition of constraints.

Each design problem has constraints that must be met, such as service, strength, and available resources constraints. Therefore, as the final step in the problem formulation phase, all the constraints on the system need to be identified and defined. Usually constraint expressions involve both the dependent and independent design variables as well as explicit bounds on the variables. Examples of the quantities to be constrained are stress, buckling load, natural frequency, and deflection. An evaluation of the constraint functions requires response analysis of the system for all applied loading cases.

In the following paragraphs, the problem formulation procedure is illustrated for the plate girder design problem.

Design Variables

For the plate girder shown in Fig. 1-1, the independent design variables are identified as

h = web height, m
b = flange width, m
t_f = flange thickness, m
t_w = web thickness, m

Design Parameters/Data

Additional parameters, material data, and loading data for the plate girder are defined as follows.

Span:	$L = 25$, m
Modulus of elasticity:	$E = 210$, GPa
Yield stress:	$\sigma_y = 262$, MPa
Allowable bending stress:	$\sigma_a = 0.55\sigma_y = 144.1$, MPa (Table 10.32.1A, AASHTO 1992)
Allowable shear stress;	$\tau_a = 0.33\sigma_y = 86.46$, MPa (Table 10.32.1A, AASHTO 1992)
Allowable fatigue stress:	$\sigma_t = 255$, MPa (Table 10.3.1A, AASHTO 1992)
Allowable deflection:	$D_a = \dfrac{L}{800}$, m (Article 10.6.2, AASHTO 1992)
Concentrated load for moment:	$P_m = 104$, kN
Concentrated load for shear:	$P_s = 155$, kN
Live load impact factor:	$1 + \dfrac{50}{(L + 125)}$

Note that the impact factor for the live load, according to AASHTO Specifications, depends on the span length L. For $L = 25$, this factor is calculated as 1.33. The loads P_m and P_s already contain the impact factor of 1.33. Note also that the same loads are used in all the numerical solutions presented later.

Dependent Variables

The dependent variables for the problem are identified as follows (these can be evaluated once the design of the girder is specified).

Cross sectional area: $$A = (ht_w + 2bt_f),\ \text{m}^2 \tag{1-1}$$

Moment of inertia: $$I = \frac{1}{12} t_w h^3 + \frac{2}{3} bt_f^3 + \frac{1}{2} bt_f h(h + 2t_f),\ \text{m}^4 \tag{1-2}$$

Uniform load: $$w = (19 + 77A),\ \text{kN/m} \tag{1-3}$$

Bending moment: $$M = \frac{L}{8}(2P_m + wL),\ \text{kN} - \text{m} \tag{1-4}$$

Bending stress: $$\sigma = \frac{M}{1000I}(0.5h + t_f),\ \text{MPa} \tag{1-5}$$

Flange buckling stress limit: $$\sigma_f = 72{,}845 \left(\frac{t_f}{b}\right)^2,\ \text{MPa} \tag{1-6}$$

(Article 10.34.2.1.3, AASHTO 1992)

Web crippling stress limit: $$\sigma_w = 3{,}648{,}276 \left(\frac{t_w}{h}\right)^2,\ \text{MPa} \tag{1-7}$$

(Article 10.34.3.1.1, AASHTO 1992)

Shear force: $$S = 0.5(P_s + wL),\ \text{kN} \tag{1-8}$$

Deflection: $$D = \frac{L^3}{384 \times 10^6 EI}(8P_m + 5wL),\ \text{m} \tag{1-9}$$

Average shear stress: $$\tau = \frac{S}{1000ht_w},\ \text{MPa} \tag{1-10}$$

Objective Function

The objective is to minimize the total mass or, equivalently, the cross-sectional area A of the girder. Therefore a feasible design with smaller cross-sectional area is called a "better design" compared to the one with a larger area. Other cost functions that include dollar costs for material, fabrication, welding, erection, and transportation, can also be formulated if these cost factors are known.

Constraints

The following constraints for the plate girder are defined as follows.

Bending stress: $\sigma \leq \sigma_a$ (1-11)

Flange buckling: $\sigma \leq \sigma_f$ (1-12)

Web crippling: $\sigma \leq \sigma_w$ (1-13)

Shear stress: $\tau \leq \tau_a$ (1-14)

Deflection: $D \leq D_a$ (1-15)

Fatigue stress: $\sigma \leq \frac{1}{2}\sigma_t$ (1-16)

Size constraints: $0.30 \leq h \leq 2.5$; $0.30 \leq b \leq 2.5$

$0.01 \leq t_f \leq 0.10$; $0.01 \leq t_w \leq 0.10$ (1-17)

Note that the lower and upper limits on the design variables in Eqs. (1-17) have been specified arbitrarily in the present example. In practice, appropriate values for the given design problem will have to be specified. It is important to note that the final solution will most likely depend on these limits, since some of the design variables are likely to attain these limits.

Optimization Problem Statement

Mathematically, the plate girder design optimization problem is stated as

- compute the design variables h, b, t_f, and t_w to minimize the cross sectional area, A
- subject to the constraints on bending stress, flange buckling, web crippling, deflection, shear stress, fatigue stress, and size of the girder given in Eqs. (1-11)–(1-17).

It is noted here that the foregoing formulation may not be appropriate for some applications. For example, it may be more appropriate to use composite girder construction for highway bridges. Such an application is discussed in Chapters 3 and 11, where all the constraints of AASHTO Specifications are imposed and the use of the web stiffeners is discussed. The foregoing formulation is used, nevertheless, to illustrate and discuss basic concepts of optimization and corresponding computational algorithms.

It is important to note that constraints of Eqs. (1-11)–(1-16) can be written explicitly in terms of the design variables h, b, t_f, and t_w by substituting into them expressions for all the dependent variables.

However, there are many applications where it is not possible to eliminate the dependent variables, such as stresses and deflections, to obtain explicit expressions for all the functions of the optimization problem in terms of the design variables. In such cases, the dependent variables must be kept in the problem formulation and treated in the solution process. Such methods have been developed for various applications and are available for use in practical applications. Chapters 2, 3, and several other chapters of this Guide also illustrate the use of intermediate variables and associated equality constraints.

The interested reader may consult Kirsch (1981, 1993), Arora (1989, 1990), and many other references cited therein for a more detailed discussion of the problem formulation for design optimization.

OPTIMUM DESIGN PROCESS VERSUS CONVENTIONAL DESIGN PROCESS

It is important to contrast the optimum design process with the conventional design process. Fig. 1-2 summarizes basic steps of the two processes. Although there are several steps that are common to both processes, the *optimum design process differs from the conventional design process* in the following ways.

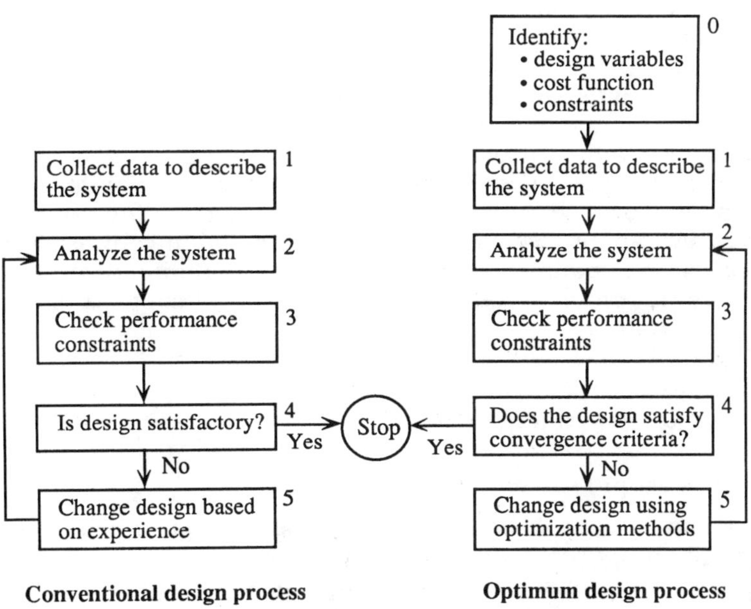

FIG. 1-2. Conventional Versus Optimum Design Process.

1. There is an additional initial Step 0 where the design problem is formulated as an optimization problem, as explained in the previous section.
2. In Step 5, the design is improved using an optimization method (rather than based on experience as in the conventional approach).
3. The stopping criterion in Step 4 is based on whether the current design is the best (the conventional approach accepts any feasible design).

It can be seen that the optimum design process is more formal than the conventional design process. The designer must clearly and precisely identify *design variables*, a design *performance measure* (the cost function) by which the relative merits of various designs can be categorized, and all the *performance constraints* that the final design must satisfy.

The conventional design process for the plate girder design problem is to select some initial values for the design variables (this is the same as the first step of the optimum design process). Then a constraint that is likely to govern the final design is guessed and used to adjust the initial design. For example, assume the bending stress to be the governing constraint, which is calculated and checked against the limiting value. If the limiting value is exceeded, then the design variables are adjusted and the process is repeated until the calculated stress is smaller than the limiting value. Then the other constraints are checked one by one until an acceptable design is obtained. This process usually results in a conservative design of the system. In contrast, the optimum design process takes into account all the constraints simultaneously, and iteratively improves the design, while minimizing the cost function. This can result in better designs. In addition, if problem parameters are changed (e.g., girder span, material properties, limit values for the constraints, etc.), the conventional process must start over again, demanding repetition of all the calculations, which can be quite tedious. In contrast, new solutions can be obtained very rapidly with the optimum design process once the problem has been formulated and coupled to an optimization program.

OPTIMUM SOLUTION

Graphical Approach

Any two-variable design problem can be solved using the graphical optimization approach. In this approach, each constraint function is plotted on a graph sheet. This gives a curve that divides the design space into two parts. One side of the curve represents all the design points that satisfy the constraint (feasible designs) and the other side represents the designs that violate the constraint (infeasible designs). The infeasible side of the constraint is identified by hatching it out. This way, all the constraint functions are plotted and a feasible region for the problem is identified. The *feasible region* represents a set of designs that satisfy all the constraints of the problem. This is called the *feasible set* of designs or simply, the *constraint set*. Any design in the feasible region is an acceptable design; that is, it satisfies all the design requirements.

Next, our job is to locate the best possible design in the feasible region. In order to do that, a few *iso-cost curves* are drawn through the feasible region and a point that gives least value to the cost function is identified as the optimum solution. The coordinates for the optimum point are read directly from the graph. The constraints that govern this design are called *tight, active, critical,* or *binding,* whereas others are called *inactive, noncritical,* or *nonbinding*. Any point outside the feasible region is called *infeasible* because some constraints are *violated* at that point.

In order to convert the foregoing plate girder design problem to a two-variable problem, it is assumed that the flange width is a fraction of the web height; that is, $b = c_1 h$ and the flange thickness is proportional to the web thickness; that is, $t_f = c_2 t_w$. Now the problem is expressed in terms of only two variables, the web height and web thickness. For the example problem, $c_1 = 0.30$ and $c_2 = 1.30$ are selected. A graphical representation of the problem is shown in Fig. 1-3 where all the constraints are plotted and the infeasible region for each one is identified by hatching it out. Thus the feasible region for the design problem is identified. Two iso-cost curves with values of 0.085 and 0.075 for the cross-sectional area are also plotted through the feasible region. By observing the trend of the iso-cost curves, it is determined that the feasible point P gives the least value for the cost function; therefore this point is identified as the optimum solution. Observe that if one makes any feasible moves away from the point P, the cost function increases; thus the point is indeed a minimum for the cost function. The coordinates of the point P are: $h = 1.633\ m$, $t_w = 0.01868\ m$, and the cross-sectional area is calculated as $A = 0.0543\ m^2$. The deflection and fatigue stress constraints are active at the final design and all other constraints are inactive (the solution is actually obtained by solving for the roots of the active constraints).

OPTIMUM DESIGNS USING NUMERICAL TECHNIQUES

Numerical Optimization Techniques

Closed form analytical and/or graphical solutions for practical optimization problems are difficult to obtain if the number of design variables is more than two and/or the constraint expressions are complex. Therefore numerical methods and computers must be used to

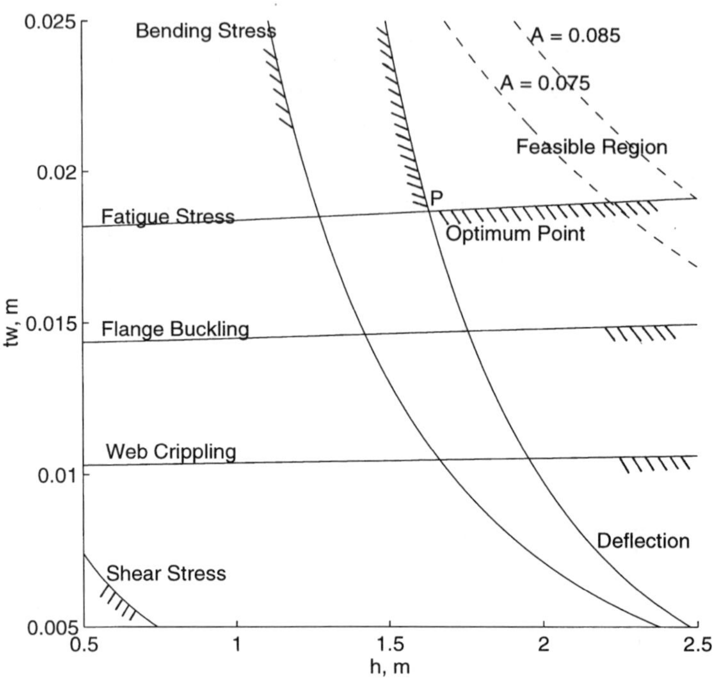

FIG. 1-3. Graphical Optimization of Plate Girder Design Problem.

solve most optimization problems. In these methods, an initial design for the system is selected which is iteratively improved until no further improvements are possible without violating any of the constraints. The stopping criteria for the methods are usually based on certain optimality conditions that are explained in Appendix A (Arora 1989, 1990). The basic concepts related to the iterative numerical optimization methods are explained in more detail in Appendices B–F.

In this section, use of two optimization programs is demonstrated for solution of the plate girder design problem. The first one is a general-purpose design optimization software called IDESIGN (Arora and Tseng 1988; Arora 1989) and the second one is a widely available spreadsheet program. Note, however, that any other similar optimization program can also be used to solve the problem. It is also important to note that use of any optimization program requires time and effort to gain familiarity with it and to properly use it. Therefore the designer should be prepared to spend some time and resources for this task as well.

One of the advantages of using optimization techniques and associated programs is that once the problem has been properly formulated and defined for the program, it is quite easy to solve it for a variety of different conditions and requirements. Many "what if" type studies can be carried out and a tradeoff between various parameters and conditions can be better analyzed.

In most practical applications, the design variables cannot have arbitrary values due to manufacturing and fabrication limitations. For example, the plate thickness and width must be selected from the available ones, the number of bolts used must be an integer, the number of rebars must be an integer and their size must be selected from those available, and so on. Design problems with such variables are called *discrete variable optimization problems* in contrast to the *continuous variable problems* where design variables can have any value within the specified limits. To solve discrete variable problems, the optimization software must have the capability to obtain a final design for which values of the variables have been selected from a specified set. Several optimization methods have been investigated for such applications; recent articles by Arora et al. (1994) and Huang and Arora (1996) and many references cited therein can be consulted for more details. Also, Appendices E and F describe two optimization methods that are suitable for discrete variable problems.

In the following paragraphs, these types of plate girder design optimization problems are solved and the solutions compared.

1. *Continuous* design variable problems where each variable can have any value from the specified range.
2. *Discrete* design variable problems where each variable must have a value from the specified set of values.

Each type of problem is solved with two and four design variable formulations. Several cases of constraints for each formulation are considered and an optimum solution is obtained. Various comparisons are presented and discussed to study the effect of

1. number of design variables on the optimum solution,
2. omitting a constraint from the problem formulation,
3. a discrete versus a continuous solution, and
4. changes in the problem parameters.

Continuous Solutions with IDESIGN

Tables 1-1 and 1-2 summarize various optimum designs with two- and four-design variable formulations, respectively. These solutions were obtained in a matter of minutes on a workstation after the problem had been defined for the program IDESIGN. The sequential quadratic programming algorithm of nonlinear optimization was used to obtain these solutions. For the solutions presented in Tables 1-1 and 1-2, values for the parameters c_1 and c_2, the span length, and the deflection limit are varied ($c_1 = 0.3$ and $c_2 = 1.3$ are used for all cases, unless noted otherwise). Note that the optimum solution depends on the values of these parameters. In addition, the set of active

TABLE 1-1 Continuous Solutions with Two Design Variables

No.	Design conditions	Optimum design	Active constraints
1	$L = 25$ m	$A = 5.430E{-}02$ $h = 1.633E{+}00$ $t_w = 1.868E{-}02$	Deflection Fatigue stress
2	minimum $t_w = 0.02$ m; $L = 25$ m $Da = 0.05$ m	$A = 4.818E{-}02$ $h = 1.353E{+}00$ $t_w = 2.000E{-}02$	Deflection
3	$c_1 = 0.35$; $c_2 = 1.30$; $L = 25$ m	$A = 5.897E{-}02$ $h = 1.541E{+}00$ $t_w = 2.003E{-}02$	Deflection Fatigue stress
4	$c_1 = 0.40$; $c_2 = 1.40$; $L = 25$ m	$A = 6.180E{-}02$ $h = 1.470E{+}00$ $t_w = 1.983E{-}02$	Deflection Fatigue stress
5	$c_1 = 0.25$; $c_2 = 1.25$; $L = 25$ m	$A = 5.024E{-}02$ $h = 1.740E{+}00$ $t_w = 1.778E{-}02$	Deflection Fatigue stress
6	$c_1 = 0.25$; $c_2 = 1.25$; $L = 30$ m	$A = 6.533E{-}02$ $h = 2.008E{+}00$ $t_w = 2.002E{-}02$	Deflection Fatigue stress
7	$c_1 = 0.25$; $c_2 = 1.25$; $L = 35$ m	$A = 8.429E{-}02$ $h = 2.256E{+}00$ $t_w = 2.300E{-}02$	Deflection Fatigue stress
8	$c_1 = 0.25$; $c_2 = 1.25$; $L = 40$ m	$A = 1.061E{-}01$ $h = 2.500E{+}00$ $t_w = 2.611E{-}02$	Deflection

constraints may change with different values for the parameters. From the results in Table 1-1 for the problem with two design variables, it is seen that the deflection and/or the fatigue stress constraint are binding at the optimum design. From the results in Table 1-2 for the problem with four design variables, the deflection constraint is always active; in addition, either the web crippling or the bending stress constraint is active. The upper limit on the height of the web and lower limit on the width of the flange are also active in some cases. It is important to note that, even for this simple design problem, it is almost impossible to guess the set of active constraints at the final solution, and thus the conventional design process will likely be quite tedious and time consuming. The optimization method automatically determines the set of active constraints at the optimum point.

A comparison of some of the optimum solutions with the two- and four-variable formulations is given in Table 1-3 ($c_1 = 0.25$ and $c_2 = 1.25$ are used for two-variable problems). Note that the minimum value of the cost function with the four-design variable formulation

TABLE 1-2 Continuous Solutions with Four Design Variables

No.	Design conditions	Optimum design	Active constraints
1	$L = 25$ m	$A = 3.620E-02$ $h = 2.067E+00$ $b = 3.000E-01$ $t_w = 1.148E-02$ $t_f = 2.080E-02$	Web crippling Deflection
2	min t_w and $t_f = 0.02$ m $L = 25$ m $Da = 0.05$ m	$A = 4.341E-02$ $h = 1.571E+00$ $b = 3.000E-01$ $t_w = 2.000E-02$ $t_f = 2.000E-02$	Bending stress Deflection
3	$L = 30$ m	$A = 4.673E-02$ $h = 2.375E+00$ $b = 3.000E-01$ $t_w = 1.287E-02$ $t_f = 2.695E-02$	Web crippling Deflection
4	$L = 35$ m	$A = 5.905E-02$ $h = 2.500E+00$ $b = 3.000E-01$ $t_w = 1.290E-02$ $t_f = 4.467E-02$	Web crippling Deflection
5	$L = 40$ m	$A = 7.643E-02$ $h = 2.500E+00$ $b = 3.000E-01$ $t_w = 1.218E-02$ $t_f = 7.665E-02$	Web crippling Deflection

is always smaller than that with the two-variable formulation (by 30% in some cases). Although these comparisons are not very precise, this observation is true in general; that is, *the more design variables (degrees of freedom) a problem has, the smaller is the optimum cost function value.*

In order to study the effect on the optimum solution due to pitfalls in the formulation, additional cases of the plate girder are optimized with the four-design variable formulation. In the first one, the deflection constraint is omitted from the formulation, and in the second one the web crippling constraint is omitted. The idea is to study what happens if one forgets to account for a constraint in the formulation. The optimum solutions for these two cases along with the original solution are given in Table 1-4. It can be seen that the three solutions are quite different. For Solution 2 where the deflection constraint is not included in the formulation, the deflection is exceeded by 64% at the final design. For Solution 3, where the web crippling constraint is omitted from the formulation, the calculated stress exceeds its limit by 61%. These are very severe violations of the constraints and it

TABLE 1-3 Comparison of Continuous Solutions with Two and Four Design Variables

Design conditions	4 Variables	2 Variables
$L = 25$ m	$A = 3.620E-02$	$A = 5.024E-02$
	$h = 2.067E+00$	$h = 1.740E+00$
	$b = 3.000E-01$	$t_w = 1.778E-02$
	$t_w = 1.148E-02$	
	$t_f = 2.080E-02$	
$L = 30$ m	$A = 4.673E-02$	$A = 6.533E-02$
	$h = 2.375E+00$	$h = 2.008E+00$
	$b = 3.000E-01$	$t_w = 2.002E-02$
	$t_w = 1.287E-02$	
	$t_f = 2.695E-02$	
$L = 35$ m	$A = 5.905E-02$	$A = 8.429E-02$
	$h = 2.500E+00$	$h = 2.256E+00$
	$b = 3.000E-01$	$t_w = 2.300E-02$
	$t_w = 1.290E-02$	
	$t_f = 4.467E-02$	
$L = 40$ m	$A = 7.643E-02$	$A = 1.061E-01$
	$h = 2.500E+00$	$h = 2.500E+00$
	$b = 3.000E-01$	$t_w = 2.611E-02$
	$t_w = 1.218E-02$	
	$t_f = 7.665E-02$	

TABLE 1-4 Comparison of Solutions with Different Constraints

No.	Design conditions	Optimum design	Active constraints
1	$L = 25$ m	$A = 3.620E-02$	Web crippling
		$h = 2.067E+00$	Deflection
		$b = 3.000E-01$	
		$t_w = 1.148E-02$	
		$t_f = 2.080E-02$	
2	$L = 25$ m	$A = 3.099E-02$	Web crippling
	No deflection constraint	$h = 1.591E+00$	Fatigue stress
		$b = 3.456E-01$	
		$t_w = 1.000E-02$	
		$t_f = 2.181E-02$	
3	$L = 25$ m	$A = 3.243E-02$	Deflection
	No web crippling constraint	$h = 2.211E+00$	Fatigue stress
		$b = 3.000E-01$	
		$t_w = 1.000E-02$	
		$t_f = 1.721E-02$	

shows that the correct formulation of the problem is quite critical for good optimum solutions. Note, however, that the optimum cost function values (the cross-sectional area) for the two new cases are smaller than the value for the case where all the constraints are included in the formulation. This shows that there may be some penalty for including more constraint in the formulation.

Discrete Solutions with IDESIGN

Tables 1-5 and 1-6 summarize optimum designs with two and four discrete variable formulations, respectively. Active or nearly active constraints are also depicted. For these discrete solutions (obtained using a genetic algorithm in IDESIGN), it is assumed that the members are available in increments of

(i) 2 cm in width or height, and
(ii) 2 mm in thickness.

TABLE 1-5 Discrete Solutions with Two Design Variables

No.	Design conditions	Optimum design	Active constraints
1	$L = 25$ m	$A = 5.696E{-}02$ $h = 1.600E{+}00$ $t_w = 2.000E{-}02$	Deflection
2	minimum $t_w = 0.02$ m; $L = 25$ m $Da = 0.05$ m	$A = 4.842E{-}02$ $h = 1.360E{+}00$ $t_w = 2.000E{-}02$	Deflection
3	$c_1 = 0.35$; $c_2 = 1.30$; $L = 25$ m	$A = 5.883E{-}02$ $h = 1.540E{+}00$ $t_w = 2.000E{-}02$	Deflection Fatigue stress
4	$c_1 = 0.40$; $c_2 = 1.40$; $L = 25$ m	$A = 6.275E{-}02$ $h = 1.480E{+}00$ $t_w = 2.000E{-}02$	Deflection
5	$c_1 = 0.25$; $c_2 = 1.25$; $L = 25$ m	$A = 5.090E{-}02$ $h = 1.740E{+}00$ $t_w = 1.800E{-}02$	Deflection
6	$c_1 = 0.25$; $c_2 = 1.25$; $L = 30$ m	$A = 6.565E{-}01$ $h = 2.020E{+}00$ $t_w = 2.000E{-}02$	Deflection Fatigue stress
7	$c_1 = 0.25$; $c_2 = 1.25$; $L = 35$ m	$A = 8.736E{-}02$ $h = 2.240E{+}00$ $t_w = 2.400E{-}02$	Deflection
8	$c_1 = 0.25$; $c_2 = 1.25$; $L = 40$ m	$A = 1.056E{-}01$ $h = 2.500E{+}00$ $t_w = 2.600E{-}02$	Deflection

TABLE 1-6 Discrete Solutions with Four Design Variables

No.	Design conditions	Optimum design	Active constraints
1	$L = 25$ m	$A = 3.648E{-}02$ $h = 2.140E{+}00$ $b = 3.000E{-}01$ $t_w = 1.200E{-}02$ $t_f = 1.800E{-}02$	Web crippling Deflection
2	min t_w and $t_f = 0.02$ m $L = 25$ m $Da = 0.05$ m	$A = 4.360E{-}02$ $h = 1.580E{+}00$ $b = 3.000E{-}01$ $t_w = 2.000E{-}02$ $t_f = 2.000E{-}02$	Deflection
3	$L = 30$ m	$A = 4.752E{-}02$ $h = 2.280E{+}00$ $b = 4.200E{-}01$ $t_w = 1.200E{-}02$ $t_f = 2.400E{-}02$	Web crippling Deflection
4	$L = 35$ m	$A = 6.096E{-}02$ $h = 2.380E{+}00$ $b = 3.000E{-}01$ $t_w = 1.200E{-}02$ $t_f = 5.400E{-}02$	Web crippling Deflection
5	$L = 40$ m	$A = 7.776E{-}02$ $h = 2.480E{+}00$ $b = 3.000E{-}01$ $t_w = 1.200E{-}02$ $t_f = 8.000E{-}02$	Web crippling Deflection

Although material selection may be treated as a design variable, this aspect is not studied here. Comparing solutions with two and four design variables, the same trend is observed again as for the continuous variable solutions; that is, *the minimum cost function value with four design variables is always smaller than that with two design variables.*

A comparison of continuous and discrete solutions with four design variables is given in Table 1-7. It can be observed that the optimum cost for the discrete solutions is always higher than the corresponding cost for the continuous solutions. This is an expected result because the requirement of having to assign discrete values to the design variables imposes additional constraints on the problem. *Whenever more constraints are imposed on the system, the optimum cost function value can be expected to increase.*

Solution by a Spreadsheet Program

Many spreadsheet programs have the capability of solving constrained nonlinear optimization problems, such as the plate girder

TABLE 1-7 Comparison of Continuous and Discrete Solutions with Four Design Variables

No.	Design conditions	Continuous solution	Discrete solution
1	$L = 25$ m	$A = 3.620E{-02}$	$A = 3.648E{-02}$
		$h = 2.067E{+00}$	$h = 2.140E{+00}$
		$b = 3.000E{-01}$	$b = 3.000E{-01}$
		$t_w = 1.148E{-02}$	$t_w = 1.200E{-02}$
		$t_f = 2.080E{-02}$	$t_f = 1.800E{-02}$
2	min t_w and $t_f = 0.02$ m	$A = 4.341E{-02}$	$A = 4.360E{-02}$
	$L = 25$ m	$h = 1.571E{+00}$	$h = 1.580E{+00}$
	$Da = 0.05$ m	$b = 3.000E{-01}$	$b = 3.000E{-01}$
		$t_w = 2.000E{-02}$	$t_w = 2.000E{-02}$
		$t_f = 2.000E{-02}$	$t_f = 2.000E{-02}$
3	$L = 30$ m	$A = 4.673E{-02}$	$A = 4.752E{-02}$
		$h = 2.375E{+00}$	$h = 2.280E{+00}$
		$b = 3.000E{-01}$	$b = 4.200E{-01}$
		$t_w = 1.287E{-02}$	$t_w = 1.200E{-02}$
		$t_f = 2.695E{-02}$	$t_f = 2.400E{-02}$
4	$L = 35$ m	$A = 5.905E{-02}$	$A = 6.096E{-02}$
		$h = 2.500E{+00}$	$h = 2.380E{+00}$
		$b = 3.000E{-01}$	$b = 3.000E{-01}$
		$t_w = 1.290E{-02}$	$t_w = 1.200E{-02}$
		$t_f = 4.467E{-02}$	$t_f = 5.400E{-02}$
5	$L = 40$ m	$A = 7.643E{-02}$	$A = 7.776E{-02}$
		$h = 2.500E{+00}$	$h = 2.480E{+00}$
		$b = 3.000E{-01}$	$b = 3.000E{-01}$
		$t_w = 1.218E{-02}$	$t_w = 1.200E{-02}$
		$t_f = 7.665E{-02}$	$t_f = 8.000E{-02}$

design problem formulated previously. This section describes the use of Microsoft Excel™ and its Solver tool to solve the problem.

The solver is an "Add-in" to Microsoft Excel, which is an optional module for adding optimization capabilities to the spreadsheet program. Depending on how Excel was initially installed, the Solver may appear under the Tools menu in Excel. If it does not, it can be installed by selecting the "Add-Ins..." menu item under the Tools menu. This action will produce a dialogue box containing all available "Add-Ins," including the Solver, among others. "Selecting it" will add it to the Tools menu.

Fig. 1-4 presents the layout of the problem in a spreadsheet format (LB and UB refer to the lower and upper bounds on the variables, respectively). Fig. 1-5 illustrates the same spreadsheet, displaying all the formulas used in the solution process. The reader should note that this format is a matter of user preference; the spreadsheet can be organized in any desired way. Main requirements are that one cell

	A	B	C	D	E	F
1	***Plate Girder Design***					
2	**Independent Variables name**	**LB**	**symbol**	**value**	**UB**	**units**
3	web height	0.30	h	2.067	2.5	m
4	flange width	0.30	b	0.3000	2.5	m
5	flange thickness	0.01	tf	0.0208	0.10	m
6	web thickness	0.01	tw	0.0115	0.10	m
7						
8	**Parameters name**	**symbol**	**value**	**units**		
9	Span length	L	25	m		
10	Modulus of elasticity	E	210	GPa		
11	Yield stress	sigma_y	262	MPa		
12	Allowable fatigue stress	sigma_t	255	MPa		
13	Concentrated load for moment	Pm	104	kN		
14	Concentrated load for shear	Ps	155	kN		
15	Live load impact factor	LLIF	1.33	none		
16						
17	**Dependent Variables name**	**symbol**	**eqn**	**units**		
18	Cross sectional area	A	0.03620	m^2		
19	Moment of inertia	I	0.0220452	m^4		
20	Uniform load	w	21.788	kN/m		
21	Bending moment	M	2352.16	kN-m		
22	Bending stress	sigma	112.49	MPa		
23	Shear force	S	349.85	kN		
24	Deflection	D	0.0312	m		
25	Average shear stress	tau	14.745	MPa		
26						
27	**Objective Function name**	**symbol**	**eqn**	**units**		
28	Volume of steel	Vol	0.905	m^3		
29						
30	**Constraints name**	**value/eqn**	**</>/=**	**value/eqn**	**name**	
31	Bending stress	112.49	<	144.1	Allowable bending stress	
32	Bending stress	112.49	<	349.9	Flange buckling limit	
33	Bending stress	112.49	<	112.5	Web crippling limit	
34	Shear stress	14.745	<	86.46	Allowable shear stress	
35	Deflection	0.0312	<	0.03125	Allowable deflection	
36	Bending stress	112.49	<	127.5	Allowable fatigue stress	

FIG. 1-4. Layout of Spreadsheet for Plate Girder Design Problem.

contain the objective function formula (cell C28 in Fig. 1-5), and the independent variables appear somewhere else (cells D3–D6 in Fig. 1-5).

The spreadsheet has been organized so that the independent variables are grouped in one area. This area contains the name, symbol, current value, units, and upper and lower bounds on each variable. The symbol is associated with the variable cell through the "insert/name" menu selection. This permits the variables to be referenced by their symbols in the subsequent formulas which greatly improves readability and makes debugging the formulation easier.

The next section of the spreadsheet contains the design parameters, including their names, symbols, values, and units. These values typically are held fixed during the solution process, but can be changed from one solution to the next to investigate "what if" scenarios.

	A	B	C	D	E	F
1	***Plate Girder Design***					
2	**Independent Variables name**	**LB**	**symbol**	**value**	**UB**	**units**
3	web height	0.3	h	2.067	2.5	m
4	flange width	0.3	b	0.3	2.5	m
5	flange thickness	0.01	tf	0.02079	0.1	m
6	web thickness	0.01	tw	0.01148	0.1	m
7						
8	**Parameters name**	**symbol**	**value**	**units**		
9	Span length	L	25	m		
10	Modulus of elasticity	E	210	GPa		
11	Yield stress	sigma_y	262	MPa		
12	Allowable fatigue stress	sigma_t	255	MPa		
13	Concentrated load for moment	Pm	104	kN		
14	Concentrated load for shear	Ps	155	kN		
15	Live load impact factor	LLIF	1.33	none		
16						
17	**Dependent Variables name**	**symbol**	**eqn**	**units**		
18	Cross sectional area	A	=h*tw+2*b*tf	m^2		
19	Moment of inertia	I	=(1/12)*tw*h^3+(2/3)*b*tf^3+(1/2)*b*tf*h*(h+2*tf)	m^4		
20	Uniform load	w	=19+77*A	kN/m		
21	Bending moment	M	=L*(2*Pm+w*L)/8	kN-m		
22	Bending stress	sigma	=M*(h/2+tf)/(1000*I)	MPa		
23	Shear force	S	=(Ps+w*L)/2	kN		
24	Deflection	D	=L^3*(8*Pm+5*w*L)/(384*E*I*10^6)	m		
25	Average shear stress	tau	=S/(1000*h*tw)	MPa		
26						
27	**Objective Function name**	**symbol**	**eqn**	**units**		
28	Volume of steel	Vol	=A*L	m^3		
29						
30	**Constraints name**	**value/eqn**	**</>/=**	**value/eqn**	**name**	
31	Bending stress	=sigma	<	=0.55*sigma_y	Allowable bending stress	
32	Bending stress	=sigma	<	=72845*(tf/b)^2	Flange buckling limit	
33	Bending stress	=sigma	<	=3648276*(tw/h)^2	Web crippling limit	
34	Shear stress	=tau	<	=0.33*sigma_y	Allowable shear stress	
35	Deflection	=D	<	=L/800	Allowable deflection	
36	Bending stress	=sigma	<	=sigma_t/2	Allowable fatigue stress	

FIG. 1-5. Layout of Spreadsheet with Data and Formulas Displayed for Plate Girder Design Problem.

The third section contains quantities that depend on those in the previous two sections. Although it is not necessary to include these (because they can be incorporated explicitly into the constraint and objective function formulas), they are very useful when included. First they simplify the formulation of the constraint expressions, reducing algebraic manipulation errors. Second, they provide a check of these intermediate quantities for debugging purposes and for information feedback.

The objective function and constraints appear at the bottom of the spreadsheet. Constraints are implemented in Excel by relating two cells through an inequality (≤ or ≥) or an equality (=) relationship. This is defined in the Solver dialogue box which is described next. Although many of the quantities appearing in the constraint section also appear elsewhere in the spreadsheet, they are simply references to other cells in the variables and parameters sections of the spreadsheet (see formulas in Fig. 1-5). This way, the only cells that need to be modified during a "what if" analysis are those in the independent variable or parameters sections. The constraints are automatically updated to reflect the changes.

Once the spreadsheet has been created, the next step is to define the optimization problem in the Solver Add-In. Fig. 1-6 shows a screen dump of the Solver dialogue box. The objective function cell is entered as the "Target Cell" that is to be minimized. The independent design variables are identified next under the "By Changing Cells:" heading. A range of cells has been entered here, but individual cells, separated by commas, could be entered instead. Finally the constraints are entered under the "Subject to the Constraints" heading. The constraints include not only those identified in the constraints section of the spreadsheet but also the bounds on the independent design variables. Clicking on the "Solve" button initiates the optimization process. The solution obtained is shown in Fig. 1-4 which is the same as the first solution in Table 1-2. By changing various parameters and conditions, other solutions can be obtained. All the solutions shown in Tables 1-1 and 1-2 for continuous variable formulations were also obtained with Excel.

GENERAL DESIGN OPTIMIZATION MODEL AND CONCEPTS

It is customary to use some standard notation to describe the problem formulation in order to describe more general concepts and methods related to design optimization problems. The reader is introduced to such a notation in this section. All design optimization problems when transcribed into a standard form look alike. Thus the concepts and methods developed for this standard form can be applied to any design optimization problem.

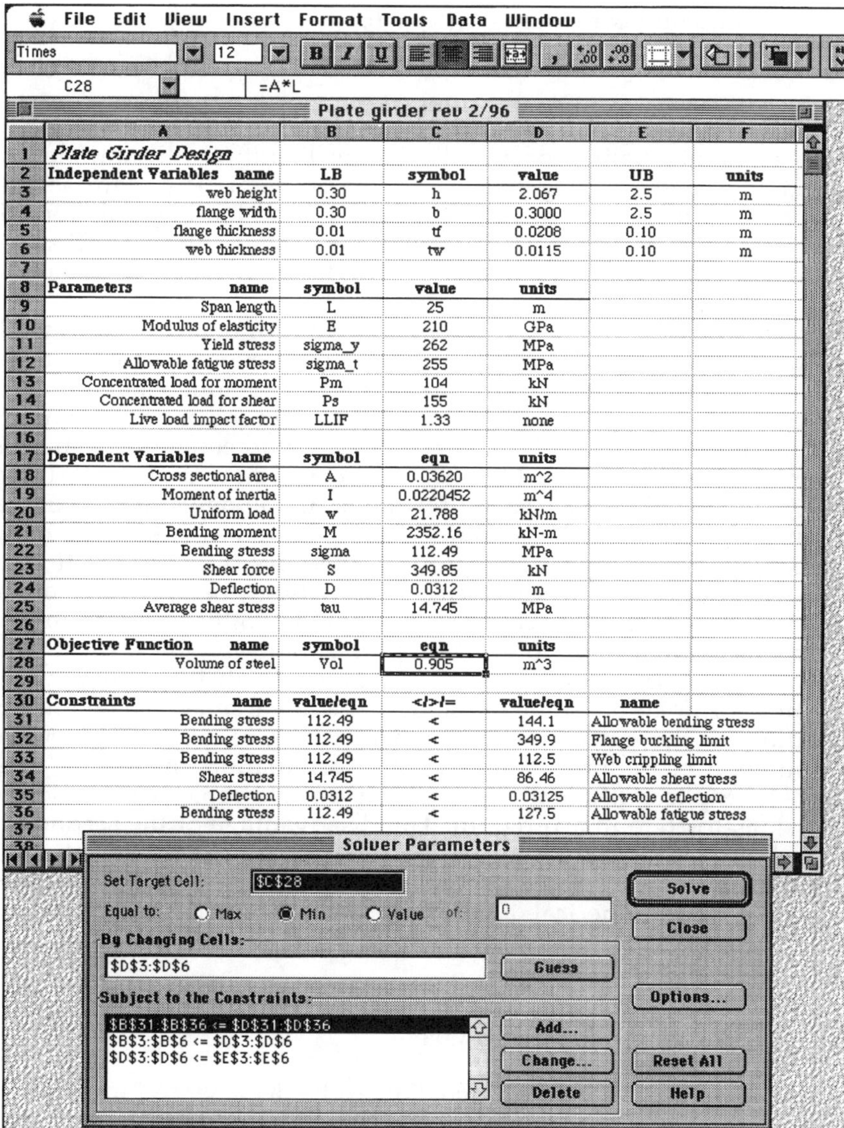

	A	B	C	D	E	F
1	*Plate Girder Design*					
2	**Independent Variables name**	**LB**	**symbol**	**value**	**UB**	**units**
3	web height	0.30	h	2.067	2.5	m
4	flange width	0.30	b	0.3000	2.5	m
5	flange thickness	0.01	tf	0.0208	0.10	m
6	web thickness	0.01	tw	0.0115	0.10	m
7						
8	**Parameters name**	**symbol**	**value**	**units**		
9	Span length	L	25	m		
10	Modulus of elasticity	E	210	GPa		
11	Yield stress	sigma_y	262	MPa		
12	Allowable fatigue stress	sigma_t	255	MPa		
13	Concentrated load for moment	Pm	104	kN		
14	Concentrated load for shear	Ps	155	kN		
15	Live load impact factor	LLIF	1.33	none		
16						
17	**Dependent Variables name**	**symbol**	**eqn**	**units**		
18	Cross sectional area	A	0.03620	m^2		
19	Moment of inertia	I	0.0220452	m^4		
20	Uniform load	w	21.788	kN/m		
21	Bending moment	M	2352.16	kN-m		
22	Bending stress	sigma	112.49	MPa		
23	Shear force	S	349.85	kN		
24	Deflection	D	0.0312	m		
25	Average shear stress	tau	14.745	MPa		
26						
27	**Objective Function name**	**symbol**	**eqn**	**units**		
28	Volume of steel	Vol	0.905	m^3		
29						
30	**Constraints name**	**value/eqn**	**</>/=**	**value/eqn**	**name**	
31	Bending stress	112.49	<	144.1	Allowable bending stress	
32	Bending stress	112.49	<	349.9	Flange buckling limit	
33	Bending stress	112.49	<	112.5	Web crippling limit	
34	Shear stress	14.745	<	86.46	Allowable shear stress	
35	Deflection	0.0312	<	0.03125	Allowable deflection	
36	Bending stress	112.49	<	127.5	Allowable fatigue stress	

FIG. 1-6. *Screen Dump of Solver Dialogue Box for Plate Girder Design Problem.*

Let **X** represent an n-dimensional design variable vector. Then any design optimization problem can be stated as follows: find **X** to

$$\text{minimize a cost function } f(\mathbf{X}) \tag{1-18}$$

subject to

$$\text{Equality constraints: } g_i(\mathbf{X}) = 0, \qquad i = 1 \text{ to } p \tag{1-19}$$

$$\text{Inequality constraints: } g_i(\mathbf{X}) \leq 0, \qquad i = p + 1 \text{ to } m \tag{1-20}$$

where p is the number of equality constraints and $(m - p)$ is the number of inequality constraints. In the literature, this problem is called the *nonlinear programming (NLP)* problem. There are several textbooks and journal articles that discuss solution methods for this problem, such as Arora (1989, 1990), Gill et al. (1981), Haftka and Gurdal (1992), Papalambros and Wilde (1988), Reklaitis et al. (1983), and Vanderplaats (1984).

For the plate girder design problem, two formulations are possible to transcribe the problem into the foregoing standard form. In the first formulation, h, b, t_f, and t_w are treated as the independent design variables; thus there are only four unknowns in the optimization procedure. In this case $\mathbf{X} = (h, b, t_f, t_w)$; there are no equality constraints, so $p = 0$; and there are 14 inequality constraints in Eqs. (1-11) to (1-17), so $m = 14$. In the second formulation, the dependent variables defined in Eqs. (1-1) to (1-10) may also be treated as unknowns in the optimization procedure. Therefore, in this case, $n = 14$, $p = 10$, and $m = 24$. Implications of these two formulations, besides sizes of the two problems, are in the calculation of the derivatives of the problem functions with respect to all the optimization variables. Most optimization algorithms need these derivatives in their calculations. In the first formulation, some of the dependent variables can be implicit functions of the independent design variables, requiring implicit differentiation procedures. In the second formulation, there is no need for this implicit differentiation, giving much simpler derivative expressions to code in a program. The disadvantage of the second approach is that the dimension of the problem for the optimization algorithm becomes much larger. More details about optimization concepts and numerical algorithms are presented in Appendices A–F.

SUMMARY AND DISCUSSION

Use of optimization techniques forces the designer to formalize the process of designing engineering systems. The design problem must be defined clearly and precisely by identifying design variables, dependent variables, a cost function (that assigns relative values to different designs), and constraints that need to be met for safe performance of the system. Since computational algorithms and computers must be used to solve such design optimization problems, computerization of the entire design process is required. In this way, modern computers can be utilized to the fullest extent to design complex engineering systems. Once a design problem has been properly formulated and coupled to an optimization software, variations in the operating environment and other conditions for the problem can be investigated in a very short time. "What if" type questions can be investigated and insights about the behavior of the system can be gained. Several "what if" studies are conducted for the plate girder design problem, and the effect of the following conditions on the optimum solution is discussed.

- number of design degrees of freedom (more design variables versus fewer design variables);
- discrete solutions as opposed to continuous solutions;
- pitfalls in the formulation (omission of a constraint); and
- inclusion of more design constraints in the formulation.

In summary, optimization techniques offer a designer the flexibility of studying different designs in a relatively short time, thus producing better and cost effective designs more efficiently. This chapter has demonstrated the use of optimization techniques on a simple plate girder design problem. The remaining chapters of the Guide present many other interesting applications of this new design methodology for structural engineering problems.

REFERENCES

AASHTO (1992). Standard specifications for highway bridges. 15th ed. Washington, DC.

Arora, J. S. (1989). Introduction to optimum design. McGraw-Hill, New York.

Arora, J. S. (1990). "Computational design optimization: A review and future directions." *Structural Safety* 7, 131–148.

Arora, J. S. and Tseng, C. H. (1988). "Interactive design optimization." *Engng. Optimization* 13, 173–188.

Arora, J. S., Huang, M. W., and Hsieh, C. C. (1994). "Methods for optimization of nonlinear problems with discrete variables: A review." *Structural Optimization* 8, 69–85.

Gill, P. E., Murray, W., and Wright, M. H. (1981). *Practical optimization*. Academic Press, New York.

Haftka, R. T. and Gurdal, Z. (1992). *Elements of structural optimization*. Kluwer Academic, Norwell, MA.

Huang, M. W. and Arora, J. S. (1996). "Optimal design with discrete variables: Some numerical experiments." *Int. J. Numerical Methods in Engng*. (in press).

Kirsch, U. (1981). *Optimum structural design*. McGraw-Hill, New York.

Kirsch, U. (1993). *Structural optimization*. Springer-Verlag, New York.

Microsoft EXCEL. Version 5.0, Microsoft Corporation, Redmond, WA.

Papalambros, P. Y. and Wilde, D. J. (1988). *Principles of optimal design*. Cambridge University Press, New York.

Reklaitis, G. V., Ravindran, A., and Ragsdell, K. M. (1983). *Engineering optimization: Methods and applications*. John Wiley and Sons, New York.

Vanderplaats, G. N. (1984). *Numerical optimization techniques for engineering design: With applications*. McGraw-Hill, New York.

Chapter 2

HOW TO OPTIMIZE LIGHT-GAUGE, COLD-FORMED STEEL BEAMS

A light-gauge, cold-formed steel beam is optimized to maximize bending strength while maintaining a fixed volume of steel and section depth. A mathematical program is formulated, based on the AISI specification for cold-formed steel members, and is solved using the optimization features of the Solver Add-in in Microsoft Excel™.

INTRODUCTION

Cold-formed steel structural members are used in a wide variety of ways, such as in building wall systems and automobile frames. Relatively simple forming operations (brake pressing, stamping, or roll forming) can produce a wide variety of structural shapes and sizes. Cold-formed sections can be very economical, particularly if production costs can be spread over a large number of identical units manufactured.

We are concerned here with the optimal design of the cross-sectional dimensions of cold-formed steel beams. Previously, Seaburg and Salmon (1971) investigated the minimum weight design of cold-formed flexural members. Here we approach the somewhat different problem of maximizing the bending strength of the beam while maintaining a fixed volume of steel and section depth.

PROBLEM STATEMENT

The *Illustrative Examples* section of the *AISI Cold-Formed Steel Design Manual* (AISI 1989) presents the analysis of a hat section (see Fig.

Scott A. Burns, Associate Professor, Department of General Engineering, University of Illinois at Urbana-Champaign, 104 S. Mathews #117, Urbana, IL 61801, email: ⟨s-burns@uiuc.edu⟩.

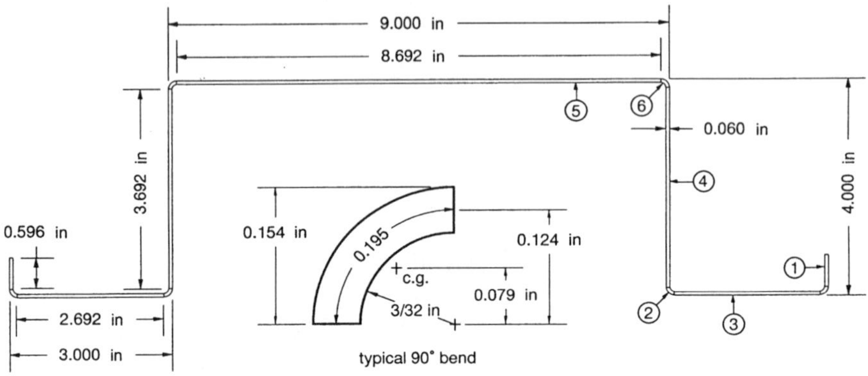

FIG. 2-1. Example #5 of Illustrative Examples Section of AISI Cold-Formed Steel Design Manual. (Bending Moment Capacity = 52.0 Kip·in; Cross-Sectional Area = 1.43 in.^{2}; Overall Depth = 4 in.)

2-1). Our goal is to optimize this hat section to maximize the allowable bending moment that it can safely support while maintaining the existing cross-sectional area (1.43 in.^{2} or 921 mm^2) and overall section depth (4 in. or 102 mm).

PROCEDURE

The optimization problem is posed as a mathematical program comprising an objective function and constraints. The constraints are based on the AISI (1986) design specification with the 1989 addendum. This mathematical program treats key dimensions of the hat section (e.g., flange width, web height, sheet thickness, etc.) explicitly as design variables that appear within constraint expressions reflecting the rules of the AISI specification. An objective function that maximizes the allowable bending moment completes the formulation.

Since optimization is typically an iterative procedure that requires the selection of a starting point, the cross-sectional dimensions of the AISI example are used to initiate the solution process. The mathematical program is solved using an existing optimization computer program (in this case, the Solver add-in of Microsoft Excel). The solution is compared to the original section to assess the increase in moment capacity achieved by the optimization process.

FORMULATION

One of the most difficult aspects of designing cold-formed steel beams is how to properly treat the effects of local buckling. Light-

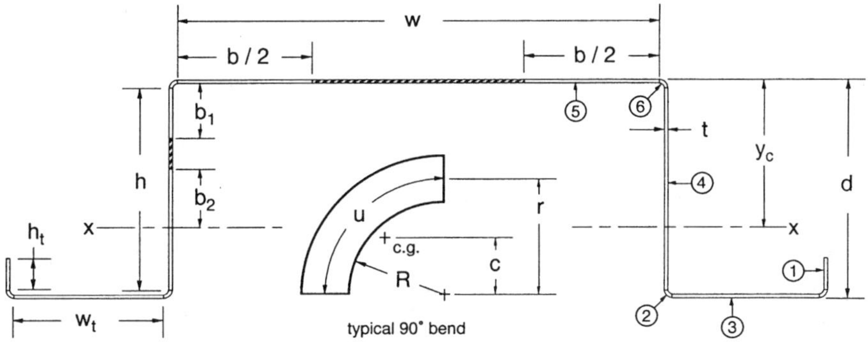

FIG. 2-2. Effective Section (Shaded Areas Removed to Account for Local Buckling).

gauge steel members are very likely to have compression elements with large width-to-thickness ratios that are susceptible to local buckling. In many cases, the local buckling of a compression element does not cause global failure. If the element is stiffened, that is, if both edges of the element parallel to the direction of stress are reinforced by some type of stiffener such as a web or a lip, then the section can carry an additional load beyond that causing the first buckling. The local buckling causes a redistribution of stress toward the stiffeners, and overall failure does not occur. This phenomenon is known as "post-buckling strength."

The AISI specification treats local buckling by eliminating a portion from the center of each stiffened compression element in the modeled cross-section. Each resulting compression element has an "effective width" that is used instead of the actual width for calculating the section properties. Fig. 2-2 shows the modeled cross-section of the hat section, where both the compression flange and the compression portion of the webs have been modified to account for local buckling. The magnitude of the effective width of each element depends on several factors, including the actual stress level in the element. This can make the design process tricky, since calculating the actual stress in the element requires knowing the section properties, but the section properties are dependent upon the effective widths of the compression elements, which in turn depend on the actual stress in the elements. Thus a simple analysis of a given cross-section can sometimes require an iterative process.

Design Variables

Fig. 2-2 shows the five independent design variables:

w, the flat width of the compression flange
h, the flat height of the web

t, the nominal thickness
w_t, the flat width of each tension flange
h_t, the flat height of the lip

The inside bend radius R is not an independent variable since it usually depends on the sheet thickness. It is specified to be twice the sheet thickness through an equality constraint in the formulation. Likewise, the effective width of the compression flange b, the overall depth d, and the distance from the neutral axis to the extreme compression fiber y_c, are also dependent variables that are specified as functions of the five independent variables in the formulation. Other dependent variables are introduced later to simplify the constraint expressions in the formulation.

Five quantities are specified as "design parameters." These quantities are assigned fixed values during the optimization process, yet are represented explicitly in the constraint expressions. They are:

m, the inside bend radius multiplier
A_{max}, the maximum cross-sectional area
d_{max}, the maximum overall depth of the section
E, the modulus of elasticity
F_y, the yield strength

By representing these parameters explicitly in the formulation, we are able to extend the applicability of the formulation to a wider range of specific cases more easily. For example, by solving the optimization problem with a series of different maximum overall depth values, we may observe how the maximum depth requirement has an impact on the optimal design.

Section Properties

The AISI Design Manual uses a tabular procedure to calculate the section properties, based on a line idealization of the cross-section. The table contains six columns, as shown in Table 2-1. This table presents the section properties for the AISI example problem shown in Fig. 2-1. Note that the length of element 5, the compression flange, has been given a reduced effective length instead of its actual length in this table to account for local buckling. Section 4.3 discusses how the effective length is calculated. Also note that the web is assumed to be fully effective (no portion is removed for local buckling effects). This assumption must, of course, be checked at the end of the design process to assure that it was justified. The distance from the neutral axis to the extreme compression fiber is calculated as the ratio of two column totals: $y_c = \Sigma(Ly)/\Sigma(L) = 43.54/17.70 = 2.46$ in. (62.5 mm). The moment of inertia is also found from the column totals using the parallel axis theorem: $I_x = t[\Sigma(Ly^2) + \Sigma(I_1) - \Sigma(L)y_c^2] = 0.06[141.4 + 8.43 - 17.7(2.46)^2] = 2.56$ in.$^4 \cdot$(107 cm^4).

TABLE 2-1 Section Properties Table for Example in Fig. 2-1

Element	Effective Length, L (in.*)	Distance from top fiber, y (in.)	Ly (in.^{2})	Ly^2 (in.^{3})	I_1' about own axis (in.^{3})
1	1.192	3.548	4.229	15.005	0.035
2	0.780	3.925	3.062	12.016	negl.
3	5.384	3.970	21.375	84.857	negl.
4	7.384	2.000	14.768	29.536	8.388
5	2.573	0.030	0.077	0.002	negl.
6	0.390	0.075	0.029	0.002	negl.
sum	17.70		43.540	141.418	8.423

$y_c = 43.540/17.70 = 2.46$ in.
$I_x = 0.06[141.418 + 8.423 - 17.70(2.46)^2] = 2.56$ in.^{4}
*Note: 1 in. = 25.4 mm.

To pose a mathematical program that reflects the effect that the design variables have on the behavior of the beam, the section properties must be expressed in terms of these design variables. Table 2-2 presents the modified section properties table. To simplify the expressions, five new dependent variables have been defined (d_1 through d_5). The following set of equations defines the location of the neutral axis (y_c) and the moment of inertia (I_x):

$$R = mt$$
$$r = R + t/2$$

TABLE 2-2 Section Properties Table for Mathematical Program

Element	Effective Length, L	Distance from top fiber, y		Ly	Ly^2	I_1' about own axis
1	$2h_t$	$t/2 + r + h - h_t/2$	$(=d_1)$	$2h_t d_1$	$2h_t d_1^2$	$2h_t^3/12$
2	$4u$	$t/2 + r + h + c$	$(=d_2)$	$4ud_2$	$4ud_2^2$	negl.
3	$2w_t$	$t/2 + h + 2r$	$(=d_3)$	$2w_t d_3$	$2w_t d_3^2$	negl.
4	$2h$	$t/2 + r + h/2$	$(=d_4)$	$2hd_4$	$2hd_4^2$	$2h^3/12$
5	b	$t/2$		$bt/2$	$bt^2/4$	negl.
6	$2u$	$r + t/2 - c$	$(=d_5)$	$2ud_5$	$2ud_5^2$	negl.
sum	$\Sigma(L)$			$\Sigma(Ly)$	$\Sigma(Ly^2)$	

$y_c = \Sigma(Ly)/\Sigma(L)$.
$I_x = t[\Sigma(Ly^2) + 2h_t^3/12 + 2h^3/12 - \Sigma(L)y_c^2]$.

$$
\begin{aligned}
u &= 1.57r \\
c &= 0.637r \\
d_1 &= t/2 + r + h - h_t/2 \\
d_2 &= t/2 + r + h + c \\
d_3 &= t/2 + h + 2r \\
d_4 &= t/2 + r + h/2 \\
d_5 &= r + t/2 - c \\
\Sigma(L) &= 2h_t + 6u + 2w_t + 2h + b \\
\Sigma(Ly) &= 2h_t d_1 + 4ud_2 + 2w_t d_3 + 2hd_4 + bt/2 + 2ud_5 \\
\Sigma(Ly^2) &= 2h_t d_1^2 + 4ud_2^2 + 2w_t d_3^2 + 2hd_4^2 + bt^2/4 + 2ud_5^2 \\
y_c &= \Sigma(Ly)/\Sigma(L) \\
I_x &= t[\Sigma(Ly^2) + 2h_t^3/12 + 2h^3/12 - \Sigma(L)y_c^2]
\end{aligned}
$$

Note that by introducing the dependent variables (R, r, u, c, d_1, d_2, d_3, d_4, d_5, $\Sigma(L)$, $\Sigma(Ly)$, $\Sigma(Ly^2)$, b, y_c, and I_x), we increase the dimensionality of the problem, but minimize the algebraic manipulations that we must perform. Imagine how complex the moment of inertia equation would be if dependent variables were not used! We reduce the chance of making algebraic errors and make the problem easier to formulate at the expense of shifting more of the computational effort to the optimization computer program.

Effective Width

The effective width of the compression flange b is the only quantity in the preceding set of equations that has not yet been defined in terms of the independent variables w, h, t, w_t, and h_t. If the w/t ratio of the compression flange is small enough, then the effective width equals the actual width because local buckling will not occur. In this case, the compression flange is termed "fully effective." The maximum value of w/t for which the flange is fully effective is expressed through a quantity called λ in the AISI specification:

$$\lambda = \frac{1.052}{\sqrt{k}} \left(\frac{w}{t}\right) \sqrt{\frac{f_c}{E}}$$

$$
\begin{aligned}
&\text{If } \lambda \leq 0.673, \quad \text{then } \rho = 1 \\
&\text{If } \lambda > 0.673, \quad \text{then } \rho = (1 - 0.22/\lambda)/\lambda \\
&b = \rho w
\end{aligned}
$$

The value of k in this case is 4. The quantity f_c is defined as the *actual* stress in the compression flange when the section first yields. Since the first yielding can occur in the tension flange, f_c might be less than F_y. Normally, f_c is initially assumed to equal F_y, and once the analysis of the section is complete, this assumption is checked. When this assumption is used and the tension flange yields first, then f_c must be adjusted, which leads to an iterative, trial-and-error procedure. We can avoid this iteration in our mathematical program if we define another dependent variable, M_n, the nominal bending moment that causes the first yield of the section. Then f_c can be defined as $f_c = M_n y_c / I_x$ using an additional equality constraint in the mathematical program. This allows us to use f_c directly in the expression for λ without the need for iteration. The constraints establishing the dependent variable M_n are developed in the following section.

Inequality Constraints

The beam is designed so that the nominal bending moment causes the first yielding at one of the extreme fibers of the section. The allowable bending moment results from dividing the nominal bending moment by a factor of safety (=1.67 for the AISI specifications). We may determine the nominal bending moment in terms of two inequality constraints: one for the compression side and the other for the tension side:

$$M_n \leq F_y I_x / y_c \quad \text{and} \quad M_n \leq F_y I_x / y_t$$

Here, we define $y_t = d - y_c$, where $d = h + 2r + t$. Since our objective is to maximize the allowable bending moment, $M_n/1.67$, one of the two inequality constraints will be forced to become active (become a strict equality) during the optimization process. We need not be concerned with which flange yields first; this will be established automatically by the optimization process.

The original problem statement was to maximize allowable bending moment while maintaining the same cross-sectional area and section depth of the AISI illustrative example. This leads to two additional inequality constraints:

$$t(2h_t + 6u + 2w_t + 2h + w) \leq A_{\max}$$

$$d \leq d_{\max}$$

Complete Formulation

The design parameters for this example, which are fixed during the optimization process, are:

m	$= 2$	the inside bend radius multiplier
$A_{\max}$	$= 1.43$ in.^{2} (921 mm^2)	the maximum cross-sectional area

d_{max}	= 4.00 in. (102 mm)	the maximum overall depth of the section
E	= 29,000 ksi (200,000 MPa)	the modulus of elasticity
F_y	= 50 ksi (345 MPa)	the yield strength

The design variables are:

w	flat width of the compression flange (independent)
h	flat height of the web (independent)
t	nominal thickness (independent)
w_t	flat width of each tension flange (independent)
h_t	flat height of the lip (independent)
R	inside bend radius
b	effective width of the compression flange
d	overall section depth
y_c	distance from the neutral axis to the extreme fiber in compression flange
r	distance from the center of radius of the bend to the centerline of the bend
u	length of the bend centerline arc
c	distance from the center of radius to the center of gravity of the bend
d_1	distance from top fiber to c.g. of lip
d_2	distance from top fiber to c.g. of lower bends
d_3	distance from top fiber to c.g. of tension flanges
d_4	distance from top fiber to c.g. of webs
d_5	distance from top fiber to c.g. of upper bends
$\Sigma(L)$	sum of column 2 of Table 2-2
$\Sigma(Ly)$	sum of column 4 of Table 2-2
$\Sigma(Ly^2)$	sum of column 5 of Table 2-2
I_x	moment of inertia of the section
λ	effective width cutoff parameter
ρ	effective width multiplier
f_c	actual stress in compression flange when first yielding occurs on either flange
M_n	nominal bending moment causing first yielding
y_t	distance from the neutral axis to the extreme fiber in tension flange

The mathematical program contains an objective function and a set of constraints:

Maximize

$$M_n/1.67$$

Subject to

$$
\begin{aligned}
R &= mt \\
r &= R + t/2 \\
u &= 1.57r \\
c &= 0.637r \\
d_1 &= t/2 + r + h - h_t/2 \\
d_2 &= t/2 + r + h + c \\
d_3 &= t/2 + h + 2r \\
d_4 &= t/2 + r + h/2 \\
d_5 &= r + t/2 - c \\
\Sigma(L) &= 2h_t + 6u + 2w_t + 2h + b \\
\Sigma(Ly) &= 2h_t d_1 + 4ud_2 + 2w_t d_3 + 2hd_4 + bt/2 + 2ud_5 \\
\Sigma(Ly^2) &= 2h_t d_1^2 + 4ud_2^2 + 2w_t d_3^2 + 2hd_4^2 + bt^2/4 + 2ud_5^2 \\
y_c &= \Sigma(Ly)/\Sigma(L) \\
I_x &= t[\Sigma(Ly^2) + 2h_t^3/12 + 2h^3/12 - \Sigma(L)y_c^2] \\
\lambda &= \frac{1.052}{\sqrt{k}}\left(\frac{w}{t}\right)\sqrt{\frac{f_c}{E}}
\end{aligned}
$$

if $\lambda \leq 0.673$, then $\rho = 1$, otherwise $\rho = (1 - 0.22/\lambda)/\lambda$

$$
\begin{aligned}
b &= \rho w \\
f_c &= M_n y_c / I_x \\
M_n &\leq F_y I_x / y_c \\
M_n &\leq F_y I_x / y_t \\
y_t &= d - y_c \\
d &= h + 2r + t
\end{aligned}
$$

$$t(2h_t + 6u + 2w_t + 2h + w) \leq A_{\max}$$

$$d \leq d_{\max}$$

SOLUTION

This mathematical program was solved using the "Solver" Add-In in Microsoft Excel. Convergence to a solution was achieved in about 10 seconds of computation time on a Macintosh Quadra 700 computer. The starting point and optimal section are presented in the following.

Variable	Starting Value	Optimal Section	Final Solution
w	8.69 in. (221 mm)	3.04 in. (77.3 mm)	2.62 in. (66.5 mm)
h	3.69 in. (93.8 mm)	3.41 in. (86.6 mm)	3.37 in. (85.7 mm)
t	0.06 in. (1.52 mm)	0.0987 in. (2.51 mm)	0.1046 in. (2.66 mm)
w_t	2.69 in. (68.3 mm)	1.15 in. (29.2 mm)	0.918 in. (23.3 mm)
h_t	0.596 in. (15.1 mm)	0.00 in. (0.00 mm)	0.00 in. (0.00 mm)
R	0.0938 in. (2.38 mm)	0.197 in. (5.00 mm)	0.209 in. (5.31 mm)

Variable	Starting Value	Optimal Section	Final Solution
b	2.57 in. (65.3 mm)	3.04 in. (77.2 mm)	2.62 in. (66.5 mm)
d	4.00 in. (102 mm)	4.00 in. (102 mm)	4.00 in. (102 mm)
y_c	2.46 in. (62.5 mm)	2.00 in. (50.8 mm)	2.00 in. (50.8 mm)
r	0.124 in. (3.15 mm)	0.247 in. (6.27 mm)	0.262 in. (6.65 mm)
u	0.195 in. (4.95 mm)	0.387 in. (9.83 mm)	0.411 in. (10.4 mm)
c	0.079 in. (2.01 mm)	0.157 in. (3.99 mm)	0.167 in. (4.24 mm)
d_1	3.55 in. (90.2 mm)	3.70 in. (94.0 mm)	3.69 in. (93.7 mm)
d_2	3.93 in. (99.8 mm)	3.86 in. (98.0 mm)	3.85 in. (97.8 mm)
d_3	3.97 in. (101 mm)	3.95 in. (100 mm)	3.95 in. (100 mm)
d_4	2.00 in. (50.8 mm)	2.00 in. (50.8 mm)	2.00 in. (50.8 mm)
d_5	0.075 in. (1.91 mm)	0.139 in. (3.53 mm)	0.147 in. (3.73 mm)
$\Sigma(L)$	17.7 in. (450 mm)	14.5 in. (368 mm)	13.7 in. (348 mm)
$\Sigma(Ly)$	43.5 in.^{2} (281 cm^2)	29.0 in.^{2} (187 cm^2)	27.3 in.^{2} (176 cm^2)
$\Sigma(Ly^2)$	141 in.^{3} (2311 cm^3)	86.3 in.^{3} (1414 cm^3)	80.0 in.^{3} (1311 cm^3)
I_x	2.56 in.^{4} (107 cm^4)	3.45 in.^{4} (144 cm^4)	3.32 in.^{4} (138 cm^4)
λ	3.14	0.673	0.547
ρ	0.296	1.000	1.00
f_c	50.0 ksi (345 MPa)	50.0 ksi (345 MPa)	50.0 ksi (345 MPa)
M_n	52.0 Kip·in (5.88 kN m)	86.3 Kip·in (9.75 kN m)	83.0 Kip·in (9.38 kN m)
y_t	1.54 in. (39.1 mm)	2.00 in. (50.8 mm)	2.00 in. (50.8 mm)

The allowable bending moment was increased 66% from the initial starting design without an increase in volume of steel or depth of section. Note that the optimized section has simultaneous yielding in both flanges at M_n. Also the compression flange is as slender as it can be without having to sacrifice material to a reduced effective width. The lip has disappeared in the optimal design.

The optimal design specifies a sheet thickness that does not match a standard gauge of available sheets. To remedy this, the variable t was set equal to the nearest standard value (12 gauge or 0.1046 in or 2.66 mm), and the optimization process was repeated. The final solution is shown in the preceding and is presented in Fig. 2-3. Note that the variables change considerably to simultaneously accommodate the new sheet thickness and the maximum permitted cross-sectional area. The compression flange moves away from the slenderness cutoff value ($\lambda = 0.673$) in order to take full advantage of the maximum permissible section depth. The flanges still yield simultaneously at failure. The nominal bending moment decreases 4% as a result of the additional restriction placed on the problem (the forced sheet thickness).

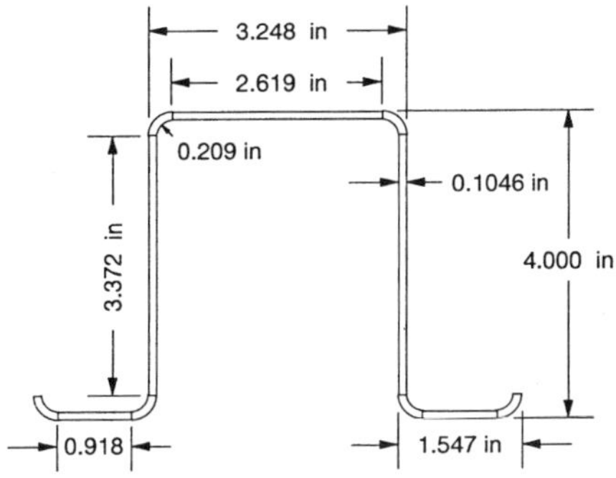

FIG. 2-3. Final Optimized Section. (Bending Moment Capacity = 83.0 Kip·in; Cross-Sectional Area = 1.43 in.^{2}; Overall Depth = 4 in.)

There are a number of final checks that must be made. It was assumed that the web would be fully effective. Checking section B2.3 of the AISI specification confirms that the optimal design maintains a fully effective web. Other checks that must be performed include maximum width-to-thickness ratios of each flange element (AISI section B1.1), maximum width-to-thickness ratios of each web (AISI section B1.2), maximum allowable shearing force (AISI section C3.2), web crippling (AISI section C3.4), and combined bending and web crippling (AISI section C3.5). The optimal solution passes all of these checks.

PARAMETER VARIATIONS

Now that the formulation has been proposed in a general form, using explicit parameters in the constraint expressions, it is a simple matter to perform a parameter study. For example, we may wish to assess how the optimal section responds to a change in maximum allowable depth while holding the maximum allowable area constant. Fig. 2-4 presents the optimal solutions for four different allowable depths. Fig. 2-5 presents a different study; the depth was held constant and the allowable cross-sectional area was varied. Finally, Fig. 2-6 presents a variation in yield strength of the sheet metal. The higher strength steels are usually much more difficult to bend, and a larger bend radius multiplier must be specified.

SUMMARY AND DISCUSSION

A cold-formed steel beam has been optimized to maximize its bending strength while maintaining a fixed volume of steel and sec-

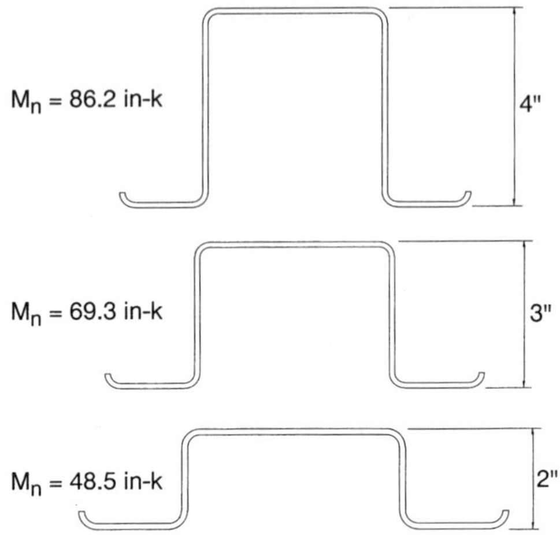

FIG. 2-4. Optimal Sections for Various Allowable Depths, Holding Cross-Sectional Area Constant.

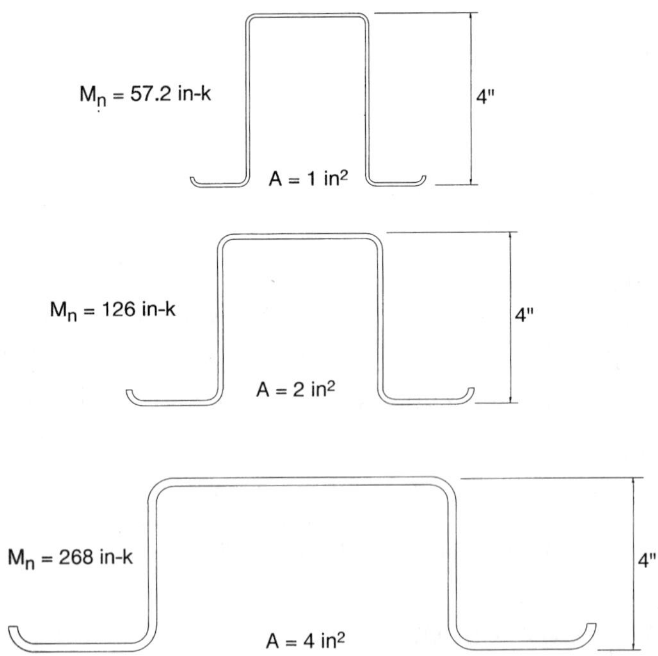

FIG. 2-5. Optimal Sections for Various Allowable Cross-Sectional Areas, Holding Section Depth Constant.

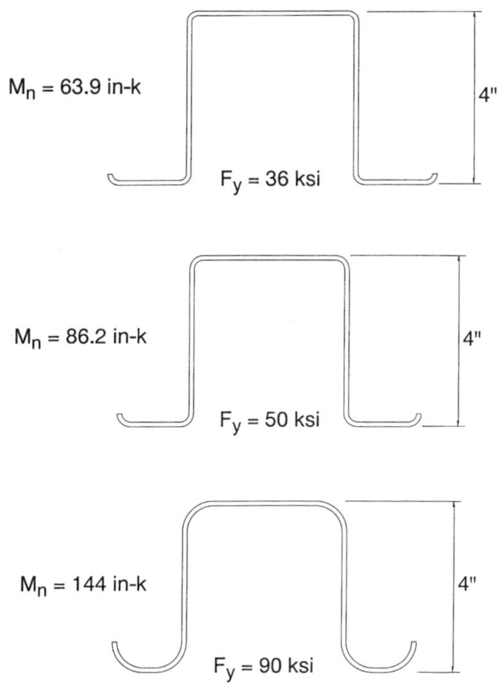

FIG. 2-6. Optimal Sections for Various Yield Strengths, Holding Cross-Sectional Area and Section Depth Constant.

tion depth. A mathematical program was formulated for a general hat section, which can be reused for a variety of different applications (i.e., different steel strength, allowable section depth, allowable cross-sectional area, etc.). With very little modification, the formulation could be changed to minimize the volume of steel for a given applied bending moment to support, or to maximize the shear strength of the beam instead of the bending moment capacity. Other design considerations could be added to the formulation to customize it for specific applications, such as constraints that determine and limit the maximum displacement of the beam.

REFERENCES

American Iron and Steel Institute. (1986). "Specification for the design of cold-formed steel structural members." August 19 Edition with December 11, 1989 Addendum, Washington, DC.

American Iron and Steel Institute. (1989). "AISI cold-formed steel design manual." Washington, DC.

Microsoft Corporation, Microsoft EXCEL, version 4.0, Redmond, WA.

Seaburg, P. A. and Salmon, C. G. (1971). "Minimum weight design of light gage steel members." ASCE J. Struct. Div., 97 (ST1), Jan., 203–222.

Chapter 3

HOW TO OPTIMIZE COMPOSITE MEMBERS

Composite construction is commonly used in civil engineering structures for its particular economic advantages. In steel-concrete construction, the use of composite sections in beams, girders, columns, and deck slabs results in significant savings. Additional savings can be realized in design and material costs for some members by automating design approaches based on optimization techniques.

A composite member consisting of a concrete slab on a built-up girder is frequently used in bridge construction. Traditionally, a trial-and-error method is used for its design. The optimization approach offers additional advantages for such members, especially when using higher strength steel for some elements of the girder.

A mathematical formulation is presented for optimization of the composite hybrid plate girders. It is based on the load factor design method of the American Association of State Highway and Transportation Officials (AASHTO) specification for highway bridges and uses *generalized geometric programming (GGP)* technique. An example is presented to illustrate the traditional and optimum approaches to the composite hybrid plate girder design.

INTRODUCTION

Composite construction involves concrete and structural steel components interconnected and acting as an integral unit. It takes effective advantage of the concrete strength in compression and steel strength in tension to result in an efficient construction.

Balaur S. Dhillon, Fellow, ASCE, Associate Professor, Department of Civil Engineering, Widener University, Chester, PA 19013, email: ⟨Balaur.S.Dhillon@widener.edu⟩ (contact for correspondence); and Chen-Hsing Kuo, Member, ASCE, Structural Engineer, Urban Engineers, Inc., Philadelphia, PA 19106.

The composite construction has been used in buildings and bridges for over fifty years. During this period, design standards and specifications have evolved for the composite member design. The common composite members in structures include composite beams and girders, composite columns, composite deck slabs, and composite truss members.

The composite construction results in considerable savings that in steel-concrete construction may range from 15 to 40%. Thus it has become a common practice to utilize the composite action in all situations where steel and concrete are in contact.

The most common construction in buildings is a composite beam consisting of a concrete slab on a steel beam connected through shear connectors, which are welded to the top of the beam. A composite floor offers an advantage in that it puts the concrete slab in compression and the entire steel beam or a large part of it in tension, and results in lighter steel beams. Additional savings come from smaller beam depths resulting in reduced building heights and lighter beam weights, effecting savings in columns and footings. Besides, composite floors provide greater stiffness with smaller deflections and greater overload capacity.

The composite beam results in an efficient construction for simply supported spans over 25 feet. The American Institute of Steel Construction (AISC/LRFD) has recognized this and provided design aids in the form of composite-beam tables. Similar tables are also provided for composite columns. The composite-beam tables are for wide-flange sections supporting concrete slabs connected through shear connectors. However, for simply supported beams, an unsymmetric section with a larger tension flange will result in greater savings in weight. The same effect can be achieved by coverplated beams, but this is offset by additional welding costs.

In bridges, the composite construction is economical for simple spans over 40 feet and continuous spans over 60 feet. In the composite-bridge girder, a concrete slab is used in combination with a built-up girder with a larger area in the tension flange. A hybrid section with higher strength steel for the tension flange results in a more reasonable size for the tension flange. The hybrid-plate girders are economical when strength governs the design and more so in composite construction as concrete provides a large compression capacity. The availability of higher strength steels makes the use of the hybrid girder more common in highway bridges.

In common practice, composite plate girders are designed by a trial-and-error approach due to the complexity of the design rules. The design of a composite hybrid-plate girder is a tedious and time-consuming job for the designer, and it is highly desirable to have design aids for its design. The use of the built-up section for the plate girder makes it more adaptable to optimization techniques (Abuyounes and Adeli 1987; Burns and Ramamurthy 1983; Dhillon and Kuo 1991), and offers large savings in composite construction. For this reason, the optimization of a composite hybrid-plate girder is selected to illustrate the procedure to optimize composite members.

This chapter presents a formulation for an automated approach to the design of composite-girder sections using optimization techniques. An effective method has been developed for the optimum design of composite hybrid-plate girders using GGP. It is based on the load factor design method of the AASHTO specification for bridges. A mathematical formulation is developed for the application of the GGP technique to find the optimum design. The traditional approach to the design of composite-plate girders is discussed. An example is undertaken to demonstrate the validity of the mathematical model and to illustrate the difference between the traditional and optimum design solutions.

MATHEMATICAL FORMULATION OF COMPOSITE HYBRID-PLATE GIRDER PROBLEM

The design of a simply supported, unstiffened or stiffened, composite, welded, hybrid-plate girder of uniform cross-section is formulated on the basis of the load factor design method of the 1992 AASHTO specification. The objective function and design constraints are expressed in a suitable form for the application of GGP technique for finding the optimum design.

In this formulation, concrete deck slab, diaphragm, bearing stiffeners, welded connections, and shear connectors are not taken into account. The top flange, web, and stiffener steel are assumed to be of the same specified minimum yield strength (ASTM A36), and the bottom flange is assumed to be of higher strength steel (ASTM A517). For the stiffened-plate girder, intermediate transverse stiffeners are assumed to be equally spaced along the span of the girder and placed on one side only.

The optimum design is generally obtained by considering either the total cost or total weight of the girder as the objective function. Because of the variations in the cost of the materials and the difficulty in obtaining accurate cost data, it is preferable to consider the weight of the member as the objective function. However, a modification may be made to account for different costs of steels used in the hybrid girder.

The objective function in the optimization problem is associated with a set of constraints. These constraints fall into two categories: behavior constraints that limit the maximum stress, displacement, or buckling strength; and design or side constraints imposed on design variables to put practical limits on some dimensions of elements. These categories give rise to inequality constraints. Further, to simplify the formulation of the optimization problems and improve the efficiency of the process, it becomes necessary to introduce dependent design variables. These variables give rise to additional constraints, called equality constraints.

The mathematical optimization problem of a composite hybrid-plate girder has a nonlinear objective function along with nonlinear equality and inequality constraints. Generalized geometric programming offers a powerful method that works effectively for solving such nonlinear optimization problems.

Generalized Geometric Problem

The generalized geometric programming problem with mixed equality and inequality constraints is expressed in the following general form (Burns 1987; Kuo 1989).

Minimize

$$G_o(x) = \sum_{j=1}^{T_o} q_{oj} C_{oj} \prod_{k=1}^{N} x_k^{a_{ojk}} \tag{3-1}$$

Subject to

$$G_i(x) = \sum_{j=1}^{Ti} q_{ij} C_{ij} \prod_{k=1}^{N} x_k^{a_{ijk}} = 0, \qquad i = 1, \ldots, M_1 \tag{3-2}$$

$$G_i(x) = \sum_{j=1}^{T_i} q_{ij} C_{ij} \prod_{k=1}^{N} x_k^{a_{ijk}} \leq 0, \qquad i = M_1 + 1, \ldots, M \tag{3-3}$$

and

$$0 < x_k^{LB} \leq x_k \leq x_k^{UB} \tag{3-4}$$

where x_k = positive design variables; x_k^{LB} = positive lower bounds on design variables; x_k^{UB} = positive upper bounds on design variables; M_1 = number of equality constraints; M = total number of constraints; N = number of design variables; T_o = number of terms in the objective function; T_i = number of terms in the ith constraint; q_{oj} and q_{ij} are equal to ± 1; C_{oj} and C_{ij} are positive coefficients; and a_{ojk} and a_{ijk} are arbitrary real constants.

The functions $G_i(x)$ are referred to as generalized polynomials. This problem is a posynomial or signomial problem, depending on the positive or negative sign of the signum functions. The optimum design of a hybrid-plate girder is treated as a signomial problem.

Design Variables

Design variables are the quantities that describe the problem and are determined in the solution process. Design variables can be clas-

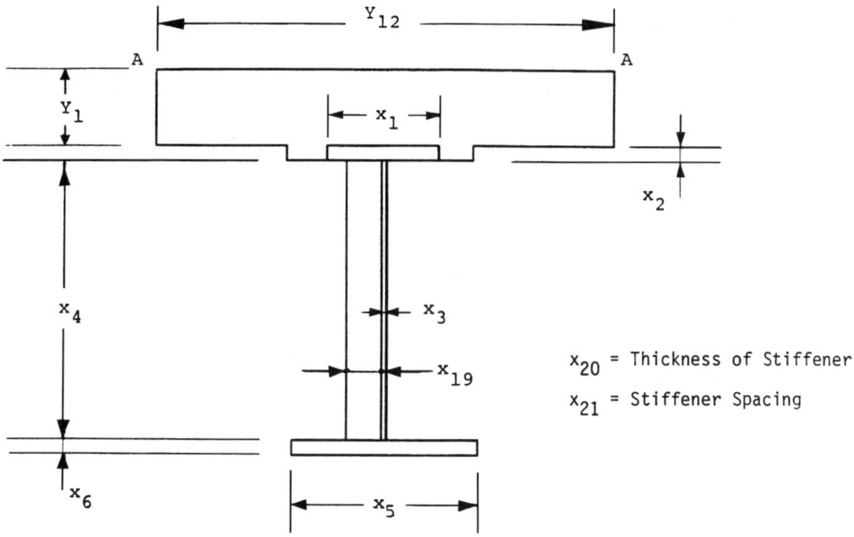

FIG. 3-1. Independent Design Variables for Composite Hybrid-Plate Girder.

sified into two categories: independent variables and dependent variables. The independent variables are sufficient to completely describe the problem. The dependent variables are functions of other variables and are introduced to simplify the problem formulation.

The independent design variables in proportioning a composite, hybrid-plate girder are the width of the compression flange x_1, the thickness of the compression flange x_2, the thickness of the web x_3, the height of the web x_4, the width of the tension flange x_5, the thickness of the tension flange x_6, the width of the transverse stiffener x_{19}, the thickness of the transverse stiffener x_{20}, and the spacing of the transverse stiffener x_{21}. These variables are shown in Fig. 3-1.

The dependent variables used in the formulation are x_7 to x_{18}, x_{22}, and x_{23}. These are defined and equations of equality constraints are provided in the following (with reference to Fig. 3-1 and Appendix 3-A for design parameters Y_1 to Y_{13}).

1. Distance from AA to centroid of compression flange

$$x_7 = Y_1 + 0.5x_2 \tag{3-5}$$

2. Distance from AA to centroid of web

$$x_8 = Y_1 + x_2 + 0.5x_4 \tag{3-6}$$

3. Distance from AA to centroid of tension flange

$$x_9 = Y_1 + x_2 + x_4 + 0.5x_6 \tag{3-7}$$

4. Distance from AA to centroid of composite section

$$x_{10} = \frac{0.5Y_1(Y_1Y_{12}/Y_4) + x_7(x_1x_2) + x_8(x_3x_4) + x_9(x_5x_6)}{(Y_1Y_{12}/Y_4) + x_1x_2 + x_3x_4 + x_5x_6} \tag{3-8}$$

5. Moment of inertia of composite section

$$\begin{aligned} x_{11} = (1/12)(x_1x_2^3 + x_3x_4^3 + x_5x_6^3 + Y_{12}Y_1^3/Y_4) + x_7^2(x_1x_2) \\ + x_8^2(x_3x_4) + x_9^2(x_5x_6) + (Y_1^2/4)(Y_1Y_{12}/Y_4) \\ - x_{10}^2(x_1x_2 + x_3x_4 + x_5x_6 + Y_1Y_{12}/Y_4) \end{aligned} \tag{3-9}$$

6. Distance from AA to centroid of plate girder section alone

$$x_{12} = \frac{x_7(x_1x_2) + x_8(x_3x_4) + x_9(x_5x_6)}{x_1x_2 + x_3x_4 + x_5x_6} \tag{3-10}$$

7. Moment of inertia plate girder alone

$$\begin{aligned} x_{13} = \frac{1}{12}(x_1x_2^3 + x_3x_4^3 + x_5x_6^3) + x_7^2(x_1x_2) + x_8^2(x_3x_4) \\ + x_9^2(x_5x_6) - x_{12}^2(x_1x_2 + x_3x_4 + x_5x_6) \end{aligned} \tag{3-11}$$

8. Factored dead-load moment

$$\begin{aligned} x_{14} = (1.3/8)[Y_5 + (x_1x_2 + x_3x_4 + x_5x_6 + x_4x_{19}x_{20}/x_{21}) \\ \cdot(0.49/144)]Y_3^2 \end{aligned} \tag{3-12}$$

9. Parameter

$$\psi = x_{15} = (Y_1 + x_2 + x_4 + x_6 - x_{10})/(x_2 + x_4 + x_6) \tag{3-13}$$

10. Capacity reduction factor

$$\begin{aligned} R &= x_{16} \\ &= 1 - \left[\frac{(x_3x_4/(x_5x_6))(x_{15})(1 - Y_9/Y_8)^2(3 - x_{15} + (x_{15}Y_9/Y_8))}{6 + (x_3x_4/(x_5x_6))x_{15}(3 - x_{15})}\right] \end{aligned} \tag{3-14}$$

11. Top flange compression stress caused by noncomposite dead load

$$x_{17} = 12x_{14}(x_{12} - Y_1)/x_{13} \tag{3-15}$$

12. Shear constant $C = x_{18}$
   (a) For unstiffened girders and for stiffened girders when (x_{21}/x_4) exceeds 3 or $[260/(x_4/x_3)]^2$

   For $(x_4/x_3) < 424.3/\sqrt{Y_9}$, $\quad x_{18} = 1.0$

   For $424.3/\sqrt{Y_9} \leq (x_4/x_3) \leq 530.3/\sqrt{Y_9}$, $\quad x_{18} = 70.7(x_3/x_4)$ (3-16)

   For $(x_4/x_3) > 530.3/\sqrt{Y_9}$, $\quad x_{18} = 6250(x_3/x_4)^2$ (3-17)

   (b) For stiffened girders with $(x_{21}/x_4) \leq 3$ or $[260/(x_4/x_3)]^2$

   For $(x_4/x_3) < 189.7\sqrt{x_{23}/Y_9}$, $\quad x_{18} = 1.0$

   For $189.7\sqrt{x_{23}/Y_9} \leq (x_4/x_3) \leq 237.2\sqrt{x_{23}/Y_9}$,

$$x_{18} = 189.7(x_3/x_4)\sqrt{x_{23}/Y_9} \tag{3-18}$$

   For $(x_4/x_3) > 237.2\sqrt{x_{23}/Y_9}$, $\quad x_{18} = 45000(x_3/x_4)^2 x_{23}/Y_9$ (3-19)

13. Variable J for transverse stiffener stiffness

$$x_{22} = 2.5x_4^2 x_{21}^{-2} - 2.0 \tag{3-20}$$

14. Web buckling coefficient

$$K = x_{23} = 5 + 5/(x_{21}/x_4)^2 \tag{3-21}$$

In addition, x_{18} and x_{24} are used to represent the objective functions for the unstiffened and stiffened cases, respectively.

Design Parameters

Design parameters are nonvarying positive quantities that are determined prior to the optimization process. The inclusion of these parameters in the problem formulation allows the use of one set of constraints for a large variety of problems by simply changing the values of the parameters. For simply supported highway bridges composed of composite hybrid-plate girders, the design parameters are Y_1 to Y_{13}, as defined in Appendix 3-A.

Objective Function

Since the deck slab thickness is fixed, it is assumed that the cost is related to the weight of the steel plate girder alone. Therefore the weight of the plate girder is used as the objective function with a modification to account for the difference in costs for the two types of steels used in the formulation, and is written as follows.
For unstiffened plate girder

$$x_{18} = \frac{1}{144}\rho Y_3(x_1x_2 + x_3x_4 + Y_{13}x_5x_6) \tag{3-22}$$

For stiffened plate girder

$$x_{24} = \frac{1}{144}\rho Y_3\left(x_1x_2 + x_3x_4 + Y_{13}x_5x_6 + \frac{x_4x_{19}x_{20}}{x_{21}}\right) \tag{3-23}$$

in which ρ = the unit weight of steel and is taken equal to 0.49 k/cu ft, Y_3 = the span length of the hybrid-plate girder, and Y_{13} is the ratio of fabricated cost of ASTM A514 steel to ASTM A36 steel.

Design Constraints

There are two kinds of constraints represented in the formulation. These are equality constraints and inequality constraints. The equality constraints provide a relationship between independent and dependent variables. The relations for the equality constraints for the composite hybrid-plate girder were listed earlier. The inequality constraints provide limitations on behavioral quantities, such as stresses and displacements. The inequality constraints in the composite hybrid-plate girder design result from the requirements of ultimate flexural strength, ultimate shear strength, concrete compressive stress, flange local buckling, and aspect ratio. For practical considerations, it is also desirable to set some geometric constraints to facilitate the process of optimization.

Ultimate Flexural Strength. According to AASHTO Section 10.53.2, the maximum flexural strength of the composite hybrid-plate girders shall be the moment at first yielding of the flanges multiplied by the reduction factor R, that is,

$$M_u \leq F_{yf}SR \tag{3-24}$$

in which F_{yf} = specified minimum yield strength of tension flange; S = section modulus of the composite section;

$$R = 1 - \left[\frac{\beta\psi(1 - \delta)^2(3 - \psi + \delta\psi)}{6 + \beta\psi(3 - \psi)}\right] \tag{3-25}$$

where $\delta = F_{yw}/F_{yf}$; $\beta = A_w/A_f$; ψ = distance from outer fiber of tension flange to neutral axis divided by depth of steel section; F_{yw} = yield strength of web steel; A_w = area of web, sq in.; and A_f = area of bottom flange, sq. in.

When the temporary supports are not provided during the placing of dead loads, the stresses are calculated with factored dead load moment acting on the steel girder alone and the moments caused by factored superimposed and live loads including impact resisted by the composite section (AASHTO 10.50.1.2.2). Thus, using the equality constraints, Eq. (3-24) may be written in the form of maximum stress in the bottom flange as follows.

$$\frac{x_{14}(x_9 + 0.5x_6 - x_{12})(12)}{x_{13}} + \frac{\left[Y_{11} + (1.3)\frac{5}{3}Y_6\right](x_9 + 0.5x_6 - x_{10})(12)}{x_{11}} \le x_{16}Y_8 \qquad (3\text{-}26)$$

Similarly for the compression flange stress, we can write

$$\frac{x_{14}(x_{12} - Y_1)(12)}{x_{13}} + \frac{\left[Y_{11} + (1.3)\frac{5}{3}Y_6\right](x_{10} - Y_1)(12)}{x_{11}} \le Y_2 \qquad (3\text{-}27)$$

Ultimate Shear Strength. For an unstiffened or stiffened hybrid-plate girder, maximum factored shear force V_u (AASHTO 10.48.8) is limited to

$$V_u \le 0.58\, CF_{yw}Dt_w \qquad (3\text{-}28)$$

where constant C is defined as in equality constraint item 12. Equation (3-28) reduces to the following inequality constraint.

$$4Y_3^{-1}x_{14} + 4Y_3^{-1}Y_{11} + 2.167Y_7 - 0.58Y_9x_3x_4x_{18} \le 0 \qquad (3\text{-}29)$$

Concrete Compressive Stress. The maximum compressive stress in the concrete slab of a composite-plate girder shall not exceed $0.85\,f'_c$ (AASHTO 10.45.3), where f'_c is the compressive strength of the concrete, that is,

$$\frac{\left[Y_{11} + (1.3)\frac{5}{3}Y_6\right](12)}{Y_4x_{11}} \le 0.85Y_{10} \qquad (3\text{-}30)$$

Aspect Ratio. For the unstiffened hybrid-plate girder, the ratio of web depth to thickness must not exceed $974\sqrt{F_{yw}}$ (ksi) (AASHTO 10.48.2.1) and is written in constraint form as

$$x_3^{-1}x_4 - 974Y_9^{-0.5} \leq 0 \tag{3-31}$$

For the stiffened hybrid-plate girder, the ratio of web depth to web thickness must not exceed $1154.23/\sqrt{F_{yw}}$ (ksi) (AASHTO 10.48.5.1) and is written in the constraint form as

$$x_3^{-1}x_4 - 1{,}154.23Y_9^{-0.5} \leq 0 \tag{3-32}$$

The width-to-thickness ratio of transverse stiffeners shall not exceed $82.22/\sqrt{F_{yw}}$ (ksi) (AASHTO 10.48.5.3) and is written in constraint form as

$$x_{19}x_{20}^{-1} - 82.22Y_9^{0.5} \leq 0 \tag{3-33}$$

The moment of inertia of transverse stiffeners is given by (AASHTO 10.48.5.3)

$$I_s \geq d_o t_w^3 J \tag{3-34}$$

where

$$J = 2.5\left(\frac{D}{d_o}\right)^2 - 2 \tag{3-35}$$

Equation (3-35) was previously expressed as equality constraint item 13 [Eq. (3-20)], and Eq. (3-34) takes the following constraint form

$$3.0x_3^3x_{21}x_{22} - x_{19}^3x_{20} \leq 0 \tag{3-36}$$

Flange Local Buckling. The ratio of projecting top compression flange plate width to thickness must not exceed $61.02/\sqrt{F_{dl1}}$ (ksi) [AASHTO 10.50.(c)] where F_{dl1} is the top flange compressive stress caused by noncomposite dead load, expressed as an equality constraint item 11 [Eq. (3-15)]. The inequality constraint arising from this limitation becomes

$$0.5x_1x_2^{-1} - 61.02x_{17}^{-0.5} \leq 0 \tag{3-37}$$

Geometric Constraints. The width of the bottom flange is limited to 35% of the depth of the web and is expressed in inequality form as

$$x_5 - 0.35x_4 \leq 0 \tag{3-38}$$

SAMPLE DESIGN PROBLEM

The design of an 80 ft simple span, composite, stiffened, hybrid-plate girder is considered to show the design application of the method. The values of the design parameters are as follows.

Span length, Y_3 = 80 ft
Deck slab thickness, Y_1 = 7 in.
Effective width of deck slab, Y_{12} = 84 in.
Compressive strength of concrete, Y_{10} = 4.0 ksi
Yield strength of top steel flange, Y_2 = 36 ksi
Yield strength of bottom flange steel, Y_8 = 100 ksi
Yield strength of web steel, Y_9 = 36 ksi
Modular ratio, Y_4 = 8.0
Dead load, excluding girder weight, Y_5 = 0.84 k/ft
Maximum live-load moment including impact, Y_6 = 1094.0 k-ft
Maximum live-load shear including impact, Y_7 = 59.7 k
Superimposed dead-load moment, Y_{11} = 132.0 k-ft
Ratio of fabricated cost of A514 to A36, Y_{13} (assumed) = 1.10

Traditional Approach

In the traditional approach, it is required to obtain a suitable section that will satisfy the design constraints. In the absence of any design aids, a trial section has to be assumed; or the section selection is based on some approximate formulation combined with engineering judgment and geometric constraints. The selected section is then analyzed to check all the design constraints, and adjustments are made as necessary. The iterative process is repeated until the designer is satisfied with the economy and safety of the final design.

The following calculation steps show the procedure for selecting a section based on flexural constraints.

1. Calculate maximum bending moment:
   Assume girder weight equal to 125 lb/ft
   Dead-load moment, $M_{D1} = \frac{1}{8}(0.84 + 0.125)(80)^2 = 768.0$ k-ft
   Superimposed load moment, M_{D2} = 132.0 k-ft
   Live-load moment, including impact, M_L = 1094.0 k-ft
   Factored moment, $M_u = (1.3)(768.0 + 132.0) + (2.17)(1094.0) =$ 3528.9 k-ft
2. Select trial section:
   Assume girder depth ($L/25$ to $L/20$), h = 42″ (say)
   Set x_4 = 40″ and x_3 = 1/4″
   Approximate bottom flange area

$$A_f = \frac{M_u}{0.9hF_{yf}} = \frac{(3528.9)(12)}{(.9)(42)(100)} = 11.2 \text{ in.}^2$$

Set $x_5 = 14''$ and $x_6 = 1''$
(to be conservative in view of approximation in calculating A_f)
Assume the top flange dimensions:
$x_1 = 10''$ and $x_2 = 1/2''$

3. Check flexural strength of the composite section.
For the composite section (Fig. 3-1), we have the trial section given by:

$$x_1 = 10'',\ x_2 = 1/2'',\ x_3 = 1/4'',\ x_4 = 40'',\ x_5 = 1'',\ x_6 = 14'',$$

$$Y_1 = 7'',\ \text{and}\ Y_{12} = 84''$$

Using the equality constraint equations we obtain the following quantities.

$$x_7 = 7.25'',\ x_8 = 27.5'',\ x_9 = 48'',\ x_{10} = 12.1''$$

$$x_{11} = 27609.0\ \text{in.}^4,\ R = 0.915,\ S = 758.5\ \text{in.}^3$$

$$M_u = RF_{yf}\,S = (0.915)(100)(758.5)\,\frac{1}{12} = 5783.6\ \text{k-ft} > 3528.9\ \text{k-ft}$$

OK

The trial section seems to be overdesigned. The section area is reduced, more likely in the flanges, and the next iteration is made. When a suitable section is obtained based on flexural considerations, stiffeners are designed for shear and all the other constraints are checked. The resulting section would form a reasonable, but not optimum, solution to the design problem.

Optimum Design Solution

The optimum design algorithm developed for the composite hybrid-plate girder requires input of initial and lower bound values of the design variables. The preceding trial section shown with a slight modification forms a good estimate for the initial values.

The iterative solutions for independent variables in the optimization process are presented in Table 3-1. This table also shows the iteration summary of the objective function. The convergence of the objective function occurs smoothly in four iterations. For practical use considerations, the optimum solution is rounded to a discrete solution as shown in the table, which forms the final design section. The stresses on the design section are checked to satisfy the AASHTO specification.

TABLE 3-1 Iterative Solution Sample Design Problem

Iteration	x_1 (in.)	x_2 (in.)	x_3 (in.)	x_4 (in.)	x_5 (in.)	x_6 (in.)	x_{19} (in.)	x_{20} (in.)	x_{21} (in.)	Obj. Function x_{24}
1	1.348	4.000	.2809	51.86	7.645	.7919	3.246	.2367	35.36	7.523
2	4.311	1.163	.2540	48.87	10.17	.6619	2.844	.2075	35.68	6.947
3	9.159	.5844	.2520	48.48	16.97	.3864	2.848	.2078	34.71	6.941
4*	11.03	.5079	.2470	47.52	16.63	.4062	2.836	.2070	33.11	6.940
Discrete solution	11	1/2	1/4	48	17	7/16	3	1/4	34	

*Optimum solution.

SUMMARY AND DISCUSSION

The composite action in members built up of different materials results in savings in construction costs. These savings can be further advanced by employing optimization techniques in the design of composite members.

The design of composite members usually involves a trial-and-error procedure. This makes the use of design aids highly desirable. Optimization techniques can be used effectively in an automated approach to the design of a composite member. This has been shown in this chapter for the design of composite hybrid-plate girders using the generalized geometric programming technique. In the case of composite hybrid girders, the use of higher strength steel for bottom flange materials gives a more reasonable size for members as well as savings in costs.

APPENDIX 3-A. NOTATION

The following symbols are used in this chapter.

A_f = area of bottom flange, sq in.
A_w = area of web, sq in.
a_{ijk} = exponent of design variable
C = shear constant
C_{ij} = positive coefficient
F_{yf} = specified minimum yield strength of flange, ksi
F_{yw} = specified minimum yield strength of web, ksi
f'_c = compressive strength of concrete, ksi
f_{dl1} = top flange compression stress caused by noncomposite dead load, ksi
K = web buckling coefficient
M = total number of constraints
M_1 = number of equality constraints
N = number of design variables
n = modular ratio
q_{ij} = equal to ± 1
R = reduction factor
S = section modulus of composite section, cu in.
T_o = number of terms in objective function
T_i = number of terms in the ith constraint
x_k = positive design variable
x_k^{LB} = positive lower bounds on design variable
x_k^{UB} = positive upper bounds on design variable
x_1 = width of top flange, in.

x_2 = thickness of top flange, in.
x_3 = thickness of web, in.
x_4 = depth of web, in.
x_5 = width of bottom flange, in.
x_6 = thickness of bottom flange, in.
x_7 = equal to $Y_1 + 0.5x_2$, in.
x_8 = equal to $Y_1 + x_2 + 0.5x_4$, in.
x_9 = equal to $Y_1 + x_2 + x_4 + 0.5x_6$, in.
x_{10} = distance from top of slab to neutral axis of composite section, in.
x_{11} = moment of inertia of composite girder, in.^{4}
x_{12} = distance from top of slab to neutral axis of steel girder section, in.
x_{13} = moment of inertia of steel girder, in.
x_{14} = dead-load moment including weight of steel girder, kips-ft
x_{15} = equal to ψ
x_{16} = reduction factor
x_{17} = top flange compression stress caused by loads on non-composite section, ksi
x_{18} = web buckling coefficient for stiffened girder or objective function for unstiffened girder case, kips
x_{19} = width of stiffener, in.
x_{20} = thickness of stiffener, in.
x_{21} = spacing of stiffeners, in.
x_{22} = equal to $2.5(x_4/x_{21})^2 - 2$
x_{23} = equal to $5 + 5/(x_{21}/x_4)^2$
x_{24} = objective function for stiffened girder case, kips
Y_1 = deck slab thickness, in.
Y_2 = yield strength of top flange steel, ksi
Y_3 = girder span length, ft
Y_4 = modular ratio
Y_5 = dead load excluding girder weight, kips/ft
Y_6 = maximum live-load moment including impact, kips-ft
Y_7 = maximum live-load shear including impact, kips
Y_8 = yield strength of bottom flange steel, ksi
Y_9 = yield strength of web steel, ksi
Y_{10} = compressive strength of concrete, ksi
Y_{11} = superimposed dead-load moment, kips-ft
Y_{12} = effective width of deck slab, in.
Y_{13} = ratio of fabricated cost of A514 steel to A36 steel
ρ = unit weight of steel, 0.49 kip/cu ft
δ = F_{yw}/F_{yf}
β = A_w/A_f
ψ = distance from outer fiber of tension flange to neutral axis of transformed section divided by depth of steel section

APPENDIX 3-B. CONVERSION TO SI UNITS

$$
\begin{aligned}
1 \text{ in.} &= 0.0254 \text{ M} \\
1 \text{ sq in.} &= 0.00064516 \text{ M}^2 \\
1 \text{ lb} &= 4.44822 \text{ N} \\
1 \text{ ksi} &= 6.89476 \text{ MPa.}
\end{aligned}
$$

REFERENCES

AASHTO (1992). *Standard specifications for highway bridges.* 15th ed., Washington, DC.

Abuyounes, S. and Adeli, H. (1987). "Optimization of hybrid steel plate girders." *Comput. Struct.* 27(5), 575–582.

AISC (1986). *Load and resistance factor design specification for structural steel buildings,* Chicago, IL.

Burns, S. A. (1987). "Generalized geometric programming with many equality constraints." *Int. J. Numer. Methods Engrg.,* 24 (4), 725–741.

Burns, S. A. and Ramamurthy, S. (1983). "Mathematical programming in structural design." *J. Struct. Div.,* ASCE, 109(7), 1669–1679.

Cook, J. P. (1977). *Composite construction methods.* John Wiley & Sons, Inc., New York.

Dhillon, B. S. and Kuo, C. H. (1991). "Optimum design of composite hybrid plate girders." *J. Struct. Div.,* ASCE 117 (7), 2088–2098.

Kuo, C. H. (1989). Optimum design of hybrid plate girders using generalized geometric programming. Thesis submitted to Widener University, Chester, PA, in partial fulfillment of the requirements for the degree of Master of Engineering.

McCormac, J. C. (1989). *Structural steel design: LRFD method.* Harper & Row, New York.

Salmon, C. G. and Johnson, J. E. (1990). *Steel structures: Design and behavior,* 3rd ed., Harper & Row, New York.

Chapter 4

HOW TO OPTIMIZE A REINFORCED CONCRETE COLUMN OR BEAM

In this chapter we consider the optimization of a single reinforced concrete column of circular or rectangular cross-section as shown in Fig. 4-1. Assume that factored axial loads P_u and moments M_{uy} and M_{uz} are specified for each loading case. Assume that the specified unbraced length is short enough so that length effects can be neglected. We also consider the optimization of a single multispan continuous beam as shown in Fig. 4-2 subjected to specified factored moment ($M_{u\max}$ and $M_{u\min}$) and shear (V_u) envelopes. Assume that span lengths and support widths are also specified. The beam may be rectangular in cross-section or it may be part of a monolithic beam-slab system with the cross-section shown in Fig. 4-3. The optimization problem for a single beam or a single column is to determine the cross-sectional dimensions and reinforcement design that minimize cost and satisfy the constraints prescribed by a design code such as the ACI Specification (1989).

Since the number and size of reinforcing bars are discrete design variables, the optimization problem is treated as a discrete optimization problem. Even the cross-sectional dimensions, b (or b_w) and h, are treated discretely since it is recommended in practice that these be rounded to the nearest whole centimeter (or inch). Because the analysis of a single column or beam is fast, and because the search space is relatively small, the discrete optimization algorithm recommended for these problems is exhaustive search. Since the exhaustive search method is relatively straightforward, the bulk of this chapter is devoted to problem formulation, method of analysis, and generalization of results from example problems. Nevertheless, some com-

Richard J. Balling, Professor, Civil Engineering, email:⟨rick@et.byu.edu⟩ (contact for correspondence), Steve S. Gallup, and Clinton S. McGrath, Department of Civil and Environmental Engineering, Brigham Young University, Provo, Utah 84602.

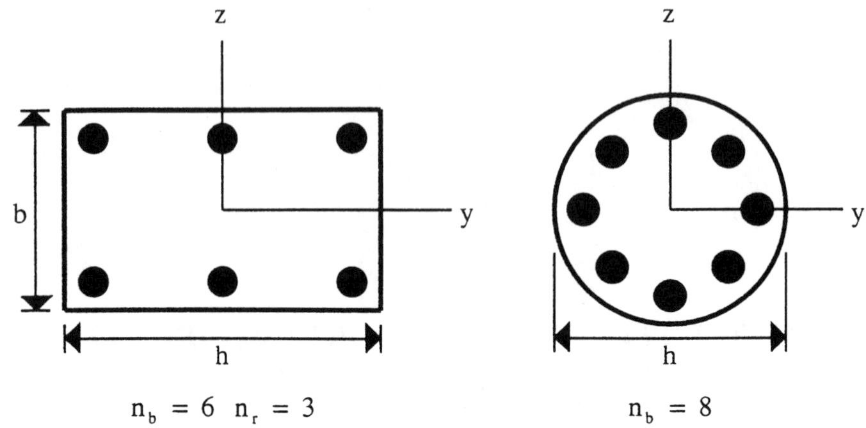

FIG. 4-1. Column Cross-Sections.

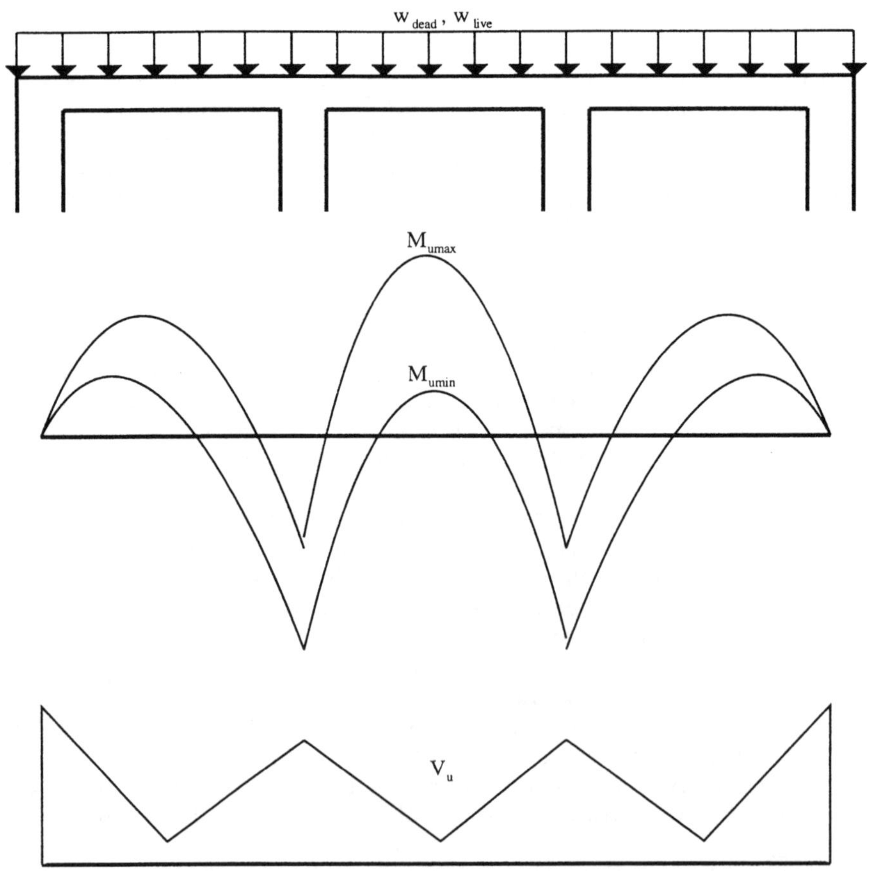

FIG. 4-2. Multispan Continuous Beam.

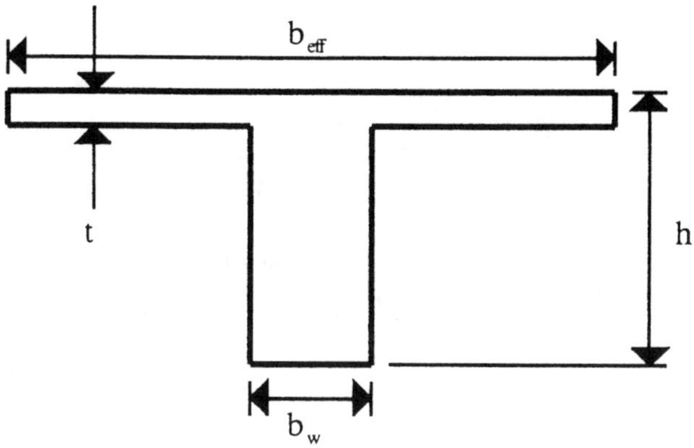

FIG. 4-3. Beam-Slab Cross-Section.

ments are made on how exhaustive search should be organized and implemented to maximize computational efficiency.

SINGLE COLUMN: PROBLEM FORMULATION

It is assumed that all longitudinal reinforcing bars in the column are the same size. It is also assumed that in circular columns, the bars are evenly spaced around the perimeter. In rectangular columns, the bars are placed around the perimeter in evenly spaced rows. For rectangular columns, there are five discrete design variables (see Fig. 4-1).

1. h = depth
2. b = width
3. d_b = reinforcing bar diameter
4. n_b = number of reinforcing bars
5. n_r = number of rows parallel to width

For a circular column, there are three discrete design variables (h, d_b, and n_b). Constraints are imposed according to the ACI Specification (1989) as follows.

1. Strength for each loading case: $P_u < \phi P_n$
2. Minimum reinforcement ratio: $\rho > 0.01$
3. Maximum reinforcement ratio: $\rho < 0.08$
4. Spacing: $b > b_{min}$ and $h > h_{min}$

where

P_n = column strength
ϕ = strength reduction factor
$\rho = (n_b \pi d_b^2)/4bh$ (rectangular) or $\rho = (n_b d_b^2)/h^2$ (circular)
b_{min}, h_{min} = smallest dimensions with bars at minimum spacing

The objective is to minimize the total cost of the column. For example, the cost of a rectangular tied column might be calculated as:

$$C_{conc}bh + (\gamma_{steel}C_{steel} - C_{conc})(n_b d_b^2 L + n_{tie} d_{tie}^2 L_{tie}) \frac{\pi}{4} + C_{bar}n_b + C_{tie}n_{tie}$$

where

L = column length
n_{tie} = number of ties
d_{tie} = tie diameter
L_{tie} = length of a single tie
C_{conc} = concrete cost per volume
C_{steel} = steel cost per weight
γ_{steel} = unit weight of steel
C_{bar} = placement cost per bar
C_{tie} = placement cost per tie

SINGLE COLUMN: METHOD OF ANALYSIS

Column analysis consists of calculating the cost and constraints from given values of the design variables. Calculation of the reinforcement ratio and spacing constraints is trivial, but calculation of the strength P_n for each loading case is not. Consider the column shown in Fig. 4-4 with steel yield stress f_y = 414 MPa (60 ksi) and concrete strength f'_c = 27.6 MPa (4 ksi) subjected to the single loading case:

P_u = 2225 kN (500 kip)
M_{uy} = 226 kN − m (2000 kip-in.)
M_{uz} = 452 kN − m (4000 kip-in.)

This loading case is equivalent to placing the load P_u at the Point A with eccentricities $e_y = M_{uz}/P_u$ = 20.3 cm (8 in.) and $e_z = -M_{uy}/P_u$ = −10.2 cm (−4 in.) as shown in Fig. 4-4. The analysis requires the determination of the two unknowns x_h and x_b which are the distances to where the neutral axis intersects y and z axes through the point of maximum compressive stress (lower left corner). The two equations used to determine these unknowns are the sums to zero of the mo-

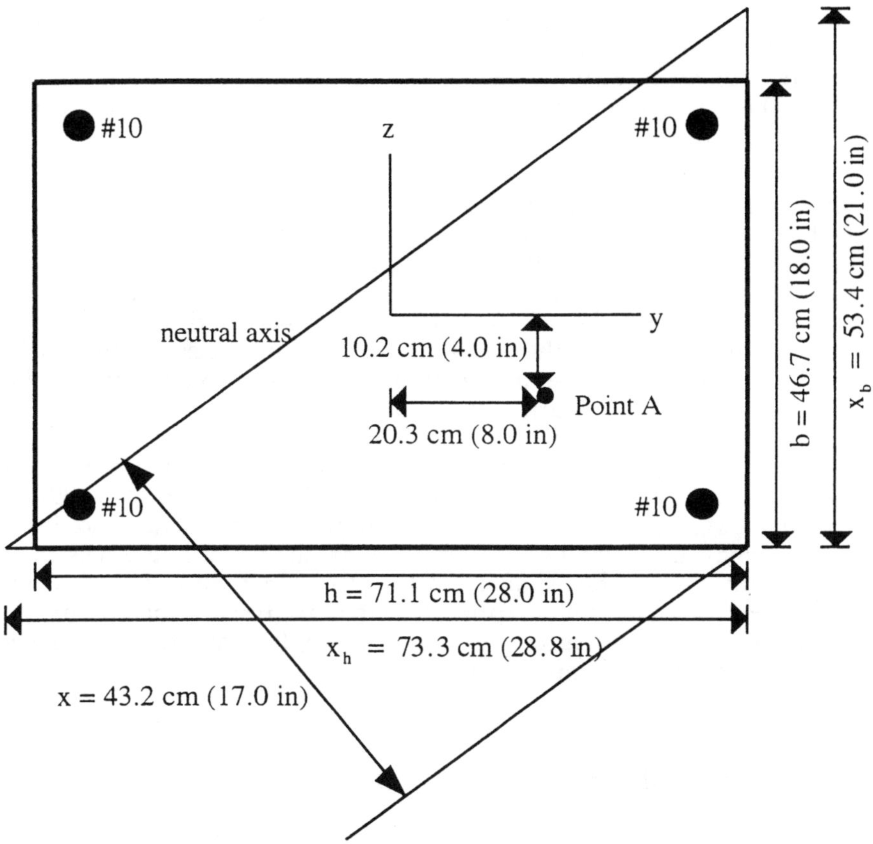

FIG. 4-4. Example Column for Analysis.

ments of the resistive concrete and steel force resultants about the y and z axes through Point A. Given trial values for x_h and x_b, the resistive force resultants for the concrete and each reinforcing bar can be determined from the standard assumptions of the ACI Specification (1989):

1. Strain in the extreme compression concrete fiber is 0.003 at ultimate strength;
2. Strain is such that the cross-sectional plane remains planar, and a good bond exists between concrete and reinforcing bars;
3. The relation between stress and strain for steel reinforcing bars is elastic perfectly plastic;
4. The tension region of the concrete is cracked and the stress is zero;

5. The compression region of the concrete has a constant stress of $0.85f'_c$ and extends inward from the extreme compression fiber for a distance β_x where $\beta < 1$ is a specified function of f'_c, and x is the perpendicular distance from the neutral axis to the point of maximum compressive stress.

Iterative methods such as the Newton-Raphson method may be used to determine the values of x_h and x_b satisfying the two moment equations (Sun and Lu 1992). The correct values of x_h and x_b for the example are shown in Fig. 4-4. Once the values of x_h and x_b are known, the resistive force resultants may be summed to get the column strength P_n. As an alternative to iterative methods, approximate formulas and graphs are also available for rapid evaluation of the strength (Bresler 1960; Parme et al. 1966).

Calculation of the column cost requires that ties or spirals be designed as part of the analysis. Formulas for the vertical spacing and size of ties and spirals are given by the ACI Specification (1989) as functions of b, h, d_b, and n_b. For tied columns, every corner and alternate bar shall be laterally supported by an included angle less than 135 degrees, and no bar shall be farther than 15.2 cm (6 in.) clear on either side from a laterally supported bar. Once the number, diameter, and length of ties and spirals are determined, the material and placement costs of the column can be summed.

SINGLE COLUMN: METHOD OF OPTIMIZATION

The exhaustive search algorithm may be organized as a set of nested loops. The outermost loops are on the dimensions b and h which span over a reasonable range of whole centimeter (or inch) values. Next comes the loop over d_b which ranges over the discrete bar sizes. Next comes the loop over n_b. One can exploit the fact that the spacing and reinforcement ratio constraints are easy to evaluate, and the fact that for fixed values of b, h, and d_b, the cost and strength are monotonic functions of n_b. Thus, for a rectangular tied column, the n_b loop begins at the maximum of four and the smallest even integer greater than $(0.01bh4)/(\pi d_b^2)$. The loop increments by two and stops when either

1. the spacing constraints cannot be satisfied;
2. n_b is greater than $(0.08bh4)/(\pi d_b^2)$; or
3. a column satisfying the strength constraints is found.

For rectangular columns, the innermost loop is on n_r which begins at two and increments by one until $n_r = n_b/2$. The exhaustive search returns the lowest cost column satisfying all constraints.

SINGLE COLUMN: OPTIMUM RESULTS

Several columns were optimized with varying values of loads and unit costs. The material properties were held constant at f_y = 414 MPa (60 ksi) and f'_c = 27.6 MPa (4 ksi). Fig. 4-5 shows the effect of varying the ratio of $\gamma_{steel}C_{steel}/C_{conc}$ on the optimum design of concentrically loaded columns with

$$P_u = 2225 \text{ kN (500 kips)}$$
$$M_{uy} = M_{uz} = 0$$
$$C_{bar} = C_{tie} = C_{spiral} = 0$$

The spacing, reinforcement ratio, and strength constraints for the

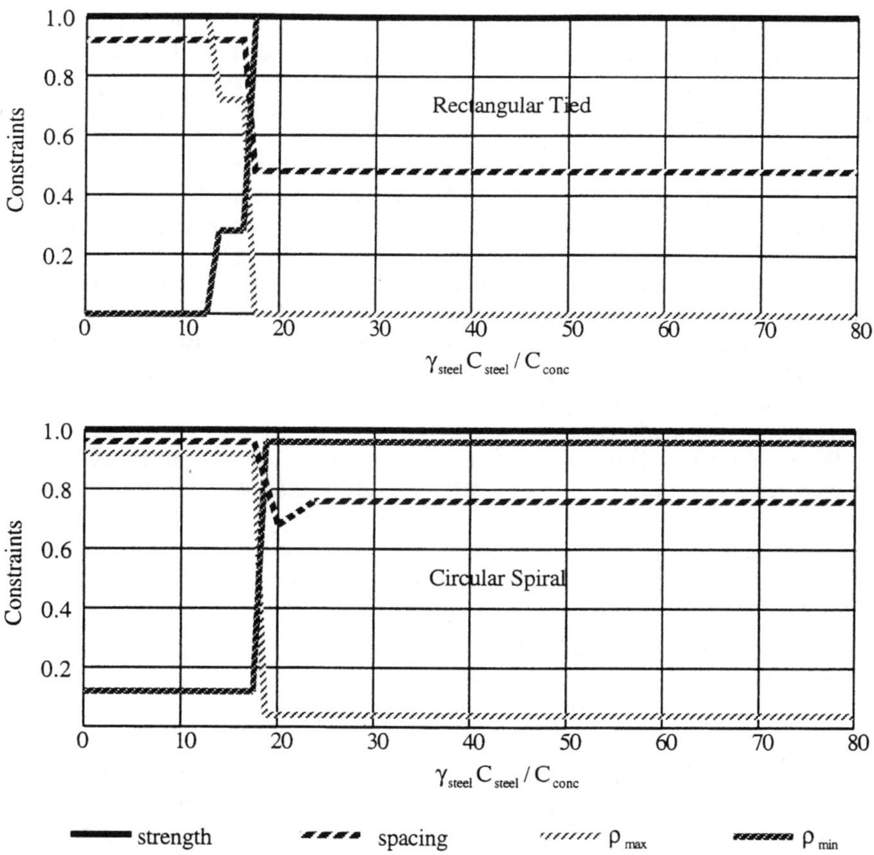

FIG. 4-5. Concentrically Loaded Columns.

various optima are plotted. The constraints have been normalized to be satisfied when less than one. As one would expect, the strength constraint controls for all optima. For low values of $\gamma_{steel}C_{steel}/C_{conc}$, either the ρ_{max} constraint or the spacing constraint controls, whereas for high values of $\gamma_{steel}C_{steel}/C_{conc}$, the ρ_{min} constraint controls. The spread in the transition zone between the ρ_{max}-controlling region and the ρ_{min}-controlling region is due to the material cost of the ties and spirals. If the material cost of ties and spirals is neglected, one can prove that the transition occurs sharply at

$$\frac{\gamma_{steel}C_{steel}}{C_{conc}} = \frac{f_y}{.85f'_c} = 17.65$$

for this example.

For typical market prices, $\gamma_{steel}C_{steel}/C_{conc}$ is usually well into the ρ_{min}-controlling region. Thus one would expect $\rho = 0.01$ for optimally designed, concentrically loaded, single columns. This may not be true for columns in a frame where the stiffness and weight of a column affects the loads placed on other members of the frame. Architectural limits on b and h may also lead to optimal columns with $\rho > 0.01$.

Figs. 4-6 and 4-7 show results for eccentrically loaded, rectangular, optimized columns subject to uniaxial bending and biaxial bending, respectively, with $C_{bar} = C_{tie} = 0$. Note that as the eccentricity increases, the transition zone widens into the ρ_{min}-controlling region. In fact, for typical market prices, the optimum design often falls within the transition zone rather than within the ρ_{min}-controlling region. The widening of the transition zone for high eccentricity can be explained by the fact that large moments put a significant portion of the concrete into tension, at which point it becomes more efficient to add steel rather than to add concrete.

Fig. 4-8 shows the effect of increasing load on the optimal values of design variables for a concentrically loaded, rectangular, tied column with

$$\begin{aligned}
M_{uy} &= M_{uz} = 0 \\
C_{bar} &= C_{tie} = 0 \\
C_{steel} &= \$0.056/\text{N} \ (\$0.25/\text{lb}) \\
\gamma_{steel} &= 77.0 \ \text{KN/m}^3 \ (490 \ \text{lb/ft}^3) \\
C_{conc} &= \$0.000122/\text{cm}^3 \ (\$93/\text{yd}^3)
\end{aligned}$$

Note that as P_u increases, both $h = b$ and A_s increase proportionately to maintain $\rho = \rho_{min} = 0.01$. For low values of increasing P_u, the optimal n_b is constant at 4 bars whereas the optimal d_b steadily increases. This can be explained by the fact that the material cost of ties is lowest when there are only four bars. When d_b reaches the diameter of a #10

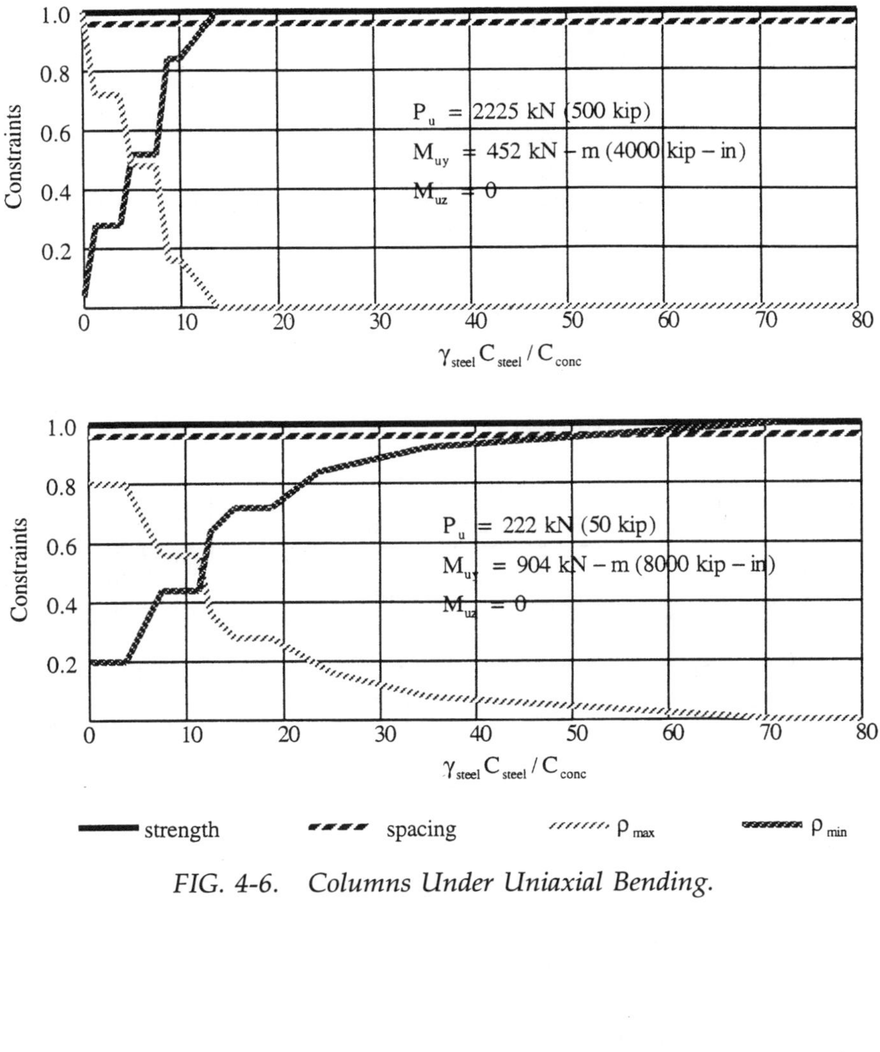

FIG. 4-6. Columns Under Uniaxial Bending.

bar, things begin to change. The ACI Specification (1989) requires that for #10 bars and smaller, #3 ties may be used, whereas for #11 bars and larger, #4 ties must be used. The optimal values of n_b and d_b change erratically with further increases in P_u indicating the truly discrete nature of the problem.

Finally, consider the effect of the unit placement costs, C_{bar}, C_{tie}, and C_{spiral}. As unit placement costs are increased, the optimal value of n_b generally decreases, which necessitates a corresponding increase in the optimal value of d_b. Furthermore, the transition zone between the ρ_{max}-controlling region and the ρ_{min}-controlling region shifts at a lower value of $\gamma_{steel}C_{steel}/C_{conc}$ since these placement costs effectively add to the cost of steel rather than to the cost of concrete.

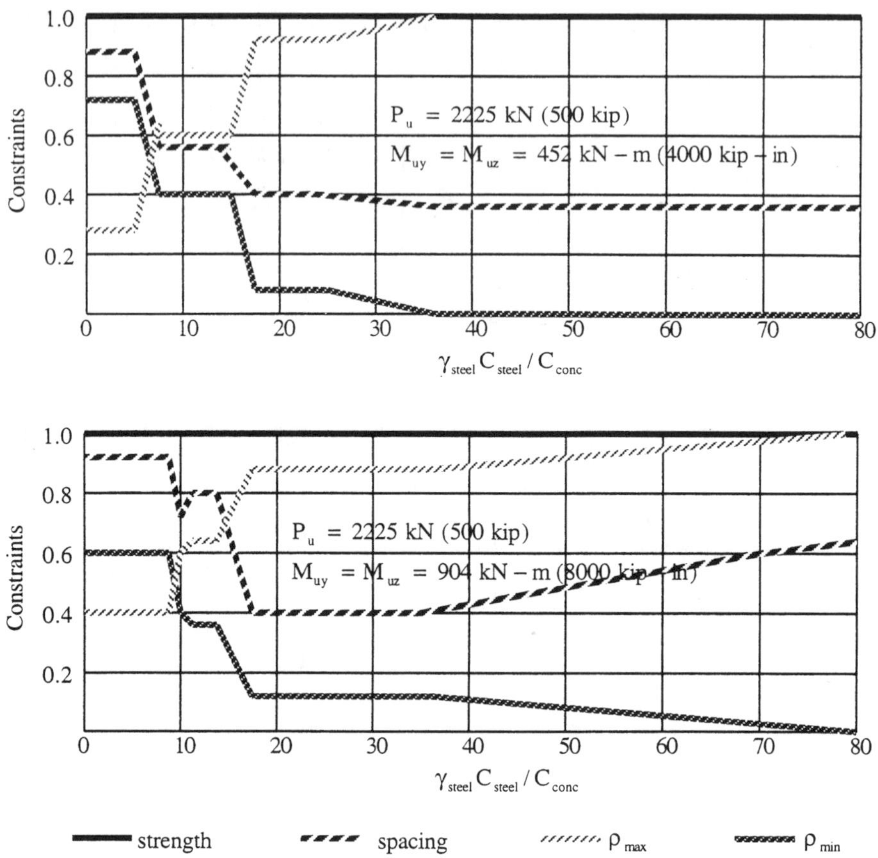

FIG. 4-7. Columns under Biaxial Bending.

SINGLE BEAM: PROBLEM FORMULATION

The design variables for a single beam include the cross-sectional dimensions. For a rectangular beam, these are simply the depth h and the width b. For the monolithic beam-slab system shown in Fig. 4-3, assume that the slab has already been designed so that t and b_{eff} are fixed, and the cross-sectional design variables are h and b_w.

Reinforcement may be arranged in different configurations or topologies. Three different topologies for the same beam are shown in Fig. 4-9. A reinforcement topology consists of a set of bar groups. Each bar group is a set of same-diameter bars that have the same cutoff and bend points. For a given topology, the design variables include the following for each bar group.

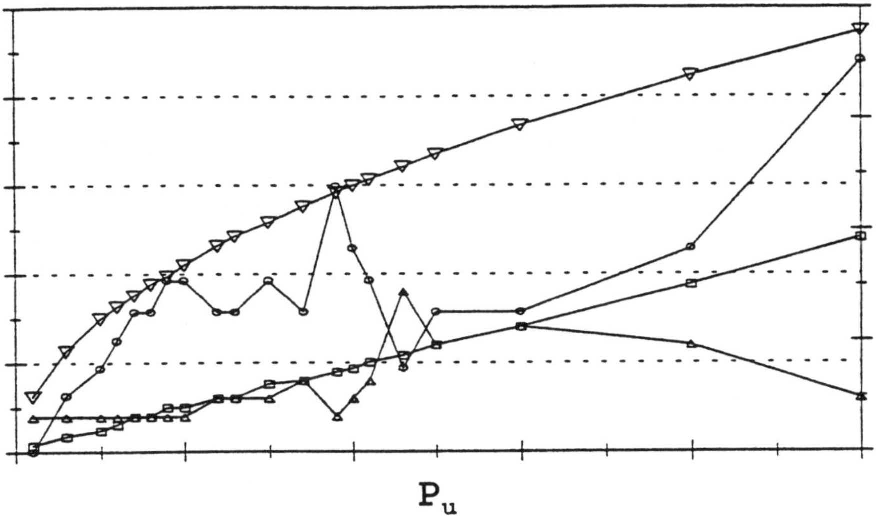

FIG. 4-8. Increasing Column Load.

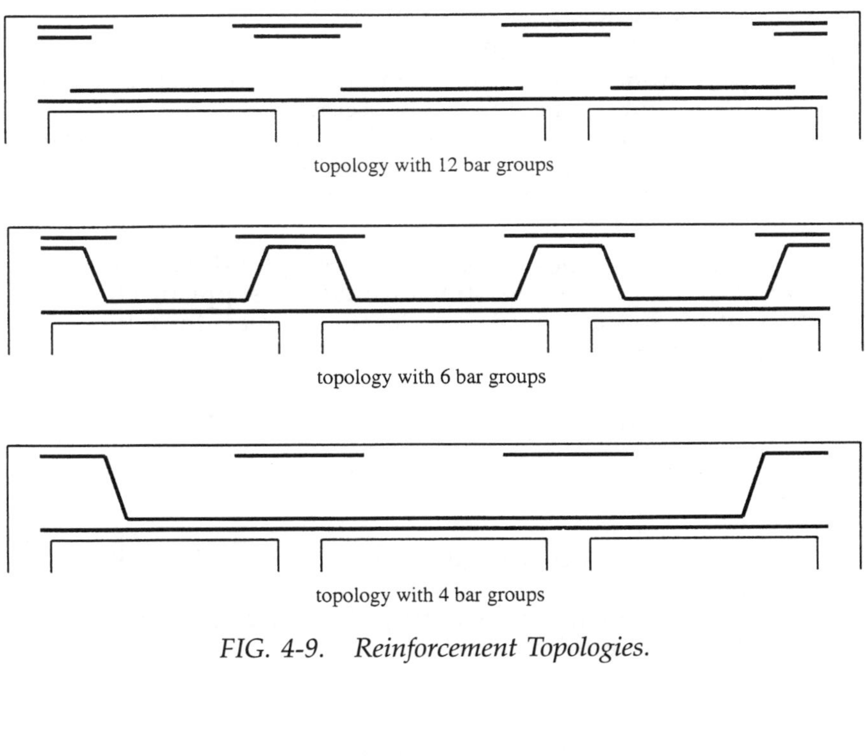

FIG. 4-9. Reinforcement Topologies.

1. d_b = reinforcing bar diameter
2. n_b = number of reinforcing bars

Constraints are imposed according to the ACI Specification (1989) at each critical location. Critical locations for positive moment include points where M_{umax} peaks positively in the interior of each span, and critical locations for negative moment occur at the face of each support. The following constraints are imposed at each positive and negative critical location.

1. Flexural strength: $M_u < \phi M_n$
2. Minimum reinforcement ratio: $\rho > \rho_{min}$
3. Balanced condition: $\rho < 0.75\rho_b$
4. Spacing: $b > b_{min}$ and $h > h_{min}$
5. Extension: $A_{sext}/A_s > 1/3$ (1/4 for positive continuous)

where

M_u = M_{max} (positive location) or $-M_{umin}$ (negative location)
M_n = flexural strength
ϕ = strength reduction factor
ρ = A_s/bd
A_s = total reinforcement area at the location
d = distance from extreme concrete compressive fiber to reinforcement centroid
ρ_{min} = 1.38 MPa/f_y (200 psi/f_y)
ρ_b = value of ρ at which simultaneous concrete failure and reinforcement yielding occurs
b_{min}, h_{min} = smallest dimensions with bars at minimum spacing
A_{sext} = area of reinforcement extending into support or past inflection points

Critical locations for shear strength are located a distance d from the face of each support. Shear strength may be supplemented by the addition of stirrups at critical shear locations unless the followed constraint is violated.

6. Shear strength: $V_u < \phi(V_c + V_{smax})$

where

V_c = concrete shear strength = 166 Pa$^{1/2}$ $\sqrt{f'_c}\, bd$ (2 psi$^{1/2}$ $\sqrt{f'_c}\, bd$)
V_{smax} = maximum allowable stirrup shear strength = $4V_c$
ϕ = strength reduction factor

The objective is to minimize the total cost of the beam. For example, the cost of a rectangular beam might be calculated as:

$$C_{\text{conc}}bhL + (\gamma_{\text{steel}}C_{\text{steel}} - C_{\text{conc}})\left(\sum_{\text{bar groups}} n_b d_b^2 L_b + n_{\text{stir}} d_{\text{stir}}^2 L_{\text{stir}}\right)\frac{\pi}{4} + C_{\text{bar}}\sum_{\text{bar groups}} n_b + C_{\text{stir}} n_{\text{stir}}$$

where

L = beam length
L_b = length of bars in a bar group
n_{stir} = number of stirrups
d_{stir} = stirrup diameter
L_{stir} = length of a single stirrup
C_{conc} = concrete cost per volume
C_{steel} = steel cost per weight
γ_{steel} = unit weight of steel
C_{bar} = placement cost per bar
C_{stir} = placement cost per stirrup

SINGLE BEAM: METHOD OF ANALYSIS

Beam analysis consists of calculating the cost and constraints for a given reinforcement topology and given values of the design variables. The calculation of A_s and the extension constraint at each critical positive and negative location is trivial.

Calculation of the other constraints begins by determining the value of d at each critical positive and negative location. This calculation requires that all bar groups present at a particular location be symmetrically placed in the cross-section. Bars may have to be placed in multiple layers in order to meet spacing and cover requirements. When multiple layers are used, bars should be placed directly over each other. Once the bar placement is determined at a critical location, d can be calculated and the spacing and reinforcement ratio constraints can be evaluated. The value of d can also be used to calculate the shear strength constraints at each critical shear location.

Calculation of the flexural strength and balanced condition constraints proceeds according to the standard assumptions of the ACI Specification (1989) listed earlier for column analysis. For a rectangular singly reinforced beam, these assumptions lead to the following formulas.

$$M_n = A_s f_y \left(d - \frac{A_s f_y}{.85 f_c' b 2}\right)$$

$$\rho_b = \frac{.85f'_c\beta}{f_y\left(1 + \dfrac{f_y}{.003E_s}\right)}$$

where E_s is the modulus of elasticity of steel.

Calculation of the cost of the beam requires determination of the length of each bar group. Thus beam analysis requires that cutoff and/or bend points be determined for each bar group. The ACI Specification (1989) requires that these cutoff and bend points satisfy the following.

1. Bar groups have embedment lengths greater than the development length, l_d, beyond the point at which their capacity is required;
2. at points where bars are cut off or bent, the flexural capacity diagram is horizontally offset from the factored moment diagram by a distance equal to the maximum of d or $12d_b$;
3. the last positive bar group to be cut off or bent extends past simple supports ($C = 1.3$) or points of inflection ($C = 1.0$) by a distance $l_d - CM_n/V_u$.

For bent bars, the point where the bar crosses the mid-depth is considered to be the effective extension for Requirement 2, and the development length for Requirements 1 and 3 is considered to be the horizontal distance from the point where the bar crosses mid-depth to the point where the bar reaches its full depth. Standard hooks may have to be used if there is not enough room at the end of the beam for bar groups to develop their capacity. After cutoff, bends, and/or hooks are determined for each bar group, the bar lengths and bar costs can be calculated.

Calculation of the cost also requires determination of the number of stirrups. Stirrups are placed to satisfy the requirement everywhere that $V_u < \phi(V_c + V_s)$ where $V_s = (\pi d^2_{stir} f_y d)/(2s_{stir})$ is the strength provided by stirrups and s_{stir} is the stirrup spacing. The stirrup spacing may vary along the length of the beam according to the factored shear envelope V_u, and the spacing must be less than either d or $d/2$ according to the provisions of the ACI Specification (1989)

SINGLE BEAM: METHOD OF OPTIMIZATION

The exhaustive search algorithm is implemented as a set of nested loops in a fashion similar to that described earlier for columns. However, the algorithm is more complicated for beams than it was for columns for the following reasons.

1. Beams require a search over reinforcement topologies; and
2. beams require that d_b and n_b be found for each bar group.

The outermost loops search over reasonable ranges of whole centimeter (or inch) values for b (or b_w) and h. Next comes the reinforcement topology search followed by loops over the d_bs for the bar groups. The innermost loops are over the n_bs for the bar groups.

One possible reinforcement topology search is described in this paragraph. The starting topology consists of a single positive bar group extending across the entire beam and negative bar groups—one per negative location—which are cut off on both sides of the location. The search loops through the critical locations. At each location a second bar group is tried which is cut off on both sides of the location. If this produces a better design, a third bar group is tried. The maximum number of bar groups at a single location is three. After looping through the locations, the search goes back and tries to find a better design by combining bar groups at adjacent locations through joining and bending.

It is bad practice to design a single beam with several different bar sizes. In fact, it is recommended that beams have at most two bar sizes, and that if two sizes are used, the difference in size is less than or equal to two bar sizes. The loops over the d_bs for the bar groups should be arranged accordingly. This can be done by having an outer loop on d_b for "Size A" and an inner loop on d_b for "Size B." The outer loop ranges over all reasonable bar sizes (e.g., #5 to #11) whereas the inner loop ranges from one size to two sizes greater than the current d_b for Size A. Inside these two loops are nested loops for the bar groups that loop over two values, either Size A or Size B. It is bad practice to place larger bars over smaller bars when bars are layered in a cross-section. Wherever the topology specifies two or three bar groups at a location, the first bar group, by convention, contains the longest bars and is placed in the lowest layers followed by the second and then the third bar group. Thus the second or third bar group at a location should not be allowed to be Size B if the bar groups below it are Size A.

The exhaustive search can be accelerated in several ways. First, the symmetry of the beam should be exploited. If the span lengths, support conditions, and load envelopes are symmetric about the centerline, bar groups on the left side of the beam should be designed identically to those on the right side. A single bar group can be used to represent identical bars on both sides.

Second, it may be possible to design locations independently depending on the topology. For example, consider a topology in which two adjacent negative locations do not have any common bar groups that are joined. The loops on d_b and n_b for the first negative location need not be nested inside the loops on d_b and n_b for the second negative location. This is possible because all constraints are associated with locations, and the cost is a sum from location to location.

Third, the range of the innermost loops on n_b may be limited. This may be done by quickly estimating the minimum A_s required for strength and the maximum A_s required for the balanced condition at each location, and appropriately restricting the loops on n_b. These estimates may be determined quickly by conservatively assuming d to be h minus cover and stirrup diameter.

SINGLE BEAM: OPTIMUM RESULTS

Several beams were optimized with varying values of loads and unit costs. The material properties were held constant at f_y = 414 MPa (60 ksi) and f'_c = 27.6 MPa (4 ksi). Fig. 4-10 shows the effect of varying the ratio of $\gamma_{steel}C_{steel}/C_{conc}$ on the optimum design of a simple rectangular beam with

$$\begin{aligned} L &= 6.1 \text{ m (20 ft)} \\ w_{dead} &= 29.2 \text{ kN/m (2.0 kip/ft)} \\ w_{live} &= 36.5 \text{ kN/m (2.5 kip/ft)} \\ C_{bar} &= C_{stir} = 0 \end{aligned}$$

Note that the transition zone from the ρ_b-controlled region to the ρ_{min}-controlled region is very wide. Recall that a wide transition zone was also observed for columns with high eccentricity. For typical market prices, the optimum design usually falls in the transition zone, and is thus neither ρ_b-controlled nor ρ_{min}-controlled.

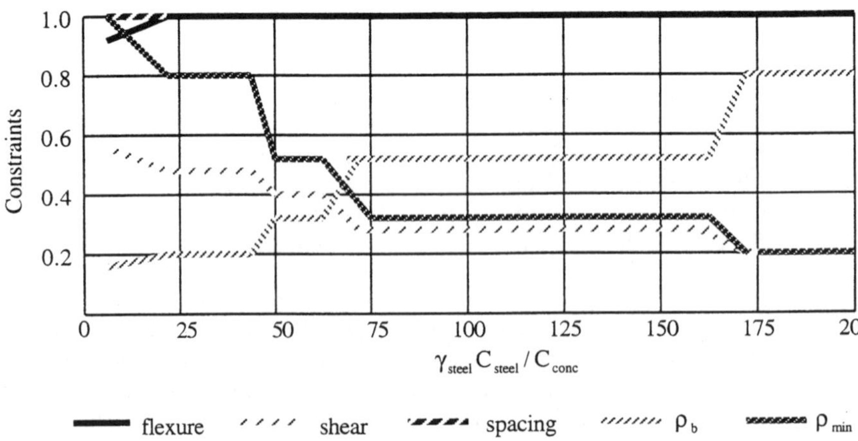

FIG. 4.10. Simple Beam.

The optimum designs of the simple beam tend to have several layers of two bars each so as to have a minimum b. This is because the most flexurally efficient use of concrete is to have large h and small b. Thin deep beams are optimal for the design of a single beam, but are not necessarily optimal for beams in a building frame. This is because increasing the depth of beams may increase the space between floors. This increases the building height and the attendant weight and cost of nonstructural materials.

The optimum designs also tend to have several small bars rather than a few large bars. This is because small bars have shorter development lengths, and several small bars can more tightly fit the moment capacity diagram to the factored moment envelopes.

When the unit placement cost per bar C_{bar} is increased from zero, n_b decreases while d_b increases as expected. As the beam widens to accommodate the larger size bars, there is a corresponding decrease in depth. When the unit placement cost per stirrup C_{stir} is increased from zero, the number of stirrups drops, and stirrup shear strength is replaced by concrete shear strength by increasing h. The increased beam depth allows a decrease in n_b.

It may be argued that many layers of two bars each is impractical because of the placement costs associated with chairs and hangers. When layer placement is added to the cost function, and the unit cost per layer per bar C_{layer} is increased from zero, bars are dropped to a single layer, and d_b is increased in order to reduce n_b and thus keep b at a minimum.

Fig. 4-11 indicates what happens as live load is increased for the simple rectangular beam with

$$
\begin{aligned}
L &= 6.1 \text{ m } (20 \text{ ft}) \\
w_{dead} &= 29.2 \text{ kN/m } (2.0 \text{ kip/ft}) \\
C_{steel} &= \$0.056/\text{N } (\$0.25/\text{lb}) \\
\gamma_{steel} &= 7.0 \text{ KN/m}^3 (490 \text{ lb/ft}^3) \\
C_{conc} &= \$0.000131/\text{cm}^3 \ (\$100/\text{yd}^3) \\
C_{bar} &= C_{stir} = C_{layer} = 0
\end{aligned}
$$

Note that the optimum area of concrete increases steadily whereas the optimum area of steel increases from plateau to plateau. Thus the optimum reinforcement ratio ρ does not stay constant as it did for concentrically loaded columns under increasing load. The same trend occurs if span length is increased rather than live load.

Multispan beams were also optimized, and results were similar to those of the simple beam. For multispan beams, the optimum topologies occasionally had joined bars between two adjacent positive locations or two adjacent negative locations. However, the optimum topologies were never observed to have bent bars between adjacent positive and negative locations regardless of how the parameters were varied. Bent bars may be optimal for slabs, but they are not generally optimal for beams.

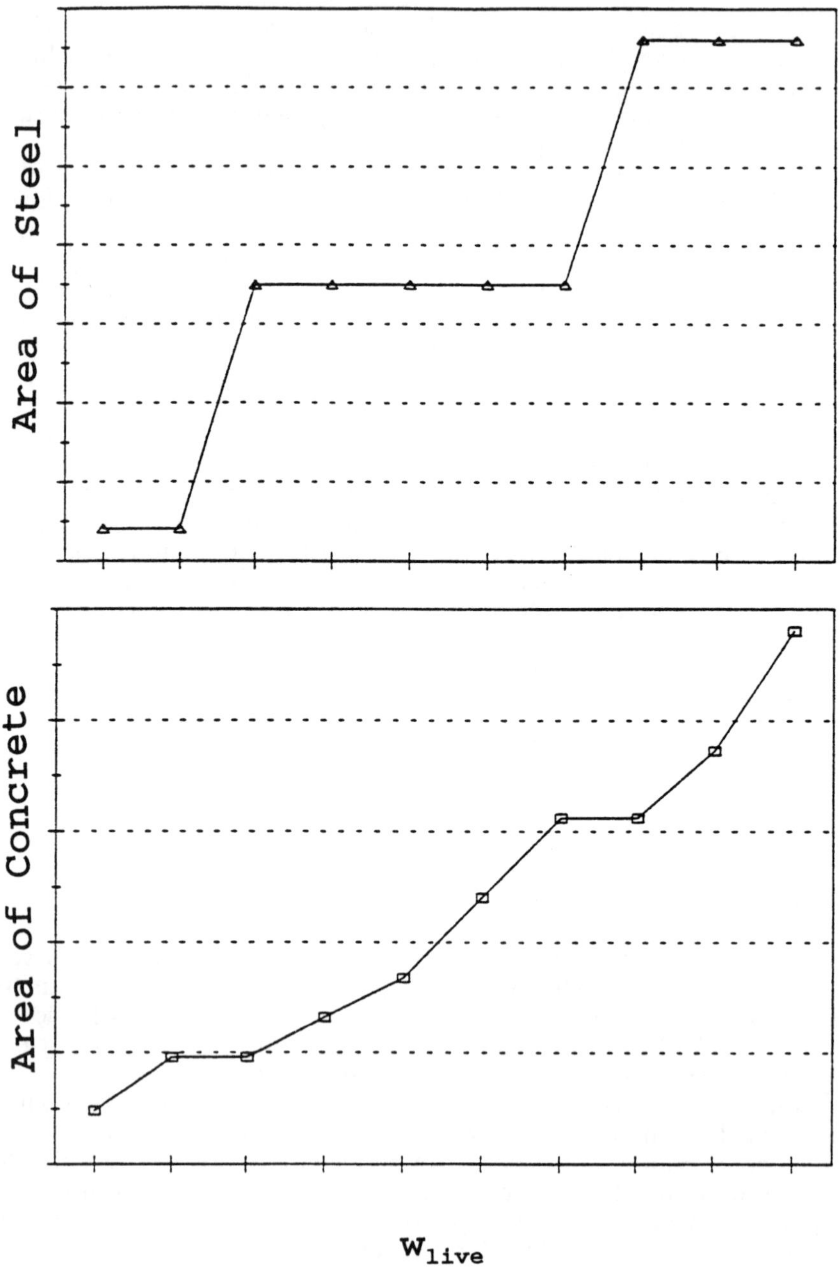

FIG. 4.11. Increasing Live Load.

SUMMARY AND DISCUSSION

The discrete optimization of single reinforced concrete columns and single reinforced concrete beams has been described and exemplified. The exhaustive search method was used. Examination of several optimally designed beams and columns reveals that the strength constraint almost always controls. In the absence of architectural or other limits, concentrically loaded columns are controlled by the ρ_{min} constraint. For beams and eccentrically loaded columns, the optimum design falls in the transition zone between ρ_{min}-controlling and ρ_{max}-controlling (or ρ_b-controlling). The optimum designs for beams tend to be very narrow and deep as one might expect in the absence of architectural or other constraints. Furthermore, several small bars are favored over a few large bars (the opposite is true for columns). Finally, it was noticed that bent bars are rarely optimal for beams.

REFERENCES

ACI (1989). *Building code requirements for reinforced concrete, ACI 318–89*. American Concrete Institute, Detroit, MI.

Bresler, B. (1960). "Design criteria for reinforced columns under axial load and biaxial bending." *ACI J.*, 63(9), 911.

Parme, A. L., Nieves, J. M., and Gouwen, A. (1966). "Capacity of reinforced concrete rectangular columns subjected to biaxial bending." *ACI J.*, 63(9), 911.

Sun, B. - J. and Lu, Z. - T. (1992). "Design aides for reinforced concrete columns." *J. Structural Engr.*, ASCE, 118(11), 2986.

Chapter 5

HOW TO OPTIMIZE PRESTRESSED CONCRETE BEAMS

Three common simplified subproblems of prestressed concrete optimization, which involve only a single structural analysis and can readily be solved by available programs, are presented: the problem of optimizing systems with uniform concrete dimensions that can be stated as an Explicit Optimization Problem (EOP); the Tendon Optimization Problem (TOP) of optimizing only the prestressing force and tendon configuration that can be stated in a linear programming form; and the explicit Concrete Optimization Problem (COP), of optimizing only the concrete dimensions. A general design procedure is then demonstrated, where solutions of the simplified problems can be used to evaluate the final optimum of practical systems. The main advantages of the solution procedure presented are: ease of implementation; useful results are obtained for various suboptimal and feasible designs; and solution of the simple subproblems is efficient.

INTRODUCTION

Considerable work has been done in recent years on optimal design of prestressed concrete structures. Despite the significant progress in this area, there are still some major obstacles in applying optimization methods to structural design:

1. The design variables are each of fundamentally different natures. Typical design variables represent, for example, the length of elements, the concrete dimensions, the prestressing force, or the

Uri Kirsch, Professor of Structural Engineering, Department of Civil Engineering, Technion—Israel Institute of Technology, Haifa, 32000, Israel, email: ⟨kirsch@tx.technion.ac.il⟩.

tendon coordinates. This may produce numerical errors in the solution process.
2. The problem size (number of variables and constraints) is relatively large even for simple structures. Typical constraints are related, for example, to stresses, displacements, and technological requirements. In general, potentially critical constraints at many points in the structure under multiple loading conditions must be considered.
3. Each redesign involves extensive calculations. The constraints are usually implicit functions of the design variables and evaluation of the constraints value for any assumed design requires the solution of simultaneous equations. In addition, calculation of the constraint derivatives with respect to design variables is often needed.
4. The number of redesigns is large. In general, the solution of optimal design problems is iterative and consists of repeated analyses followed by redesign steps. The number of redesigns and repeated analyses is often a function of the problem type and dimensionality.

As a result, the implementation of optimization algorithms is greatly complicated and the total computational effort involved in the solution of practical design problems may become prohibitive. Due to these difficulties, optimization methods are usually not used to solve the complete design problem automatically without human interaction. Rather, optimization procedures are employed to solve specific subproblems and the field of structural optimization is strongly connected with computer-aided design.

In this chapter, three common subproblems of prestressed concrete optimization that involve only a single structural analysis and can readily be solved by available programs, are presented: the problem of optimizing systems with uniform concrete dimensions that can be stated as an Explicit Optimization Problem (EOP); the Tendon Optimization Problem (TOP), of optimizing only the prestressing force and tendon configuration, that can be stated in a Linear Programming (LP) form; and the explicit Concrete Optimization Problem (COP), of optimizing only the concrete dimensions.

It is shown that solutions of these simple subproblems can be used to evaluate the final optimum of practical prestressed concrete systems. Multilevel formulation and iterative solution procedures are employed for this purpose.

GENERAL FORMULATION

Considering a typical indeterminate structure (Fig. 5-1), the problem of optimizing prestressed concrete systems can be stated as the following Nonlinear Programming (NLP) problem: find the vector of

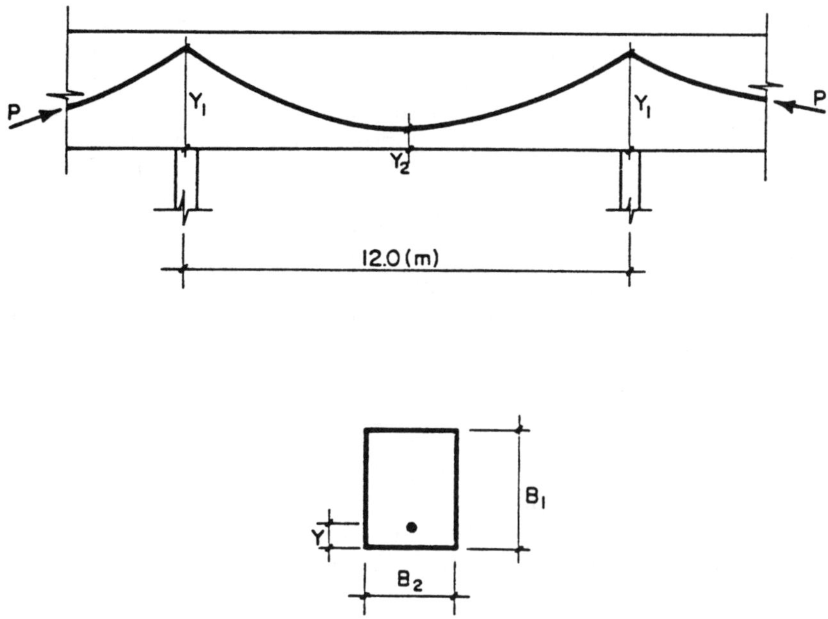

FIG. 5-1. Continuous Beam.

concrete dimensions **B**, the initial prestressing force P, and the vector of tendon coordinates **Y** such that

$$Z = C_c V(\mathbf{B}) + C_p P \rightarrow \min \quad \text{(objective function)} \tag{5-1}$$

$$\sigma^L \leq \sigma(\mathbf{B}, P, \mathbf{Y}) \leq \sigma^U \quad \text{(stress)} \tag{5-2}$$

$$\delta^L \leq \delta(\mathbf{B}, P, \mathbf{Y}) \leq \delta^U \quad \text{(deflection)} \tag{5-3}$$

$$\mathbf{B}^L \leq \mathbf{B} \leq \mathbf{B}^U \quad \text{(concrete dimensions)} \tag{5-4}$$

$$P^L \leq P \leq P^U \quad \text{(prestressing force)} \tag{5-5}$$

$$\mathbf{Y}^L \leq \mathbf{Y} \leq \mathbf{Y}^U \quad \text{(tendon coordinates)} \tag{5-6}$$

where C_c = unit cost of concrete; C_p = unit cost of prestressing force; V = volume of concrete; σ, δ = vectors of normal stresses and deflections, respectively; and L, U = superscripts denoting lower and upper bounds, respectively.

The objective function Z, to be minimized, is usually an explicit nonlinear function of the variables. For simplicity of presentation it is assumed that Z is a linear function of the volume of concrete and the prestressing force. It is instructive to note that although other components (such as the cost of forms) could be considered, their omission does not affect the optimum appreciably.

The constraints (5-2) and (5-3) are related to the structural behavior under service load conditions. Additional constraints related to ultimate bending moments, shearing stresses, and the like may be considered but are omitted here for simplicity of presentation. All potentially critical points and loading conditions should be considered in the behavior constraints. In statically indeterminate structures these constraints are often implicit functions of the design variables.

Equations (5-4)–(5-6) are the "technological constraints." These are upper and lower bounds on the variables due to functional or fabricational requirements (e.g., minimum concrete coverage).

A general solution procedure of the problem (5-1)–(5-6) usually involves the following steps.

1. Assume initial values for the design variables **B**, **Y**, and *P*, satisfying (5-4)–(5-6).
2. Analyze the structure for all loading conditions and evaluate the stresses and deflections.
3. Check the objective function and the behavior constraints. If the solution is not the optimum, modify the design variables and go to step 2.

Steps 2 and 3 are repeated until the optimum design is achieved. The main drawback of this approach is that it is necessary to repeat the analysis of the structure for each trial design. This operation involves much computational effort and, consequently, optimization of large systems might become prohibitive. Another problem is that appropriate computer programs, which combine structural analysis capabilities and optimization algorithms into an integrated system, are often not available or might not be suitable for some applications.

The simplified solution procedures presented subsequently do not involve multiple repeated analyses and, as a result, such integrated systems might not be necessary.

OPTIMIZATION OF SYSTEMS WITH UNIFORM CONCRETE DIMENSIONS

Assuming uniform concrete dimensions, the relative stiffnesses of the members are constant, and the design moments, stresses, and deflections (due to both external loadings and prestressing) are explicit functions of the design variables. It is shown in this chapter that the number of stress and deflection constraints at each point can be reduced from $2 \cdot Q$ (Q = number of loading conditions) to only two. This is a significant reduction in the number of constraints to be considered in the solution process, particularly in problems where the structure is subjected to several loading conditions.

Stress Constraints

The normal stress constraints for a given cross section can be expressed in this case as

$$\sigma_q^L \leq -\frac{\gamma_q P}{A_c(\mathbf{B})} - \frac{M_q + \gamma_q M(P, \mathbf{Y})}{W_t(\mathbf{B})} \leq \sigma_q^U \quad \text{(top fiber)}$$

$$\sigma_q^L \leq -\frac{\gamma_q P}{A_c(\mathbf{B})} + \frac{M_q + \gamma_q M(P, \mathbf{Y})}{W_b(\mathbf{B})} \leq \sigma_q^U \quad \text{(bottom fiber)} \tag{5-7}$$

where q = subscript denoting the loading condition; γ_q = given coefficient for prestressing losses, which is the ratio between the effective prestressing force after losses and the initial prestressing force ($0 < \gamma_q \leq 1$); A_c = concrete cross-sectional area; W_t, W_b = section moduli for top and bottom fibers, respectively; M_q, $M(P, \mathbf{Y})$ = moments due to the qth external loading and prestressing, respectively (moments causing tension in bottom fibers are assumed to be positive).

Define the following limits, which are functions of only **B**.

$$\sigma_{Pt}^L = \max_q \left(\sigma_q^L + \frac{M_q}{W_t(\mathbf{B})}\right) \Big/ \gamma_q \qquad \sigma_{Pt}^U = \min_q \left(\sigma_q^U + \frac{M_q}{W_t(\mathbf{B})}\right) \Big/ \gamma_q$$

$$\sigma_{Pb}^L = \max_q \left(\sigma_q^L - \frac{M_q}{W_b(\mathbf{B})}\right) \Big/ \gamma_q \qquad \sigma_{Pb}^U = \min_q \left(\sigma_q^U - \frac{M_q}{W_b(\mathbf{B})}\right) \Big/ \gamma_q \tag{5-8}$$

Substituting Eqs. (5-8) into Eqs. (5-7), the following reduced set of constraints is obtained

$$\sigma_{Pt}^L \leq -\frac{P}{A_c(\mathbf{B})} - \frac{M(P, \mathbf{Y})}{W_t(\mathbf{B})} \leq \sigma_{Pt}^U$$

$$\sigma_{Pb}^L \leq -\frac{P}{A_c(\mathbf{B})} + \frac{M(P, \mathbf{Y})}{W_b(\mathbf{B})} \leq \sigma_{Pb}^U \tag{5-9}$$

Equation (5-9) shows that for any assumed **B** only two stress constraints should be considered at each point, instead of the original $2 \cdot Q$ constraints in Eq. (5-7).

Deflection Constraints

The deflection constraints can be expressed as

$$\delta_q^L \leq \delta_q(\mathbf{B}) + \gamma_q \delta(\mathbf{B}, P, \mathbf{Y}) \leq \delta_q^U \tag{5-10}$$

in which $\delta_q(\mathbf{B})$ = deflection due to the qth external loading, and $\delta(\mathbf{B}, P, \mathbf{Y})$ = deflection due to the initial prestressing force, both being explicit functions of the variables. It should be noted that, in general, the prestressing force and the external loading act in opposite directions. In addition, prestressing losses due to creep, shrinkage, relaxation, and so on reduce the effect of prestressing. Therefore the latter losses usually increase the total deflection due to both external loading and prestressing.

Defining the limits

$$\begin{aligned} \delta_P^L &= \max_q \, [\delta_q^L - \delta_q(\mathbf{B})]/\gamma_q \\ \delta_P^U &= \min_q \, [\delta_q^U - \delta_q(\mathbf{B})]/\gamma_q \end{aligned} \tag{5-11}$$

which are functions of only **B**, then Eq. (5-10) is reduced to the form

$$\delta_P^L \leq \delta(\mathbf{B}, P, \mathbf{Y}) \leq \delta_P^U \tag{5-12}$$

That is, similar to the stress constraints, the number of deflection constraints considered at a given point is reduced from $2 \cdot Q$ to only two.

Simplified Problem Formulation

Based on Eqs. (5-9) and (5-12), the problem (5-1)–(5-6) can be stated as the following EOP (Kirsch 1993, 1985): find **B**, P, **Y**, such that

$$\begin{aligned}
&Z = C_c A_c(\mathbf{B}) + C_p P \rightarrow \min && \text{(objective function)} \\
&\left.\begin{aligned}
&\sigma_{Pt}^L \leq -\frac{P}{A_c(\mathbf{B})} - \frac{M(P, \mathbf{Y})}{W_t(\mathbf{B})} \leq \sigma_{Pt}^U \\
&\sigma_{Pb}^L \leq -\frac{P}{A_c(\mathbf{B})} + \frac{M(P, \mathbf{Y})}{W_b(\mathbf{B})} \leq \sigma_{Pb}^U
\end{aligned}\right\} && \text{(stress)} \\
&\delta_P^L \leq \delta(\mathbf{B}, P, \mathbf{Y}) \leq \delta_P^U && \text{(deflection)} \\
&\mathbf{B}^L \leq \mathbf{B} \leq \mathbf{B}^U && \text{(concrete dimensions)} \\
&P^L \leq P \leq P^U && \text{(prestressing force)} \\
&\mathbf{Y}^L \leq \mathbf{Y} \leq \mathbf{Y}^U && \text{(tendon coordinates)}
\end{aligned} \tag{5-13}$$

In this formulation the bounds σ_{Pt}^L, σ_{Pt}^U, σ_{Pb}^L, σ_{Pb}^U, δ_P^L, δ_P^U are functions of only **B**. The elimination process [Eqs. (5-8) and (5-11)] may considerably reduce the number of stress and deflection constraints in structures subjected to several loading conditions.

Example 1

Assume the typical continuous beam of a uniform cross section shown in Fig. 5-1. (all dimensions are in tons and meters). A parabolic tendon is assumed with Y_1, Y_2 being the tendon coordinate variables. Two loading conditions have been considered:

Dead load = 2.0 and P ($\gamma_1 = 1$). The allowable stresses are $\sigma^L = -1500$, $\sigma^U = 100$.
Dead + live load = 3.0 and $0.8P$ ($\gamma_2 = 0.8$). The allowable stresses are $\sigma^L = -1200$, $\sigma^U = 0$.

The concrete modulus of elasticity is $3 \cdot 10^6$ and the allowable deflections in the middle of the span are

$$\delta^U = -\delta^L = 0.04$$

The bounds on the concrete dimensions, on the tendon coordinates, and on the prestressing force are

$$B_1^L = 0.5 \qquad B_1^U = 1.0$$
$$B_2^L = 0.1 \qquad B_2^U = 0.2$$
$$Y^L = 0.05 \qquad Y^U = B_1 - 0.05$$
$$P^L = 10 \qquad P^U = 100$$

The resulting EOP of Eqs. (5-13) can be stated as follows: find B_1, B_2, P, Y_1, Y_2, such that the objective function is minimized

$$Z = C_c B_1 B_2 + C_p P \rightarrow \min$$

and the following constraints are satisfied.

Stresses at the top fiber of the support section (stresses in other sections are not considered since they are not potentially active)

$$-1500 - 144/(B_2 B_1^2) \leq - P/(B_1 B_2) - 4P(Y_1 - Y_2)/(B_2 B_1^2) \leq -270/(B_2 B_1^2)$$

Stresses at the bottom fiber of the support section

$$-1500 + 270/(B_2 B_1^2) \leq -P/(B_1 B_2) + 4P(Y_1 - Y_2)/(B_2 B_1^2) \leq 100 + 144/(B_2 B_1^2)$$

Deflection constraints at the mid-span

$$112.5 - 4167 B_2 B_1^3 \leq P(Y_1 - Y_2) \leq 36 + 3333 B_2 B_1^3$$

Technological constraints

$$0.50 \leq B_1 \leq 1.00$$
$$0.10 \leq B_2 \leq 0.20$$
$$10 \leq P \leq 100$$
$$0.05 \leq Y_1 \leq B_1 - 0.05$$
$$0.05 \leq Y_2 \leq B_1 - 0.05$$

A simplified solution procedure of the EOP is demonstrated later.

OPTIMIZATION OF THE PRESTRESSING FORCE AND TENDON CONFIGURATION

We now show that the problem of optimizing only the prestressing force and tendon configuration can be stated in an LP form. This formulation covers a wide range of problems, including:

1. Indeterminate systems with *nonuniform cross sections.*
2. Structures subjected to *multiple-load conditions* and prestressing forces.
3. Structures with a *desired tendon length and prestressing loss diagrams.*
4. Structures with a *desired tendon shape.*

Assuming the equivalent load method, the effect of prestressing is considered as that of an axial force plus a system of equivalent transverse loads, and the general problem of Eqs. (5-1)–(5-6) can be stated for any assumed **B** as the following TOP (Kirsch 1973, 1981): find P and **Y** such that:

$$\begin{aligned}
&Z = P \rightarrow \min && \text{(objective function)}\\
&\sigma_q^L \leq \sigma_q + \gamma_q P\left(a_0 + \sum_{i=1}^{n} a_j Y_j\right) \leq \sigma_q^U && \text{(stress)}\\
&\delta_q^L \leq \delta_q + \gamma_q P\left(b_0 + \sum_{j=1}^{n} b_j Y_j\right) \leq \delta_q^U && \text{(deflection)} \qquad \text{(5-14)}\\
&P^L \leq P \leq P^U && \text{(prestressing force)}\\
&Y_j^L \leq Y_j \leq Y_j^U && \text{(tendon coordinates)}
\end{aligned}$$

where a, b = constant coefficients, and n = the number of tendon coordinates.

Similar to Eqs. (5-8) and (5-11), define the constant limits

$$\sigma_P^L = \max_q (\sigma_q^L - \sigma_q)/\gamma_q \qquad \sigma_P^U = \min_q (\sigma_q^U - \sigma_q)/\gamma_q$$
$$\delta_P^L = \max_q (\delta_q^L - \delta_q)/\gamma_q \qquad \delta_P^U = \min_q (\delta_q^U - \delta_q)/\gamma_q \tag{5-15}$$

That is, only two behavior constraints should be considered at each point instead of the original $2 \cdot Q$ constraints.

To convert the NLP problem (5-14) into an LP, the following transformation of variables is used

$$X_0 = P$$
$$X_j = PY_j \qquad j = 1, \ldots, n \tag{5-16}$$

Substituting Eqs. (5-15) and (5-16) into Eqs. (5-14), the TOP can be stated in the following LP form: find X_j ($j = 0, 1, \ldots, n$) such that

$$Z = X_0 \rightarrow \min$$
$$\sigma_P^L \leq \sum_{j=0}^{n} a_j X_j \leq \sigma_P^U$$
$$\delta_P^L \leq \sum_{j=0}^{n} b_j X_j \leq \delta_P^U \tag{5-17}$$
$$P^L \leq X_0 \leq P^U$$
$$Y_j^L X_0 \leq X_j \leq Y_j^U X_0$$

Linear programming is a fundamental mathematical programming method. The special characteristic of the problem is that all the constraints and the objective function are expressed in linear terms of the variables. The method is widely used and its basics are explained in many textbooks dealing with structural optimization [e.g., Arora (1989) and Kirsch (1993)]. Some of the reasons for its popularity are:

1. The exact global optimum is reached in a finite number of steps and there are no local optima.
2. Computer programs of LP are most efficient. Systems with a relatively large number of variables and constraints can be solved in a reasonable computation time.
3. Many reliable computer programs that can be used as "black boxes" are available. Data preparation is simple and only the nonzero fixed coefficients of the constraints and the objective function must be supplied.

4. Due to the high efficiency and reliability of LP programs they are often used as subroutines in solving nonlinear programming problems.
5. Some practical nonlinear problems are often approximated by a linear formulation and solved by LP algorithms.

Example 2

Consider the beam of Example 1 and assume, for illustrative purposes, the following concrete dimensions and tendon coordinate constraints

$$B_1 = 1.00 \qquad B_2 = 0.1$$

$$(Y_1 - Y_2) \leq 0.90$$

The resulting TOP [Eqs. (5-14)] is to find P, Y_1, and Y_2 such that the objective function is minimized

$$Z = P \rightarrow \min$$

and the following constraints are satisfied.

Stresses at the top fiber of the support section

$$-2940 \leq -10P - 40P(Y_1 - Y_2) \leq -2700$$

Stresses at the bottom fiber of the support section

$$1200 \leq -10P + 40P(Y_1 - Y_2) \leq 2440$$

Deflection constraints at the mid-span

$$-304.2 \leq P(Y_1 - Y_2) \leq 369.3$$

Technological constraints

$$10 \leq P \leq 100$$

$$(Y_1 - Y_2) \leq 0.90$$

Using the transformations

$$P = X_0$$

$$P(Y_1 - Y_2) = X_1$$

the problem becomes an LP, with the design space shown in Fig. 5-2a. The optimal solution is $P = 58.7$, $(Y_1 - Y_2) = 0.90$, $Y_1 = 0.95$, $Y_2 = 0.05$.

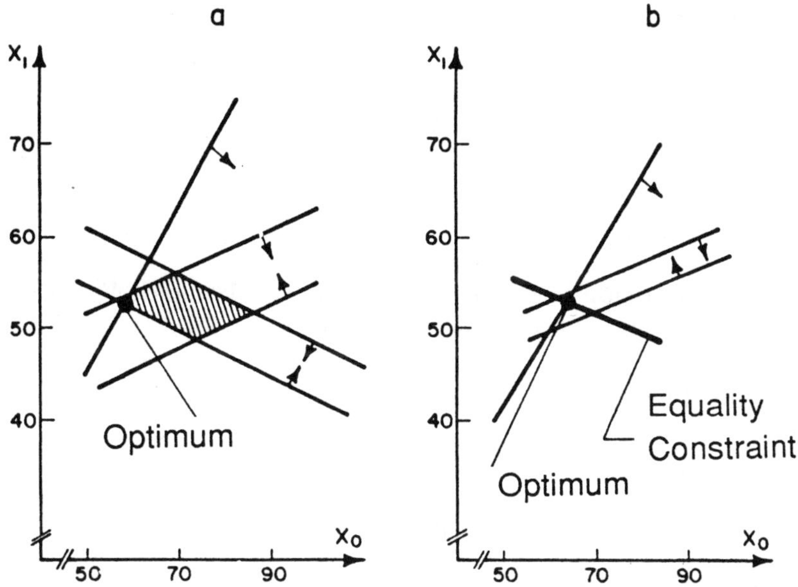

FIG. 5-2. Design Space, TOP: (a) $B_1 = 1.00$, $B_2 = 0.1$; (b) $B_1 = 0.92$, $B_2 = 0.1$.

To illustrate the effect of the concrete dimensions on the feasible region, assume $B_1 = 0.92$, $B_2 = 0.1$. The resulting TOP solution is $P = 64.5$, $(Y_1 - Y_2) = 0.82$, $Y_1 = 0.87$, $Y_2 = 0.05$. The upper and lower limits on stress constraints become equal, two of the inequality constraints become an equality, and the feasible region is reduced to the line segment shown in Fig. 5-2b. Further reduction in B_1 leads to a problem with no feasible solutions.

OPTIMIZATION OF THE CONCRETE DIMENSIONS

Stress Constraints

In cases where the object is to optimize the concrete dimension **B**, it is necessary to introduce constraints on **B** that help prevent infeasible solution of the TOP (5-17). From the stress constraints of Eq. (5-9), it is clear that necessary conditions for a feasible solution of the TOP are

$$\sigma_{Pt}^{L} \leq \sigma_{Pt}^{U} \qquad \text{(5-18)}$$
$$\sigma_{Pb}^{L} \leq \sigma_{Pb}^{U}$$

Substituting Eqs. (5-8) into Eqs. (5-18), denoting q = 1, 2, 3, 4 for the four terms in Eqs. (5-8) and rearranging yields

$$\begin{aligned} W_t(\mathbf{B}) &\geq \frac{\gamma_2 M_1 - \gamma_1 M_2}{\gamma_1 \sigma_2^U - \gamma_2 \sigma_1^L} \equiv W_t^L \\ W_b(\mathbf{B}) &\geq \frac{\gamma_4 M_3 - \gamma_3 M_4}{\gamma_3 \sigma_4^U - \gamma_4 \sigma_3^L} \equiv W_b^L \end{aligned} \tag{5-19}$$

The right-hand side terms in these equations (W_t^L and W_b^L) are constant lower bounds on the section moduli.

Deflection Constraints

From the deflection constraints of Eq. (5-12), a necessary condition for a feasible solution of the TOP (5-17) is

$$\delta_p^L \leq \delta_p^U \tag{5-20}$$

Substituting Eq. (5-11) into Eq. (5-20), denoting q = 5, 6 for the two terms in Eqs. (5-11) and rearranging yields

$$I(\mathbf{B}) \geq \frac{\gamma_5 d_6 - \gamma_6 d_5}{\gamma_5 \delta_6^U - \gamma_6 \delta_5^L} \equiv I^L \tag{5-21}$$

where $I(\mathbf{B})$ = moment of inertia of the concrete cross-section; d_5, d_6 = the deflections $\delta_5(\mathbf{B})$, $\delta_6(\mathbf{B})$, respectively, assuming $I(\mathbf{B})$ = 1.0; and I^L = constant lower bound on $I(\mathbf{B})$.

Simplified Formulation

Based on the relations (5-19) and (5-21), the problem of minimizing the concrete dimensions can be formulated as the following COP (Kirsch 1985, 1993): find **B** such that

$$\begin{aligned} &Z = V(\mathbf{B}) \rightarrow \min && \text{(objective function)} \\ &\left.\begin{aligned} W_t(\mathbf{B}) \geq W_t^L \\ W_b(\mathbf{B}) \geq W_b^L \end{aligned}\right\} && \text{(stress)} \\ &I(\mathbf{B}) \geq I^L && \text{(deflection)} \\ &\mathbf{B}^L \leq \mathbf{B} \leq \mathbf{B}^U && \text{(concrete dimensions)} \end{aligned} \tag{5-22}$$

The numbers of variables and constraints in this explicit problem are relatively small. It should be noted that, although the stress and deflection constraints (5-19) and (5-21) are necessary conditions for a feasible solution of the TOP (5-17), solution of the problem (5-22) may lead to an infeasible design due to the nature of the tendon coordinate constraints.

Example 3

Assuming the beam of Example 1, the COP [Eqs. (5-22)] can be stated as follows: find B_1 and B_2 such that the objective function is minimized

$$Z = B_1 B_2 \rightarrow \min$$

and the following constraints are satisfied.

Stresses at the top fiber of the support section

$$B_2 B_1^2 \geq 0.084$$

Stresses at the bottom fiber of the support section

$$B_2 B_1^2 \geq 0.079$$

Deflection constraints at the mid-span

$$B_2 B_1^3 \geq 0.010$$

Technological constraints

$$0.50 \leq B_1 \leq 1.00$$

$$0.10 \leq B_2 \leq 0.20$$

The two-dimensional design space and the objective function contours are shown in Fig. 5-3. The optimal solution is $\mathbf{B}_{\min}^T = \{0.92, 0.10\}$. Considering the total cost of concrete and prestressing, it is shown subsequently that the optimal solution is at some point on the line segment between the two points $\mathbf{B}_{\min}$ and $\mathbf{B}_{1\max}$.

GENERAL DESIGN PROCEDURE

It has been noted earlier that, due to some obstacles involved in the optimization of complex prestressed concrete systems, it is usually not practical to solve the complete problem automatically. Rather, optimization procedures can be used to solve simplified subproblems such as the EOP, the TOP, and the COP discussed in this chapter. Solutions of these subproblems are then used to evaluate the final optimum of practical prestressed systems.

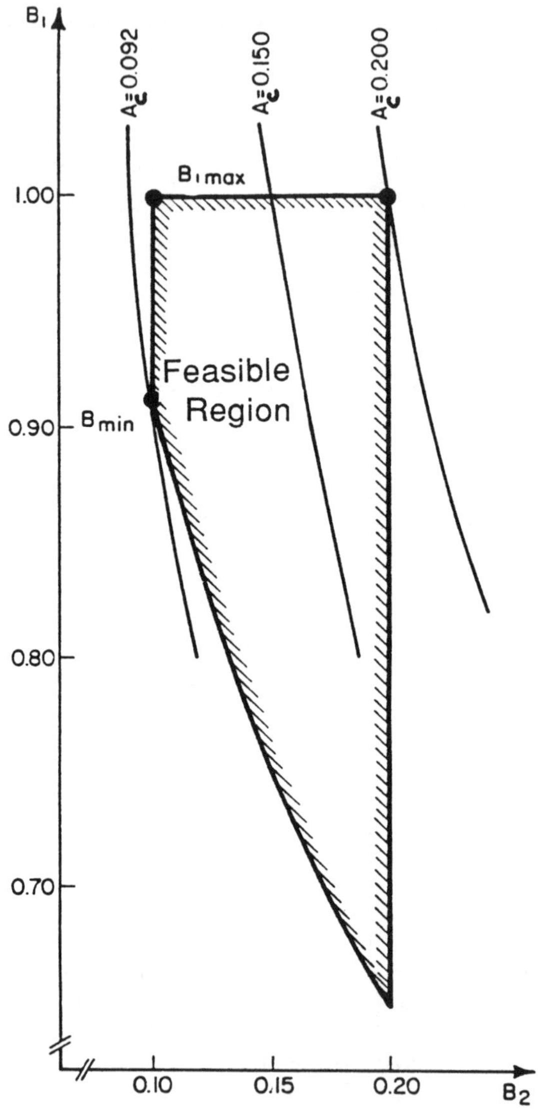

FIG. 5-3. Design Space, COP.

Based on the various problem formulations, a general design procedure has been established (Kirsch 1985, 1993). Considering the EOP [Eqs. (5-13)], the main steps of the solution process are as follows.

1. The COP [Eqs. (5-22)] is first solved. Since all the constraints are explicit functions of the variables, the analysis of the structure

need not be repeated during the solution process. In addition, the numbers of variables and constraints are relatively small, and the problem can readily be solved by available NLP methods.

2. The TOP [Eqs. (5-17)] is solved for the current concrete dimensions. Since the problem is an LP, it can be solved effectively by available programs, without repeating the analysis during the solution process. If the solution is feasible and the convergence criteria for **B** are satisfied, then the solution is optimal.
3. Modify the concrete dimensions according to some desired redesign rules and go to Step 2. It is instructive to note that changes in the depth of the cross-section are usually much more effective than changes in the width. This can greatly simplify the search for optimal concrete dimensions.

Example 4

To illustrate the design procedure, consider again the EOP of Example 1. The steps of the solution process are as follows.

1. Solving the COP, the optimal solution is (see Example 3) $\mathbf{B}_{\min}^T = \{0.92, 0.10\}$.
2. Solving the TOP for the given concrete dimensions, the optimal solution is $P = 64.5$, $Y_1 = 0.87$, $Y_2 = 0.05$ (see Example 2). It is instructive to note that the optimal solutions in Steps 1 and 2 are independent of the cost coefficients C_c and C_p.
3. It has been shown (Kirsch 1985) that in this example, depending on the ratio C_c/C_p, the final optimal solutions are as follows.

C_c/C_p	Optimal **B**	Optimal P
0 − 430	$\mathbf{B}_{1\max}^T = \{1.00, 0.1\}$	58.7
430 − 894	Between $\mathbf{B}_{1\max} - \mathbf{B}_{\min}$	58.7–64.5
≥894	$\mathbf{B}_{\min}^T = \{0.92, 0.10\}$	64.5

That is, the optimal solution is at $\mathbf{B}_{\min}$ (for high C_c/C_p ratios), at $\mathbf{B}_{1\max}$ (maximum depth and minimum width, for low C_c/C_p ratios), or on the line segment between the two points. The objective function contours for some typical C_c/C_p ratios are shown in Fig. 5-4, and design spaces of the TOP for $\mathbf{B}_{1\max}^T = \{1.00, 0.1\}$ and $\mathbf{B}_{\min}^T = \{0.92, 0.10\}$ are shown in Figs. 5-2a and 5-2b, respectively.

In cases where some practical constraints not considered in the example (such as ultimate moment requirements) are violated, the optimal solution should be modified.

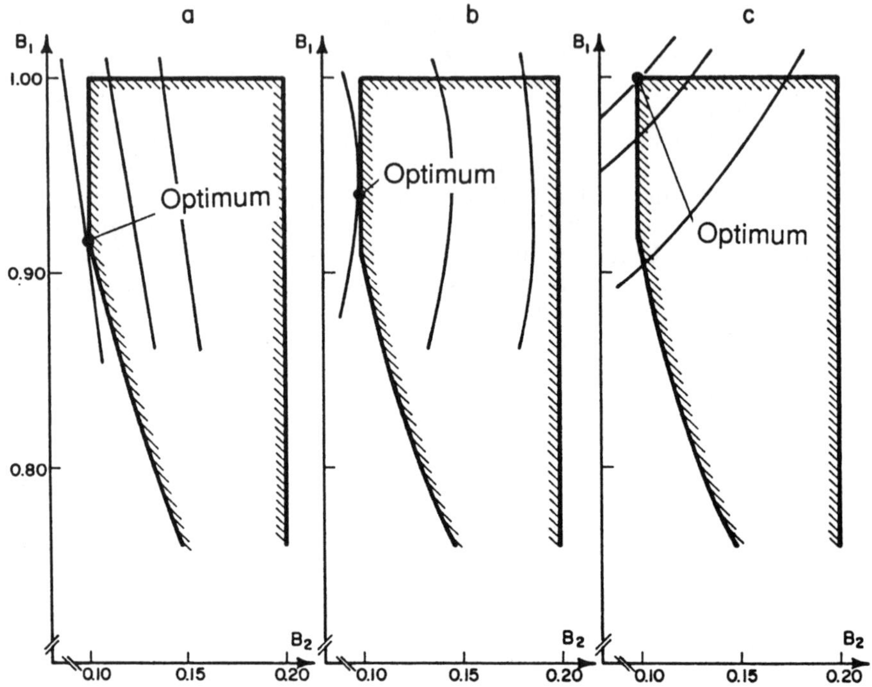

FIG. 5-4. Objective Function Contours: (a) $C_c/C_p = 1000$; (b) $C_c/C_p = 700$; (c) $C_c/C_p = 100$.

SUMMARY AND DISCUSSION

Despite the significant work that has been done in recent years on optimization of prestressed concrete systems, there are still some major obstacles in applying optimization methods to structural design. In a typical problem, the design variables are each of fundamentally different natures, the problem size is relatively large, each redesign involves extensive calculations, and the number of redesigns is large. In general, the solution of optimal design problems is iterative and consists of repeated analyses followed by redesign steps. The number of redesigns and repeated analyses is often a function of the problem type and dimensionality. As a result, the total computational effort involved in the solution of practical design problems may become prohibitive.

Optimization procedures are usually employed to solve specific subproblems. In this chapter, three simplified subproblems of prestressed concrete optimization, that involve only a single structural analysis and can readily be solved by available programs, have been presented. A general design procedure has been demonstrated, where solutions of the simplified problems can be used to evaluate the final optimum of practical systems. The main advantages of the solution procedure presented are:

1. Ease of implementation. The general problem is divided into some simplified subproblems that involve only a single structural analysis and can be solved readily by available programs.
2. Useful results are obtained for various suboptimal and feasible designs. Solutions of the EOP, the TOP, and the COP provide practical bounds on the minimum concrete dimensions and prestressing force.
3. Solution of the simple subproblems is most efficient.

APPENDIX 5-A. NOTATION

A_c = cross-sectional area of concrete
a = constant coefficient
$\mathbf{B}$ = vector of concrete dimensions
b = constant coefficient
C_c = unit cost of concrete
C_p = unit cost of prestressing force
I = moment of inertia
j = subscript of LP variables
L = superscript denoting lower bound
M = bending moment
n = number of tendon coordinates
P = initial prestressing force
Q = number of loading conditions
q = subscript denoting the loading condition
U = superscript denoting upper bound
V = volume of concrete
W_t, W_b = modulus of section for top and bottom fibers, respectively
X = variables of linear programming
$\mathbf{Y}$ = vector of tendon coordinates
Z = objective function
γ = coefficient for prestressing losses
δ = vector of deflections
δ_P^L, δ_P^U = deflections defined by Eqs. (5-11) or Eqs. (5-15)
σ = vector of normal stresses
σ_{Pt}^L, σ_{Pt}^U, σ_{Pb}^L, σ_{Pb}^U, σ_{Pt}^L, σ_{Pt}^U, σ_P^L, σ_P^U = stresses defined by Eqs. (5-8) and (5-11).

REFERENCES

Arora, J. S. (1989). *Introduction to optimum design*. McGraw-Hill, Singapore.

Kirsch, U. (1973). "Optimized prestressing by linear programming." *International J. for Numerical Methods in Engng.*, 7, 2, 125–136.

Kirsch, U. (1981). *Optimum structural design*. McGraw-Hill, New York.

Kirsch, U. (1985). "A bounding procedure for synthesis of prestressed systems." *Computers and Structures*, 20, 5, 885–895.

Kirsch, U. (1993). *Structural optimization—Fundamentals and applications*. Springer-Verlag, Berlin.

Chapter 6

SENSITIVITY ANALYSIS FOR STRUCTURAL DESIGN

This chapter presents the details of how to make explicit use of structural analysis in the structural design process in order that the overall synthesis process may more effectively establish an economical design. Specifically, the sensitivity analysis of skeletal frameworks is introduced to find the first-order sensitivities of linear-elastic displacements and stresses to changes in the cross-section area or moment of inertia of the members of a structure under static loads. The calculation of these response sensitivities involves only a nominal amount of computational effort beyond that required to conduct the baseline analysis to determine the magnitudes of the displacements and stresses for the structure. It is shown that sensitivity analysis results provide explicit information regarding the way to modify a structure so as to ensure satisfaction of governing design conditions, as well as how much to modify the structure so as to achieve an economical design.

INTRODUCTION

The primary role of structural analysis is to serve as the underlying basis for carrying out structural design computations to proportion a structure to withstand applied loading. In this chapter, we consider how to make explicit use of structural analysis in the structural design process with specific reference to the sensitivity analysis of skeletal frameworks. We see that the results of sensitivity analysis provide us with clear insights into the behavior of a structure and the way to economically modify it to satisfy design requirements. The concepts

---

Donald E. Grierson, Civil Engineering, University of Waterloo, Ontario, Canada N2L 3G1, email: <grierson@sail.uwaterloo.ca>.

are then subsequently extended in a later chapter to the computer-automated design optimization of frameworks.

We recall the fundamental aspect of much of our structural analysis considerations in practice, specifically, to deal with one-time analysis of a given structure under applied loading. (We remind ourselves that the given structure has been predetermined through a process of design involving a sequence of designer decisions and modifications, and, moreover, that further modifications may be necessary before the structure has evolved to an acceptable state). Having the results of the analysis, we can then carry out an assessment to determine the merit of the current state of the structure. Our primary concerns are for stiffness and strength, which in general terms relate to the strain and stress levels developed in the structure, respectively. If we find these levels to be acceptable, this aspect of the structural design process is complete; if not, we must modify the structure so as to bring the strains and stresses to within their acceptable limits.

To illustrate the foregoing, consider the simple pin-jointed truss loaded as shown in Fig. 6-1. We analyze this truss using the conventional Displacement (Stiffness) Method of indeterminate analysis. Through reference to the $n = 5$ degrees of freedom for nodal displacement $u_1, u_2, \ldots, u_5$, we first establish the global-axis stiffness matrices K_i ($i = 1, 2, \ldots, m$) for the $m = 6$ truss members. Then the individual K_i matrices are summed together to form the stiffness matrix for the assembled truss

$$K = \sum_{i=1}^{m} K_i \tag{6-1}$$

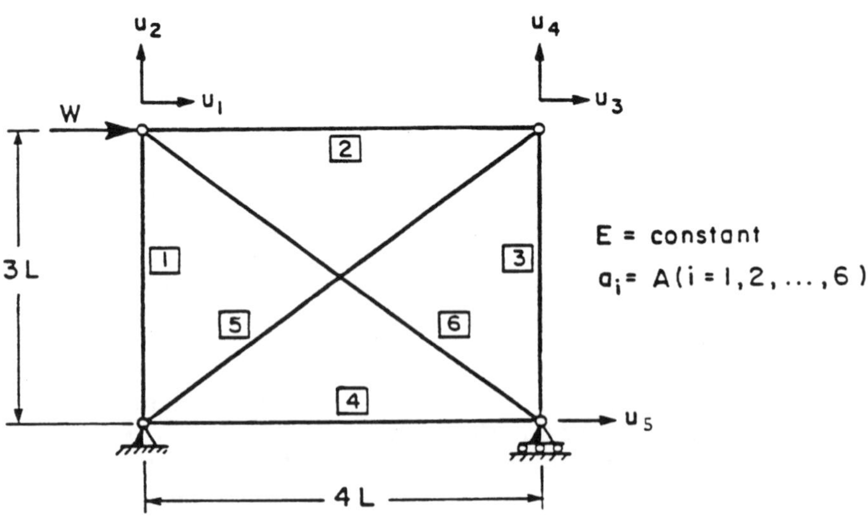

FIG. 6-1. Pin-Jointed Truss with Actual Loading.

For the vector P of applied loads, we then formulate the independent equilibrium conditions for the truss

$$Ku = P \tag{6-2}$$

and solve them to find the vector of $n = 5$ nodal displacements

$$u = K^{-1} P \tag{6-3}$$

that is,

$$\begin{bmatrix} u_1 \\ u_2 \\ u_3 \\ u_4 \\ u_5 \end{bmatrix} = \frac{L}{EA} \begin{bmatrix} 6.750 & 1.125 & 4.750 & -1.125 & 2.000 \\ 1.125 & 2.813 & 0.792 & -0.187 & -0.333 \\ 4.750 & 0.792 & 6.157 & -1.458 & 1.407 \\ -1.125 & -0.187 & -1.458 & 2.813 & -0.333 \\ 2.000 & -0.333 & 1.407 & -0.333 & 3.407 \end{bmatrix} \begin{bmatrix} W \\ 0 \\ 0 \\ 0 \\ 0 \end{bmatrix}$$

which gives

$$u^T = \frac{WL}{EA} [6.750 \quad 1.125 \quad 4.750 \quad -1.125 \quad 2.000]$$

where E is Young's material modulus and W, L, and A are applied load, member length, and cross-section area parameters, respectively. Positive displacements take place in the global-axis directions indicated in Fig. 6-1. We then identify the topological or compatibility matrix C for the truss and find the vector q of $m = 6$ axial deformations for the truss members from

$$q = Cu \tag{6-4}$$

that is,

$$\begin{bmatrix} q_1 \\ q_2 \\ q_3 \\ q_4 \\ q_5 \\ q_6 \end{bmatrix} = \begin{bmatrix} 0 & 1 & 0 & 0 & 0 \\ -1 & 0 & 1 & 0 & 0 \\ 0 & 0 & 0 & 1 & 0 \\ 0 & 0 & 0 & 0 & 1 \\ 0 & 0 & 0.8 & 0.6 & 0 \\ -0.8 & 0.6 & 0 & 0 & 0.8 \end{bmatrix} \cdot \frac{WL}{EA} \begin{bmatrix} 6.750 \\ 1.125 \\ 4.750 \\ -1.125 \\ 2.000 \end{bmatrix}$$

which gives

$$q^T = \frac{WL}{EA} [1.125 \quad -2.000 \quad -1.125 \quad 2.000 \quad 3.125 \quad -3.125]$$

where a positive axial deformation refers to tensile elongation. To complete the analysis, we find the axial stress σ_i for each truss member i from

$$\sigma_i = \frac{q_i E}{L_i} \qquad (i = 1, 2, \ldots, m) \tag{6-5}$$

that is, having the member deformations q_i $(i = 1, 2, \ldots, m)$ from Eq. (6-4), and the member lengths L_i $(i = 1, 2, \ldots, m)$ from Fig. 6-1, the vector of $m = 6$ axial stresses for the truss members is found through Eq. (6-5) to be

$$\sigma^T = \frac{W}{A}[0.375 \quad -0.500 \quad -0.375 \quad 0.500 \quad 0.625 \quad -0.625]$$

where a positive axial stress is tensile.

The translations of the joints and the axial stresses of the members for a pin-jointed truss characterize its stiffness and strength properties, respectively. In other words, if the five nodal displacements found in the foregoing are all within their acceptable limits, then the example truss has adequate stiffness for the applied loading. Similarly, if the axial stresses found for the six members are all within their acceptable limits, then the truss has adequate strength for the applied loading. On the other hand, if any of the displacements and/or stresses exceed their allowable limits under the applied loading, then we must modify the truss in a way such that the corresponding stiffness and strength requirements for the design are satisfied for the applied loading.

Before proceeding to the question of design modifications, consider as another example the flexural frame in Fig. 6-2. Applying the Displacement Method of indeterminate analysis, the three degrees-of-freedom nodal displacement vector is found through Eq. (6-3) as

$$\begin{bmatrix} u_1 \\ u_2 \\ u_3 \end{bmatrix} = \frac{L}{EI}\begin{bmatrix} 0.331 & -0.057 & 0.070L \\ -0.057 & 0.361 & 0.259L \\ 0.070L & 0.259L & 0.738L^2 \end{bmatrix}\begin{bmatrix} 0 \\ 0 \\ W \end{bmatrix}$$

which gives

$$u^T = \frac{WL^2}{EI}[0.070 \quad 0.259 \quad 0.738L]$$

where I is the cross-section moment of inertia parameter. Positive displacements take place in the global-axis directions indicated in Fig. 6-2. We then establish the transformation matrix M^1 for the frame and find the vector M of $2m = 6$ bending moments at the end-sections of

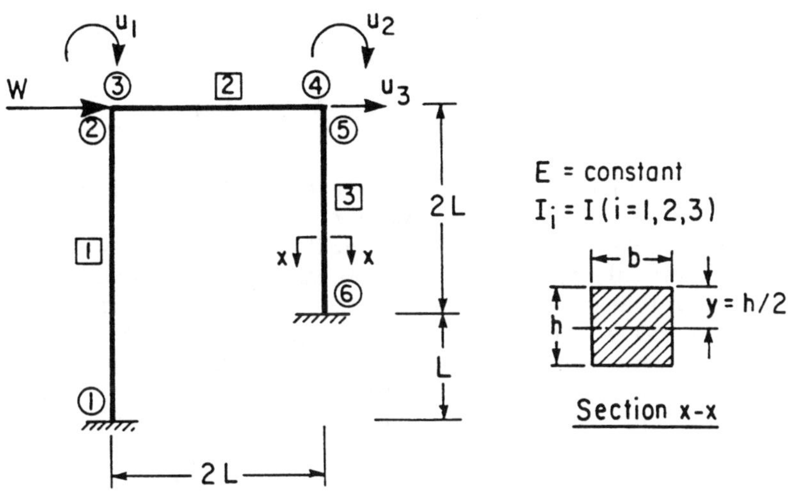

FIG. 6-2. Flexural Frame with Actual Loading.

the $m = 3$ frame members from

$$M = M^0 + M^1 u \tag{6-6}$$

where, M^0 is the vector of member-end moments due to the applied loading when all nodal displacements are constrained to be zero (i.e., fixed-end moments), and each column of M^1 is the vector of member-end moments due to unit value of a particular nodal displacement being alone imposed on the structure. For the frame in Fig. 6-2, Eq. (6-6) is

$$\begin{bmatrix} M_1 \\ M_2 \\ M_3 \\ M_4 \\ M_5 \\ M_6 \end{bmatrix} = \begin{bmatrix} 0 \\ 0 \\ 0 \\ 0 \\ 0 \\ 0 \end{bmatrix} + \frac{EI}{L} \begin{bmatrix} 2/3 & 0 & -2/3L \\ 4/3 & 0 & -2/3L \\ 2 & 1 & 0 \\ 1 & 2 & 0 \\ 0 & 2 & -3/2L \\ 0 & 1 & -3/2L \end{bmatrix} \cdot \frac{WL^2}{EI} \begin{bmatrix} 0.070 \\ 0.259 \\ 0.738L \end{bmatrix}$$

which gives

$$M^T = WL\,[-0.445 \quad -0.399 \quad 0.399 \quad 0.588 \quad -0.588 \quad -0.848]$$

where positive bending moments act clockwise at member ends. To complete the analysis, we find the bending stress σ_k at each end-section k of the m frame members from

$$\sigma_k = \frac{M_k \cdot y_k}{I_k}, \qquad (k = 1, 2, \ldots, 2m) \tag{6-7}$$

where, so as to identify the greatest stress, y_k is taken as the distance from the neutral axis to the extreme fibers of the cross-section. (For a symmetrical cross-section, y_k is one-half the depth of the section. For an unsymmetrical cross-section, y_k is measured to the extreme edge of the section that experiences tensile or compressive stress, depending on the concerns of the design). All members of the frame in Fig. 6-2 have a common cross-section that is symmetrical about its neutral axis and, as such, the distances from the neutral axis to the extreme tensile and compressive fibers for each cross-section are both equal to one-half the section depth $y_k = y$. Having bending moments M from Eq. (6-6) and the member cross-section properties from Fig. 6-2, the vector of $2m = 6$ extreme-fiber bending stresses at the end-sections of the $m = 3$ frame members is found from Eq. (6-7) to be

$$\sigma^T = \frac{WLy}{I}\,[-0.445 \quad -0.399 \quad 0.399 \quad 0.588 \quad -0.588 \quad -0.848]$$

where each stress can be viewed as being either tensile or compressive; the positive or negative sign simply indicates whether the corresponding member-end bending moment acts clockwise or counterclockwise, respectively.

Similar to our previous discussion for a truss, we note that the rotations and translations of the joints and the bending stresses of the members for a flexural frame characterize its stiffness and strength properties, respectively.

There are a number of ways that a structure can be modified in order to effect desirable changes in its response to the applied loading. We may modify the topological layout of the structure by changing the length, orientation, or even the number of members (e.g., if other design considerations permit, we may reduce the clear span of the truss in Fig. 6-1 by shortening the length of members 2 and 4, thereby also changing the length and orientation of members 5 and 6). Or, we may modify the support and joint connection details, which in some instances may result in changing the type of structure (e.g., we may change from pinned to rigid joint connections so as to convert the truss in Fig. 6-1 into a frame). We may even modify the fundamental stiffness and strength properties for the members of the structure by changing the type or grade of material from which they are fabricated (e.g., from aluminum to steel, or from mild to high-strength steel, etc.).

In most cases, the foregoing options for modifying the structure are exercised by the designer quite early on in the design process. Later, after the structure has evolved to a state of fixed topology and type and material grade, any further modifications mainly involve changing cross-sections of the members. For example, to modify the applied loading response of a truss we change the cross-section area of the members, whereas for a flexural frame we change the cross-section moment of inertia of the members. Such modifications often involve major computational effort before a satisfactory structural design is found.

At this point we note the limited value of conventional structural analysis results to the design process. Namely, although such results are the basis for checking the stiffness and strength conditions for a structure, they otherwise do not provide us with any explicit information as to how to modify a structure should we find one or more of these conditions to be violated. For example, suppose that the displacement $u_3 = 4.750\ WL/EA$ found in the foregoing for the truss in Fig. 6-1 exceeds the allowable limit for lateral translation. Also suppose that the stress $\sigma_5 = 0.625\ W/A$ found for member 5 exceeds the allowable limit for axial tensile stress. Lacking further information, we may make an educated guess that increasing the cross-section area of member 5 will have the desirable effect of decreasing the magnitudes of both u_3 and σ_5. We may perhaps draw a similar conclusion regarding member 6. But, what is the relative influence of members 5 and 6 on u_3 and σ_5? Or, by how much do we modify the cross-section area of any member in order to effect the required changes in the magnitudes of u_3 and σ_5? A related question concerning the economy of the truss design should also be asked. Namely, how do we modify the truss so as to correctly change u_3 and σ_5 while requiring the least amount of additional material for the fabrication of the members? We could also ask similar questions for the flexural frame in Fig. 6-2.

The answers to the fundamental design questions posed in the foregoing are not at all evident from the analysis results determined thus far, even though they refer to very simple structures. For the complicated structures that are generally encountered in everyday engineering practice, the answers to similar questions of design are even more difficult to comprehend. In fact, in the absence of further information, the limited nature of the analysis results that we have determined thus far poses a double dilemma for the designer. Namely, not only must the required modifications to the structure be guessed at, but also a complete re-analysis of the modified structure must be conducted before the validity of the modifications can be assessed. This trial-and-error process of design is expensive and, as shown in the remainder of this chapter, largely unnecessary.

The missing information alluded to in the foregoing concerns the response sensitivities for a structure; that is, a measure of the way that each strain or stress-related response of a structure under applied loading will vary with changes in the parameters that define the structure. Although it is possible to calculate a variety of different response sensitivities for a structure, we confine our attention here to the calculation of the sensitivities of linear-elastic displacements and stresses under static loads, which have direct relation to a large part of everyday structural engineering practice. For example, for the truss in Fig. 6-1 we would like to know how the nodal displacement u_3 will vary for unit change in the cross-section area of any particular member, or how the axial stress σ_5 for member 5 will vary for similar unit change in cross-section area. Or, for the frame in Fig. 6-2, how the lateral displacement u_3 and member bending stress σ_6 will vary for unit change in the cross-section moment of inertia of any member. Evidently, such response sensitivity information would allow us to readily answer the design questions posed in the foregoing.

We note that each response sensitivity is in reality a gradient or, more precisely, the first derivative of the response taken with respect to the cross-section area or moment of inertia of a particular member. Although it is possible to calculate second and higher-order derivatives, such first-order response information is generally all that we require to effectively carry out design computations for most structures encountered in engineering practice.

The next section of this chapter establishes the first-order sensitivities of linear-elastic displacements and stresses to changes in the cross-section area or moment of inertia of the members of framework structures under static loads. We see that the calculation of these response sensitivities involves only a nominal amount of computational effort beyond that required to conduct the baseline analysis to determine the magnitudes of the displacements and stresses for the structure. Moreover, we note that not only do sensitivity analysis results provide us with explicit information regarding the way that we should modify a structure so as to ensure its design conditions, but that they also allow us to estimate closely just how much we should modify the structure, and how to achieve an economical design.

Sensitivity analysis is conducted in the following using the, so-called, virtual-load technique. Finite-difference and pseudoload techniques for doing sensitivity analysis are discussed in Chapter 8, "How to Optimize Structural Steel Frameworks."

SENSITIVITY ANALYSIS

Displacement Sensitivity Analysis

Consider a particular displacement δ_j for a given framework structure. The displacement δ_j is related to the vector of nodal displacements u of the structure as

$$\delta_j = b_j^T u \tag{6-8}$$

where the entries of the vector b_j depend on the nature of the displacement δ_j. For example, if δ_j is the relative vertical translation of the two upper joints of the truss in Fig. 6-1, then $b_j^T = [\,0\ 1\ 0\ -1\ 0]$ and Eq. (6-8) becomes

$$\delta_j = [\,0\ 1\ 0\ -1\ 0]\cdot\begin{bmatrix} u_1 \\ u_2 \\ u_3 \\ u_4 \\ u_5 \end{bmatrix} = u_2 - u_4$$

Or, if δ_j is the horizontal translation of the frame in Fig. 6-2 then $b_j^T = [0\ 0\ 1]$ and Eq. (6-8) reduces to $\delta_j = u_3$.

Suppose now that Eqs. (6-1) to (6-3) are applied for the analysis of the given structure, and it is found from the results that the current magnitude of the displacement δ_j exceeds its allowable limit for feasible design. We must therefore modify the cross-sections of the framework members such that, under the same applied loading, the magnitude of δ_j becomes less than or equal to its allowable displacement limit. For axial truss members we modify cross-section areas, and for flexural members we modify cross-section moments of inertia.

In the following, we first calculate the first-order sensitivity of δ_j to change in the cross-section area a_i of any one member i for a truss structure (we then extend the result to a flexural structure). Taking the first-derivative of Eq. (6-8) with respect to a_i, we have

$$\frac{\partial \delta_j}{\partial a_i} = b_j^T \frac{\partial u}{\partial a_i} + u^T \frac{\partial b_j}{\partial a_i} = b_j^T \frac{\partial u}{\partial a_i} \tag{6-8a}$$

which recognizes that the vector b_j is invariant with changes in member cross-section areas; that is, $\partial b_j / \partial a_i = 0$. And taking the first-derivative of Eq. (6-2) with respect to a_i, we have

$$K \frac{\partial u}{\partial a_i} + \frac{\partial K}{\partial a_i} u = \frac{\partial P}{\partial a_i} = 0 \tag{6-8b}$$

where it is assumed that the vector of applied loads $\mathbf{P}$ is also invariant with changes in member cross-section areas; that is, $\partial \mathbf{P} / \partial a_i = 0$ (this implies that the selfweight of the members is taken to be a fixed part of the applied loading, which is often usual practice for structural design computations). Equation (6-8b) gives

$$\frac{\partial u}{\partial a_i} = -K^{-1} \frac{\partial K}{\partial a_i} u \tag{6-8c}$$

and therefore Eq. (6-8a) becomes

$$\frac{\partial \delta_j}{\partial a_i} = -b_j^T K^{-1} \frac{\partial K}{\partial a_i} u \tag{6-8d}$$

We note now from Eq. (6-1) that

$$\frac{\partial K}{\partial a_i} = \frac{\partial}{\partial a_i} (K_1 + K_2 + \cdots + K_i \cdots + K_m) = \frac{\partial K_i}{\partial a_i} \tag{6-8e}$$

which recognizes that the stiffness matrix K_i for member i is a function of the cross-section area a_i alone; that is, $\partial K_r / \partial a_i = 0$ for $r \neq i$. We recall that the stiffness matrix for a truss member is a linear function of the cross-section area, such that for member i

$$K_i = a_i K_i^* \tag{6-8f}$$

where K_i^* is a matrix of constants, and therefore

$$\frac{\partial K_i}{\partial a_i} = \frac{\partial}{\partial a_i} (a_i K_i^*) = K_i^* = \frac{K_i}{a_i} \tag{6-8g}$$

Hence Eq. (6-8d) becomes from Eqs. (6-8e) and (6-8g)

$$\frac{\partial \delta_j}{\partial a_i} = -b_j^T K^{-1} \frac{K_i}{a_i} u \tag{6-8h}$$

An alternative form of Eq. (6-8h) that is easier to interpret in the analysis context can be arrived at as follows. Imagine that the displacement δ_j in Eq. (6-8) is multiplied by a unit virtual force. Equation (6-8) then defines a condition of virtual work equivalence, and we may view b_j as a vector of virtual loads. If we now apply b_j as a load vector on the given truss, the corresponding vector of virtual nodal displacements u_j is immediately found from

$$u_j = K^{-1} b_j \tag{6-8i}$$

since the inverse stiffness matrix K^{-1} is already known from Eq. (6-3) for the baseline analysis of the truss. Therefore Eq. (6-8h) becomes from Eq. (6-8i)

$$\frac{\delta_j}{\partial a_i} = -\frac{1}{a_i} (u_j^T K_i u) \tag{6-9}$$

Equation (6-9) defines the first-order sensitivity $\partial \delta_j / \partial a_i$ of the displacement δ_j to change in the cross-section area a_i of any one member

i of the truss. We note that its evaluation involves little computation beyond that required to conduct the baseline analysis of the truss. In fact, since a_i is given and K_i and u have already been found from the basic analysis, we need only additionally compute the vector of virtual nodal displacements u_j. From Eq. (6-8i), we recognize this latter computation as being nothing more than the simple matrix operation that is usually carried out for a structure to find displacement responses due to multiple load cases (i.e., in this instance, due to the load case defined by the vector b_j).

Following a similar development it can be shown that the first-order sensitivity of a displacement δ_j to change in the cross-section moment of inertia I_i of any one member i of a flexural framework is given as

$$\frac{\partial \delta_j}{\partial I_i} = -\frac{1}{I_i}\,(u_j^T K_i u) \tag{6-10}$$

which is exactly the same in form as Eq. (6-9), with the only difference being in the nature of the stiffness properties for the different structures.

To illustrate the application of Eq. (6-9), consider the pin-jointed truss in Fig. 6-1. The results found from the baseline analysis of this truss were given earlier. We calculate the sensitivity of the horizontal displacement $\delta_j = u_3$ to change in the cross-section area of each of the six members for the truss. From Eq. (6-8), the associated vector of virtual loads is $b_j^T = [0\ 0\ 1\ 0\ 0]$. Upon applying the load vector b_j on the truss as shown in Fig. 6-3, we find from Eq. (6-8i) that the corresponding vector of five virtual nodal displacements is

$$u_j = K^{-1} b_j$$

$$= \frac{L}{EA}\begin{bmatrix} 6.750 & 1.125 & 4.750 & -1.125 & 2.000 \\ 1.125 & 2.813 & 0.792 & -0.187 & -0.333 \\ 4.750 & 0.792 & 6.157 & -1.458 & 1.407 \\ -1.125 & -0.187 & -1.458 & 2.813 & -0.333 \\ 2.000 & -0.333 & 1.407 & -0.333 & 3.407 \end{bmatrix}\begin{bmatrix} 0 \\ 0 \\ 1 \\ 0 \\ 0 \end{bmatrix}$$

$$= \frac{L}{EA}\begin{bmatrix} 4.750 \\ 0.792 \\ 6.157 \\ -1.458 \\ 1.407 \end{bmatrix}$$

where the inverse stiffness matrix K^{-1} is previously known from the results of the baseline analysis for the truss.

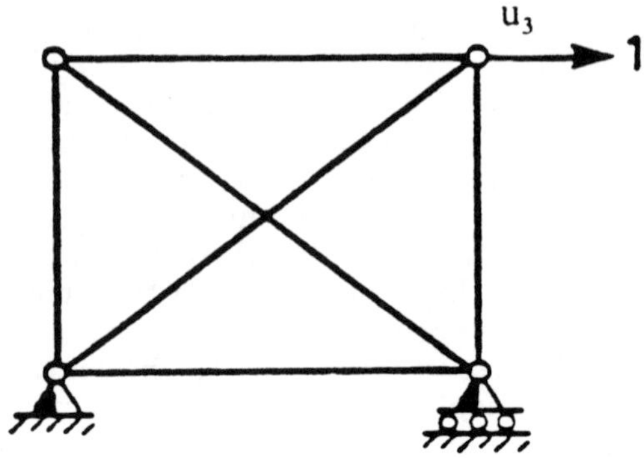

FIG. 6-3. Virtual Loading Corresponding to Displacement u_3.

Let us initially evaluate Eq. (6-9) in detail for member 5 of the given truss. From Fig. 6-1, the cross-section area for member 5 is $a_5 = A$. The global-axis stiffness matrix for member 5 is

$$K_5 = \frac{EA}{5L}\begin{bmatrix} 0 & 0 & 0 & 0 & 0 \\ 0 & 0 & 0 & 0 & 0 \\ 0 & 0 & 0.64 & 0.48 & 0 \\ 0 & 0 & 0.48 & 0.36 & 0 \\ 0 & 0 & 0 & 0 & 0 \end{bmatrix}$$

Therefore, Eq. (6-9) becomes for member 5

$$\frac{\partial \delta_j}{\partial a_5} = (u_j^T K_5 u)$$

$$= -\frac{1}{A}\cdot\frac{L}{EA}[4.750 \;\; 0.792 \;\; 6.157 \;\; -1.458 \;\; 1.407]\cdot K_5 \cdot \frac{WL}{EA}\begin{bmatrix} 6.750 \\ 1.125 \\ 4.750 \\ -1.125 \\ 2.000 \end{bmatrix}$$

$$= -\frac{1}{A}\left(2.5318\,\frac{WL}{EA}\right) = -2.5318\,\frac{WL}{EA^2}$$

where the vector of nodal displacements u is previously known from the results of the basic analysis for the truss. Hence, for the given truss in Fig. 6-1, the first-order sensitivity of the displacement $\delta_j = u_3$ to change in the cross-section area a_5 for member 5 is $\partial\delta_j/\partial a_5 = -2.5318\ WL/EA^2$. The negative sign for the sensitivity indicates that

δ_j will decrease as a_5 increases, and vice versa. The magnitude of the sensitivity is a measure of the amount by which δ_j will change for a unit change in a_5.

We note from Eq. (6-9) that the two displacement vectors u and u_j also apply for the calculation of the sensitivity of δ_j to change in the cross-section area of each of the remaining members 1, 2, 3, 4, and 6 of the truss. From Fig. 6-1, the cross-section areas for these five members are $a_i = A$ (i = 1, 2, 3, 4, 6). It then simply remains for us to identify the global-axis stiffness matrices K_i (i = 1, 2, 3, 4, 6) for these five members in order to calculate the corresponding sensitivities through Eq. (6-9) to find (the detailed calculations are left as an exercise for the reader)

$$\frac{\partial \delta_j}{\partial a_i} = -\frac{1}{a_i}(u_j^T K_1 u) = -\frac{1}{A}\left(0.2970\,\frac{WL}{EA}\right) = -0.2970\,\frac{WL}{EA^2}$$

$$\frac{\partial \delta_j}{\partial a_2} = -\frac{1}{a_2}(u_j^T K_2 u) = -\frac{1}{A}\left(-0.7035\,\frac{WL}{EA}\right) = +0.7035\,\frac{WL}{EA^2}$$

$$\frac{\partial \delta_j}{\partial a_3} = -\frac{1}{a_3}(u_j^T K_3 u) = -\frac{1}{A}\left(0.5467\,\frac{WL}{EA}\right) = -0.5467\,\frac{WL}{EA^2}$$

$$\frac{\partial \delta_j}{\partial a_4} = -\frac{1}{a_4}(u_j^T K_4 u) = -\frac{1}{A}\left(0.7035\,\frac{WL}{EA}\right) = -0.7035\,\frac{WL}{EA^2}$$

$$\frac{\partial \delta_j}{\partial a_6} = -\frac{1}{a_6}(u_j^T K_6 u) = -\frac{1}{A}\left(1.3745\,\frac{WL}{EA}\right) = -1.3745\,\frac{WL}{EA^2}$$

The calculated sensitivities provide us with a range of information concerning the behavior of the displacement $\delta_j = u_3$ for changes in the cross-section areas of the members of the truss loaded as shown in Fig. 6-1. The negative sign for the sensitivities corresponding to members 1, 3, 4, 5, and 6 indicates that δ_j will decrease as the cross-section area for each of these members increases, and vice versa. The positive sign for the sensitivity corresponding to member 2 indicates that δ_j will actually increase as a_2 increases and vice versa. (This latter phenomenon occurs because member 2 is stressed in axial compression; in fact, any change in u_3 as a consequence of changing a_2 will be exactly equal to the change in the axial shortening of member 2). That the sensitivities for members 2 and 4 have equal magnitude but opposite sign indicates that δ_j will not change at all for similar changes in a_2 and a_4. The fact that the magnitude of the sensitivities corresponding to members 5 and 6 are both much larger than those for the other four members of the truss indicates, as expected, that these two diagonal bracing struts have the greatest influence on $\delta_j = u_3$. That the magnitude of the sensitivity for member 5 is larger than that for member 6 indicates that member 5 is the dominant bracing strut affecting δ_j. (In fact, for similar changes in a_5 and a_6, member 5 will have almost twice the influence on u_3 as member 6.) Finally, the calculated sensitivities for the six members indicate that member 5 has the greatest individual influence on δ_j whereas member 1 has the least. The foregoing behavioral information provides us with valuable insight into the way the truss in Fig. 6-1 can be modified in order to effect any desired change in the magnitude of the displacement $\delta_j = u_3$.

To illustrate the application of Eq. (6-10), consider the flexural frame in Fig. 6-2. The results found from the basic analysis of this frame were given earlier. We calculate the sensitivity of the lateral translation $\delta_j = u_3$ to change in the cross-section moment of inertia of each of the three members for the frame. From Eq. (6-8), the associated vector of virtual loads is $b_j^T = [0\ 0\ 1]$. Upon applying the load vector b_j on the frame as shown in Fig. 6-4, we find from Eq. (6-8i) that the corresponding vector of three virtual nodal displacements is

$$u_j = K^{-1}b_j = \frac{L}{EA}\begin{bmatrix} 0.331 & -0.057 & 0.070L \\ -0.057 & 0.361 & 0.259L \\ 0.070L & 0.259L & 0.738L^2 \end{bmatrix}\begin{bmatrix} 0 \\ 0 \\ 1 \end{bmatrix} = \frac{L^2}{EI}\begin{bmatrix} 0.070 \\ 0.259 \\ 0.738L \end{bmatrix}$$

where the inverse stiffness matrix K^{-1} is previously known from the results of the baseline analysis for the frame.

Let us evaluate Eq. (6-10) in detail for member 1 of the given frame. From Fig. 6-2, the cross-section moment of inertia for member 1 is $I_1 = I$. The global-axis flexural stiffness matrix for member 1 is

$$K_1 = \frac{EI}{3L}\begin{bmatrix} 4 & 0 & -2/L \\ 0 & 0 & 0 \\ -2/L & 0 & 4/3L^2 \end{bmatrix}$$

Therefore Eq. (6-10) becomes for member 1

$$\begin{aligned} \frac{\partial \delta_j}{\partial I_1} &= -\frac{1}{I_1}(u_j^T K_1 u) \\ &= -\frac{1}{I}\cdot\frac{L^2}{EI}[0.070 \quad 0.259 \quad 0.738L]\cdot K_1 \cdot \frac{WL^2}{EI}\begin{bmatrix} 0.070 \\ 0.259 \\ 0.738L \end{bmatrix} \\ &= -\frac{1}{I}\left(0.1797\frac{WL^3}{EI}\right) = -0.1797\frac{WL^3}{EI^2} \end{aligned}$$

where the vector of nodal displacements u is previously known from the results of the baseline analysis for the frame. Hence, for the flexural frame in Fig. 6-2, the first-order sensitivity of the lateral trans-

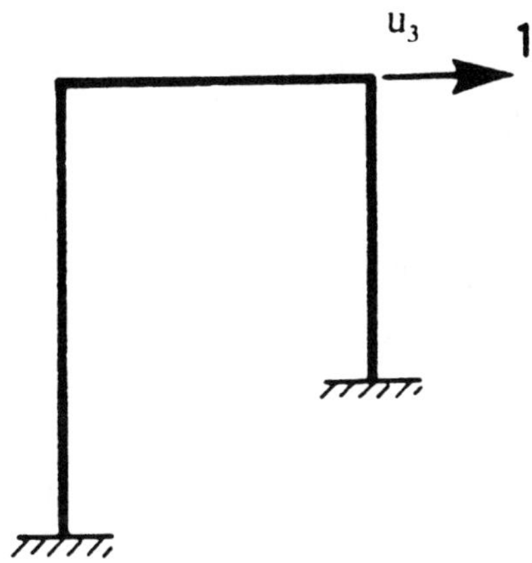

FIG. 6-4. Virtual Loading Corresponding to Displacement u_3.

lation $\delta_j = u_3$ to change in the cross-section moment of inertia I_1 for member 1 is $\partial\delta_j/\partial I_1 = -0.179WL^3/EI^2$. We note from Eq. (6-10) that the two displacement vectors u and u_j also apply for the calculation of the sensitivity of δ_j to change in the cross-section moment of inertia of each of the remaining members 2 and 3 of the frame. From Fig. 6-2, the cross-section moments of inertia for these two members are $I_i = I$ $(i = 2, 3)$. It then simply remains for us to identify the global-axis flexural stiffness matrices K_i $(i = 2, 3)$ for these two members in order to calculate the corresponding sensitivities through Eq. (6-10), to find (the detailed calculations are left as an exercise for the reader)

$$\frac{\partial \delta_j}{\partial I_2} = -\frac{1}{I_2}(u_j^T K_2 u) = -\frac{1}{I}\left(0.1802\frac{WL^3}{EI}\right) = -0.1802\frac{WL^3}{EI^2}$$

$$\frac{\partial \delta_j}{\partial I_3} = -\frac{1}{I_3}(u_j^T K_3 u) = -\frac{1}{I}\left(0.3777\frac{WL^3}{EI}\right) = -0.3777\frac{WL^3}{EI^2}$$

The calculated sensitivities provide us with a range of information concerning the behavior of the displacement $\delta_j = u_3$ for changes in the cross-section moments of inertia of the members of the flexural frame loaded as shown in Fig. 6-2. The negative sign for the sensitivities corresponding to all three members indicates that δ_j will decrease as the cross-section moment of inertia of each member increases, and vice versa. That the magnitude of the sensitivity for member 3 is larger than that for members 1 and 2 indicates that member 3 has the dominant influence on the lateral translation $\delta_j = u_3$ of the frame. (In fact, for similar changes in I_1, I_2, and I_3, member 3 will have twice the influence on u_3 as will members 1 and 2.) That the magnitude of the sensitivity for beam member 2 is greater than that for column member 1, and almost half that for column member 3, serves to illustrate the fact that horizontal members contribute significantly to the lateral stiffness of rigid frame structures.

Stress Sensitivity Analysis

Consider a particular member k of a given framework structure. The axial or bending stress σ_k for the member is related to the vector of nodal displacements u for the structure as

$$\sigma_k = t_k^T u \tag{6-11}$$

where t_k is the global-axis stress vector for member k. From Eqs. (6-4) and (6-5), we see that the stress vector t_k for the kth member of a truss is given by row k of the compatibility matrix **C** for the truss multiplied by Young's material modulus E and divided by the length of member k. For example, from the compatibility matrix **C** given previously for the truss in Fig. 6-1, the stress vector for member k = 5 is $t_5^T = [0\ 0\ 0.8\ 0.6\ 0]\ E/5L$. Similarly, assuming that there are no loads applied in the span of members [i.e., $M_k^0 = 0$ in Eq. (6-6)], from Eqs. (6-6) and (6-7) we see that the stress vector t_k is given by row k of the transformation matrix M^1 for the structure divided by the moment of inertia I_k for section k and multiplied by the distance y_k from the neutral axis to the point on the section where the stress σ_k is measured. For example, from the M^1 matrix given previously for the flexural frame in Fig. 6-2, the stress vector corresponding to stress in the extreme fibers of section k = 6 is $t_6^T = [0\ 1\ -3/2L]\ Ey/L$, where y is one-half the depth of the section.

Suppose now that it is found from the results of analysis of the structure that the current magnitude of the stress σ_k exceeds its allowable value for proper design. We must therefore modify the cross-sections of the members of the structure such that, under the same applied loading, the magnitude of σ_k becomes less than or equal to its allowable stress limit. To this end, we first calculate the first-order sensitivity of σ_k to change in the cross-section area a_i of the ith member of a truss (we later extend the result to a flexural structure). Taking the first-derivative of Eq. (6-11) with respect to a_i, we have

$$\frac{\partial \sigma_k}{\partial a_i} = t_k^T \frac{\partial u}{\partial a_i} + u^T \frac{\partial t_k}{\partial a_i} = t_k^T \frac{\partial u}{\partial a_i} \tag{6-11a}$$

which recognizes that the stress vector t_k for any member k of a pin-jointed truss is invariant with changes in the member cross-section areas; that is, $\partial t_i / \partial a_i = 0$. From Eqs. (6-8c), (6-8e), and (6-8g), we have that

$$\frac{\partial u}{\partial a_i} = -K^{-1} \frac{K_i}{a_i} u \tag{6-11b}$$

Hence, Eq. (6-11a) becomes from Eq. (6-11b)

$$\frac{\partial \sigma_k}{\partial a_i} = -t_k^T K^{-1} \frac{K_i}{a_i} u \tag{6-11c}$$

Another form of Eq. (6-11c) that is easier to interpret in the analysis context can be arrived at as follows. Imagine that the stress σ_k in Eq. (6-11) is multiplied by a unit virtual strain. Equation (6-11) then defines a condition of virtual work equivalence, and we may view t_k as a vector of virtual loads. If we now apply t_k as a load vector on the given truss, the corresponding vector of virtual nodal displacements u_k is immediately found from

$$u_k = K^{-1} t_k \tag{6-11d}$$

since the inverse stiffness matrix K^{-1} is already known from the baseline analysis of the truss. Therefore Eq. (6-11c) becomes from Eq. (6-11d)

$$\frac{\partial \sigma_k}{\partial a_i} = -\frac{1}{a_i} (u_k^T K_i u) \tag{6-12}$$

Equation (6-12) defines the first-order sensitivity $\partial \sigma_k / \partial a_i$ of the stress σ_k to change in the cross-section area a_i of any member i for the truss [Note the similarity of Eq. (6-12) for stress sensitivity analysis with Eq. (6-9) for displacement sensitivity analysis. In fact, the two types of sensitivities are calculated in exactly the same way, with a difference only in the nature of their respective virtual load vectors.] We note that the evaluation of Eq. (6-12) involves little computation beyond that required to conduct the basic analysis of the truss. In fact, since a_i is given and K_i and u have already been found from the basic analysis, we need only additionally compute the vector of virtual nodal displacements u_k. From Eq. (6-11d), we recognize this latter computation as being nothing more than the simple matrix operation that is regularly carried out for a structure to find displacement responses due to multiple load cases (i.e., in this instance, due to the load case defined by the vector t_k).

We now extend the foregoing to calculate the first-order sensitivity of a bending stress σ_k to change in the cross-section moment of inertia I_i of any member i for a flexural structure. Taking the first-derivative of Eq. (6-11) with respect to I_i, we have

$$\frac{\partial \sigma_k}{\partial I_i} = t_k^T \frac{\partial u}{\partial I_i} + u^T \frac{\partial t_k}{\partial I_i} = t_k^T \frac{\partial u}{\partial I_i} \tag{6-11e}$$

which assumes that the stress vector t_k is invariant with changes in member cross-section moment of inertia; that is, $\partial t_k / \partial I_i = 0$. In fact, this assumption is true for each member i of the structure that is not associated with the stress σ_k. On the other hand, the assumption is in error for the particular member that is associated with σ_k since t_k is a function of the neutral axis depth and, therefore, of the moment of inertia for the cross-section of the member. However, particularly for small to moderate modifications of the structure, this error is generally negligible. (In any event, as shown later, it is readily accounted for after the sensitivity analysis has been completed.) Following now the development previously carried out for a truss structure, it is readily shown that the first-order sensitivity of a flexural stress σ_k to change in the cross-section moment of inertia I_i of any one member i of a flexural framework is found to be

$$\frac{\partial \sigma_k}{\partial I_i} = -\frac{1}{I_i}\,(u_k^T K_i u) \tag{6-13}$$

Note the similarity of Eq. (6-13) for stress sensitivity analysis with Eq. (6-10) for displacement sensitivity analysis; in fact, the two types of sensitivities are calculated in exactly the same way, with the difference being only in the nature of their respective virtual load vectors. Moreover, note the similarity of Eq. (6-13) for a flexural structure with Eq. (6-12) for a pin-jointed truss; the stress sensitivity analysis of the two types of structures is carried out in exactly the same way, with a difference only in the structure stiffness properties.

To illustrate the application of Eq. (6-12), consider the pin-jointed truss in Fig. 6-1. The results found from the basic analysis of this truss were given earlier. We calculate the sensitivity of the axial stress $\sigma_k = \sigma_5$ for member 5 to change in the cross-section area of each of the six members for the truss. From Eq. (6-11), the associated vector of virtual loads is $t_k^T = [0 \;\; 0 \;\; 0.8 \;\; 0.6 \;\; 0] \; E/5L$. Upon applying the load vector t_k on the truss as shown in Fig. 6-5, we find from Eq. (6-11d) that the corresponding vector of five virtual nodal displacements is

$$u_k = K^{-1} t_k$$

$$= \frac{L}{EA}\begin{bmatrix} 6.750 & 1.125 & 4.750 & -1.125 & 2.000 \\ 1.125 & 2.813 & 0.792 & -0.187 & -0.333 \\ 4.750 & 0.792 & 6.157 & -1.458 & 1.407 \\ -1.125 & -0.187 & -1.458 & 2.813 & -0.333 \\ 2.000 & -0.333 & 1.407 & -0.333 & 3.407 \end{bmatrix} \cdot \frac{E}{5L}\begin{bmatrix} 0 \\ 0 \\ 0.8 \\ 0.6 \\ 0 \end{bmatrix}$$

$$= 1/A \begin{bmatrix} 0.625 \\ 0.104 \\ 0.810 \\ 0.104 \\ 0.185 \end{bmatrix}$$

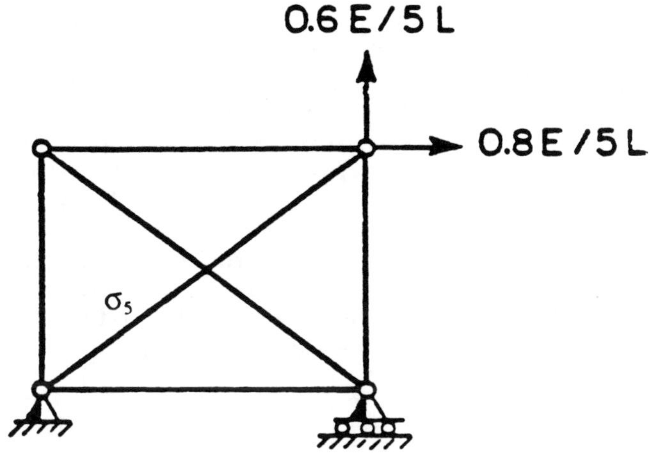

FIG. 6-5. Virtual Loading Corresponding to Axial Stress σ_5.

where the inverse stiffness matrix K^{-1} is previously known from the results of the basic analysis for the truss.

Let us initially evaluate Eq. (6-12) for member $i = 6$ of the given truss. From Fig. 6-1, the cross-section area for member 6 is $a_6 = A$. The global-axis stiffness matrix for member 6 is

$$K_6 = \frac{EA}{5L}\begin{bmatrix} 0.64 & -0.48 & 0 & 0 & -0.64 \\ -0.48 & 0.36 & 0 & 0 & 0.48 \\ 0 & 0 & 0 & 0 & 0 \\ 0 & 0 & 0 & 0 & 0 \\ -0.64 & 0.48 & 0 & 0 & 0.64 \end{bmatrix}$$

Therefore Eq. (6-12) for member 6 is written as

$$\frac{\partial \sigma_k}{\partial a_6} = -\frac{1}{a_6}\,(u_k^T K_6 u)$$

$$= -\frac{1}{A}\cdot\frac{1}{A}\,[0.625 \quad 0.104 \quad 0.810 \quad 0.104 \quad 0.185]\cdot K_6\cdot\frac{WL}{EA}\begin{bmatrix} 6.750 \\ 1.125 \\ 4.750 \\ -1.125 \\ 2.000 \end{bmatrix}$$

$$= -\frac{1}{A}\left(0.1808\,\frac{W}{A}\right) = -0.1808\,\frac{W}{A^2}$$

where the vector of nodal displacements u is previously known from the results of the basic analysis for the truss. Hence, for the given truss in Fig. 6-1, the first-order sensitivity of the axial stress $\sigma_k = \sigma_5$ for member 5 to change in the cross-section area a_6 for member 6 is $\partial\sigma_k/\partial a_6 = -0.1808\ W/A^2$. We note from Eq. (6-12) that the two displacement vectors u and u_k also apply for the calculation of the sensitivity of σ_k to change in the cross-section area of each of the remaining members 1 through 5 for the given truss. From Fig. 6-1, the cross-section areas for these five members are $a_i = A$ $(i = 1, 2, \ldots, 5)$. It then simply remains for us to identify the global-axis stiffness matrices K_i $(i = 1, 2, \ldots, 5)$ for these five members in order to calculate the corresponding sensitivities through Eq. (6-12), to find (the detailed calculations are left as an exercise for the reader)

$$\frac{\partial\sigma_k}{\partial a_1} = -\frac{1}{a_1}(u_k^T K_1 u) = -\frac{1}{A}\left(0.0391\,\frac{W}{A}\right) = -0.0391\,\frac{W}{A^2}$$

$$\frac{\partial\sigma_k}{\partial a_2} = -\frac{1}{a_2}(u_k^T K_2 u) = -\frac{1}{A}\left(-0.0926\,\frac{W}{A}\right) = +0.0926\,\frac{W}{A^2}$$

$$\frac{\partial\sigma_k}{\partial a_3} = -\frac{1}{a_3}(u_k^T K_3 u) = -\frac{1}{A}\left(-0.0391\,\frac{W}{A}\right) = +0.0391\,\frac{W}{A^2}$$

$$\frac{\partial\sigma_k}{\partial a_4} = -\frac{1}{a_4}(u_k^T K_4 u) = -\frac{1}{A}\left(0.0926\,\frac{W}{A}\right) = -0.0926\,\frac{W}{A^2}$$

$$\frac{\partial\sigma_k}{\partial a_5} = -\frac{1}{a_5}(u_k^T K_5 u) = -\frac{1}{A}\left(0.4442\,\frac{W}{A}\right) = -0.4442\,\frac{W}{A^2}$$

The calculated sensitivities provide us with a range of information concerning the behavior of the axial stress $\sigma_k = \sigma_5$ for member 5 with changes in the cross-section areas of the members of the truss loaded as shown in Fig. 6-1. The negative sign for the sensitivities corresponding to members 1, 4, 5, and 6 indicates that σ_5 will decrease as the cross-section area for each of these members increases, and vice versa. The positive sign for the sensitivities corresponding to members 2 and 3 indicates that σ_5 will actually increase as a_2 or a_3 increases, and vice versa. That the sensitivities for members 1 and 3 have equal magnitude but opposite sign indicates that σ_5 will not change at all for similar changes in a_1 and a_3. The same is true for members 2 and 4. In other words, then, a common change in the cross-section areas for members 1, 2 and 3, 4 will not affect the axial stress for member 5. That the magnitudes of the sensitivities for members 5 and 6 are both much larger than those for the other four members of the truss indicates that these two diagonal bracing struts have the dominant influence on σ_5. As expected, member 5 has the greatest influence on σ_5. The foregoing behavioral information provides us with valuable insight into the way the truss in Fig. 6-1 can be modified in order to effect any desired change in the magnitude of the axial stress $\sigma_k = \sigma_5$ for member 5.

To illustrate the application of Eq. (6-13), consider the flexural frame in Fig. 6-2. The results found from the baseline analysis of this frame were given earlier. We calculate the sensitivity of the extreme-fiber bending stress $\sigma_k = \sigma_6$ at end-section $k = 6$ of member 3, to change in the cross-section moment of inertia of each of the three members of the frame. Since member 3 has a symmetrical cross-section, σ_k may be viewed as either a tensile or compressive stress. From Fig. 6-2, end-section 6 of member 3 has moment of inertia I and neutral axis depth y. Therefore, from Eqs. (6-6) for $M^0 = 0$ and Eqs. (6-7) and (6-11) and the sixth row of the transformation matrix M^1, the virtual load vector associated with the stress $\sigma_k = \sigma_6$ is $t_k^T = t_6^T = [0\ 1\ -3/2L]\ Ey/L$. Upon applying the load vector t_k on the frame as shown in Fig. 6-6, we find from Eq. (6-11d) that the corresponding vector of three virtual nodal displacements is

$$u_k = K^{-1}t_k = \frac{L}{EI}\begin{bmatrix} 0.331 & -0.057 & 0.070L \\ -0.057 & 0.361 & 0.259L \\ 0.070L & 0.259L & 0.738L^2 \end{bmatrix} \cdot \frac{Ey}{L}\begin{bmatrix} 0 \\ 1 \\ -3/2L \end{bmatrix}$$

$$= \frac{y}{I}\begin{bmatrix} -0.162 \\ -0.028 \\ -0.848L \end{bmatrix}$$

where the inverse stiffness matrix K^{-1} is previously known from the results of the basic analysis for the frame.

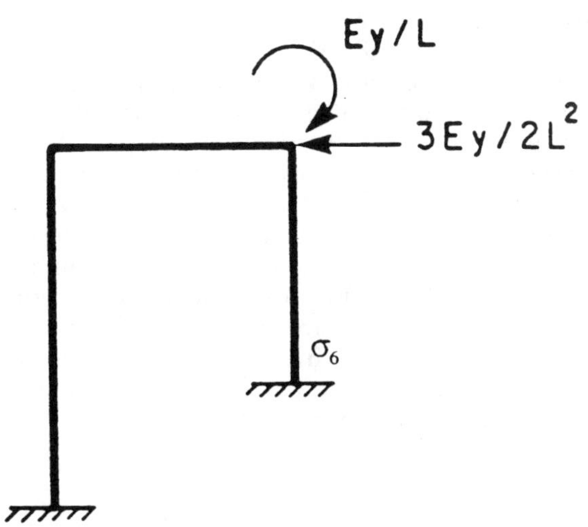

FIG. 6-6. Virtual Loading Corresponding to Bending Stress σ_6.

Let us evaluate Eq. (6-13) in detail for member 3 of the given frame. From Fig. 6-2, the cross-section moment of inertia for member 3 is $I_3 = I$. The global-axis stiffness matrix for member 3 is

$$K_3 = \frac{EI}{2L}\begin{bmatrix} 0 & 0 & 0 \\ 0 & 4 & -3/L \\ 0 & -3/L & 3/L^2 \end{bmatrix}$$

Therefore Eq. (6-13) becomes for member 3

$$\frac{\partial \sigma_k}{\partial I_3} = -\frac{1}{I_3}(u_k^T K_3 u)$$

$$= -\frac{1}{I}\cdot\frac{y}{I}[-0.162 \quad -0.028 \quad -0.848L]\cdot K_3\cdot\frac{WL^2}{EI}\begin{bmatrix} 0.070 \\ 0.259 \\ 0.738L \end{bmatrix}$$

$$= -\frac{1}{I}\left(-0.5928\,\frac{WLy}{I}\right) = +0.5928\,\frac{WLy}{I^2}$$

where the vector of nodal displacements u is previously known from the results of the basic analysis for the frame. Hence, for the given flexural frame in Fig. 6-2, the first-order sensitivity of the extreme-fiber bending stress $\sigma_k = \sigma_6$ at end-section $k = 6$ of member 3 to change in the cross-section moment of inertia I_3 for member 3 is $\partial\sigma_k/\partial I_3 = +0.5928\, WLy/I^2$. In general terms, the positive sign for the sensitivity indicates that σ_k will increase in the sense of becoming more positive as I_3 increases, and vice versa. Since we see from the earlier analysis results that $\sigma_k = \sigma_6 = -0.848\, WLy/I$ is a negative-valued stress according to the sign convention adopted for the basic analysis of the frame, this actually means that σ_k will tend to become less negative as I_3 increases. In other words, as we would expect, the absolute magnitude of σ_k decreases as I_3 increases, and vice versa.

We note from Eq. (6-13) that the two displacement vectors u and u_k also apply for the calculation of the sensitivity of σ_k to change in the cross-section moment of inertia of each of the remaining members 1 and 2 of the frame. From Fig. 6-2, the cross-section moments of inertia for these two members are $I_1 = I$. It then simply remains for us to identify the global-axis stiffness matrices K_1 and K_2 for members 1 and 2 in order to calculate the corresponding sensitivities through Eq. (6-13), to find (the detailed calculations are left as an exercise for the reader)

$$\frac{\partial \sigma_k}{\partial I_1} = -\frac{1}{I_1}(u_k^T K_1 u) = -\frac{1}{I}\left(-0.1740\,\frac{WLy}{I}\right) = +0.1740\,\frac{WLy}{I^2}$$

$$\frac{\partial \sigma_k}{\partial I_2} = -\frac{1}{I_2}(u_k^T K_2 u) = -\frac{1}{I}\left(-0.0811\,\frac{WLy}{I}\right) = +0.0811\,\frac{WLy}{I^2}$$

The calculated sensitivities provide us with a range of information concerning the behavior of the bending stress $\sigma_k = \sigma_6$ at end-section $k = 6$ of member 3 to changes in the cross-section moments of inertia of the three members of the flexural frame loaded as shown in Fig. 6-2. As discussed earlier, the positive sign for each sensitivity indicates that the absolute magnitude of σ_k will decrease as the cross-section moment of inertia of each of the three members is increased, and vice versa. The magnitudes of the sensitivities indicate the relative influence that the three members have on the bending stress $\sigma_k = \sigma_6$ at section 6 of the frame. Member 3 has the greatest influence on σ_k, and member 2 has the least.

To conclude this section of the chapter, we note that response sensitivities are easy to calculate and provide us with valuable insight into structural behavior. Moreover, they provide us with an effective means to carry out design modifications for a structure in a systematic and economical way. In the next section we present a simple sensitivity-based procedure for modifying a structure that is the forerunner of a full computer-based design optimization capability (which is covered in Chapter 8 on "How to Optimize Structural Steel Frameworks").

DESIGN MODIFICATIONS

For the sake of brevity, we confine our attention to the design of pin-jointed trusses for acceptable displacements, using the truss in Fig. 6-1 as illustration. There is sufficient information in the previous section concerning stresses for trusses, and stresses and displacements for flexural frameworks for the reader to ascertain that the principles apply equally for these design parameters as well.

We now discuss the modification of a given truss in order to effect a required change in a particular displacement δ_j. Let us define the given values of the cross-section areas for the m truss member as a_i^0 ($i = 1, 2, \ldots, m$), where the superscript ^{0} denotes quantities that apply for the current state of the truss. Moreover, let us define the new values of the member cross-section areas for the modified state of the truss as a_i^1 ($i = 1, 2, \ldots, m$). In other words,

$$a_i^0 \rightarrow a_i^1 \qquad (i = 1, 2, \ldots, m) \tag{6-14}$$

describes the modification to the truss. Upon calculating the sensitivities $(\partial\delta_j/\partial a_i)^0$ for the $i = 1, 2, \ldots, m$ members through Eq. (6-9), we may estimate the new magnitude δ_j^1 of the displacement for the modified truss from the first-order Taylor series

$$\delta_j^1 = \delta_j^0 + \sum_{i=1}^{m} \left(\frac{\partial \delta_j}{\partial a_i}\right)^0 (a_i^1 - a_i^0) \tag{6-15}$$

where δ_j^0 is the magnitude of the displacement found from the baseline analysis of the truss prior to the modification. Equation (6-15) allows us to directly estimate the new magnitude of the displacement δ_j for the modified truss without having to conduct a complete re-analysis. In the following, we develop a bounding technique that allows us to use the sensitivity analysis results to also determine the extent to which the truss is to be modified in order to effect a required change in δ_j. Since sensitivity analysis involves only a fraction of the computational effort required to conduct a complete analysis of a structure, these operations can lead to significant savings for the design process in comparison to a trial-and-error approach that requires complete re-analysis of a structure each time it is modified. Moreover, as shown, these operations allow us to make economical design changes.

To commence development of the bounding technique, we first observe that the displacements for a structure are generally inversely related to the sizes of the members; in other words, that displacements will decrease as members' sizes are increased, and vice versa. When this is the case, the relationship between a displacement δ_j and the m member cross-section areas a_i for a truss is defined by a convex function $\delta_j = f(a_1, a_2, \ldots, a_m)$. Equation (6-15) is therefore a tangent-plane approximation to a convex function and, as such, its approximate displacement value δ_j^1 will underestimate the actual magnitude of the displacement for the modified truss. Conversely, it follows that the displacement δ_j is generally directly related to the reciprocals of the member cross-section areas

$$x_i = 1/a_i \qquad (i = 1, 2, \ldots, m) \tag{6-16}$$

and, therefore, that the relationship between δ_j and the reciprocal areas x_i is a concave function $\delta_j = g(x_1, x_2, \ldots, x_m)$. As such, when this is the case, a corresponding Taylor series approximation that is similar to Eq. (6-15) but which is expressed in terms of the x_i, that is,

$$\delta_j^1 = \delta_j^0 + \sum_{i+1}^{m} \left(\frac{\partial \delta_j}{\partial x_i}\right)^0 (x_i^1 - x_i^0) \tag{6-17}$$

will yield an approximate displacement value δ_j^1 that overestimates the actual magnitude of the displacement for the modified truss. The x_i^0 and x_i^1 quantities appearing in Eq. (6-17) are simply the reciprocals of the cross-section area values a_i^0 and a_i^1 that define the modification of the truss through Eq. (6-14). Upon accounting for Eq. (6-16) in the development of Eq. (6-9), we can readily show that each first-order sensitivity $\partial\delta_j/\partial x_i$ appearing in Eq. (6-17) is given by (the detailed development is left as an exercise for the reader)

$$\frac{\partial\delta_j}{\partial x_i} = \frac{1}{x_i}\,(u_j^T K_i u) = -a_i^2 \left(\frac{\partial\delta_i}{\partial a_i}\right) \tag{6-18}$$

We see that the determination of each sensitivity $\partial\delta_j/\partial x_i$ through Eq. (6-18) involves only nominal computational effort beyond that required to determine the corresponding sensitivity $\partial\delta_j/\partial a_i$ through Eq. (6-9).

From the foregoing, we have that the actual magnitude δ_j^{actual} of the displacement for the modified truss is bounded from below and above by the displacement estimates δ_j^1 provided by Eqs. (6-15) and (6-17), respectively. In other words, that

$$\delta_j^1\big|_{\text{Eq. (6-15)}} \leq \delta_j^{\text{actual}} \leq \delta_j^1\big|_{\text{Eq. (6-17)}} \tag{6-19}$$

We now employ the bounding technique suggested by Eq. (6-19) in the context of a design problem. Suppose for the design of the truss in Fig. 6-1 that the displacement $\delta_j = u_3$ is to be not greater than 4.4 WL/EA. From the earlier analysis results we note that $u_3 = 4.750$ WL/EA for the current state of the truss indicated in Fig. 6-1. We must therefore modify the cross-section area a_i of one or more of the $i = 1, 2, \ldots, 6$ members so as to decrease the magnitude of the displacement δ_j to within its acceptable limit. Although it is possible to modify the cross-section areas of a number of members simultaneously, it is perhaps most practical to change but one member cross-section area at a time. We should be concerned with maintaining the symmetry of the truss design (e.g., members 5 and 6 should have the same cross-section area, etc.). Moreover, we should consider the economy of the truss design when we make our modifications.

From the previously calculated sensitivities $\partial\delta_j/\partial a_i$ for the truss in Fig. 6-1, we note that member 2 certainly affords the most economical modification since $\delta_j = u_3$ may be decreased by decreasing a_2. This is perhaps not a practical modification, however, since it may result in excessive compressive stress for member 2. Moreover, if we decrease a_2, we should similarly change a_4 for member 4 in order to maintain the symmetry of the truss; in which case, we recall, the magnitude of δ_j will not be changed at all from its original value. We therefore confine our attention to members 1, 3, 5, and 6, for each of which the cross-section area must be increased in order to decrease $\delta_j = u_3$. We may apply a simple rule of thumb in order to determine which of these members affords the most economical modification; namely, it is the member i of length L_i for which the absolute-value-ratio of its sensitivity $\partial\delta_j/\partial a_i$ to its current material volume $L_i a_i$ is the largest of all such ratios for the members. This member requires the smallest increase in material volume to effect a unit decrease in δ_j. Applying this rule for the previously found sensitivities $\partial\delta_j/\partial a_i$ ($i = 1, 3, 5, 6$) and the length and area values indicated in Fig. 6-1, we readily determine that member 5 affords the most economical modification to $\delta_j = u_3$ from among members 1, 3, 5, and 6, that member 6 is the next most economical, and then members 3 and 1 in that order. For example, the economy ratio for member 5 is

$$\left|\frac{\partial\delta_j/\partial a_5}{L_5 a_5}\right| = \left|\frac{-2.5318\,\dfrac{WL}{EA^2}}{5LA}\right| = 0.5064\,\frac{W}{EA^3}$$

Since member 5 affords the most economical modification, let us begin by arbitrarily increasing its cross-section area by 10 percent. In other words, a_5 is increased from A to $1.1A$. We first estimate the new magnitude δ_j^1 of the displacement $\delta_j = u_3$ for the modified truss using Eq. (6-15). Having the known values of δ_j^0 and $(\partial\delta_j/a_i)^0$ that prevail for the truss prior to the modification, we find

$$\begin{aligned}\delta_j^1\big|_{\text{Eq. (6-15)}} &= 4.750\,\frac{WL}{EA} - 2.5318\,\frac{WL}{EA^2}\,(1.1A - A)\\ &= 4.4968\ WL/EA \qquad \text{(6-19a)}\end{aligned}$$

We next use Eq. (6-17) to estimate the new displacement magnitude δ_j^1. In this case, from Eq. (6-16), the modification to the truss is defined by a decrease from $1/A$ to $1/1.1A$ for the reciprocal area x_5 of member 5. Upon noting from Eq. (6-18) that each sensitivity $\partial\delta_j/\partial x_i$ has the same numerical coefficient as the corresponding sensitivity $\partial\delta_j/\partial a_i$, but an opposite sign and a constant multiplier of WL/E rather than WL/EA^2, we find

$$\begin{aligned}\delta_j^1\big|_{\text{Eq. (6-17)}} &= 4.750\,\frac{WL}{EA} + 2.5313\,\frac{WL}{E}\left(\frac{1}{1.1} - 1\right)\cdot\frac{1}{A}\\ &= 4.5198\ WL/EA \qquad \text{(6-19b)}\end{aligned}$$

Therefore, from Eqs. (6-19a), (6-19b), and (6-19), the actual magnitude δ_j^{actual} of the displacement for the modified truss is such that

$$4.4968\,\frac{WL}{EA} \le \delta_j^{\text{actual}} \le 4.5198\,\frac{WL}{EA} \qquad \text{(6-19c)}$$

Equation (6-19c) provides us with some important items of information. First, the lower-bound value tells us that increasing the cross-section area of member 5 by 10 percent is alone not sufficient to decrease the magnitude of the displacement to within its acceptable limit of 4.4 WL/EA. In other words, we have been able to determine that further modification is required without having to conduct a complete re-analysis of the truss as modified thus far. Second, the upper-bound value tells us that increasing a_5 by 10 percent causes the magnitude of δ_j to decrease from its original value of 4.750 WL/EA to at least 4.5198 WL/EA. In other words, we have been able to determine a measure of the influence of the modification without having to re-analyze the modified truss. This latter information serves as a useful guide for further modifications.

Specifically, Eq. (6-19c) informs us that a 10 percent increase in the cross-section area of member 5 for the truss in Fig. 6-1 causes the magnitude of displacement $\delta_j = u_3$ to decrease a little more than two-thirds of the way from its original value of 4.750 WL/EA to its acceptable value of 4.4 WL/EA. Since, from the previous section, the magnitude of the sensitivity $\partial\delta_j/\partial a_6$ for member 6 is a little more than half that of the sensitivity $\partial\delta_j/\partial a_5$ for member 5, a similar 10 percent increase in the cross-section area of member 6 will therefore bring the displacement to within its acceptable limit. This is a desirable change since, we recall, member 6 affords the next most economical modification after member 5. Such further modification also serves to restore the symmetry of the truss design. For a_6 as well as a_5 increased from A to $1.1A$, we readily find that the new estimate of the displacement magnitude according to Eq. (6-15) is

$$\delta_j^1\big|_{\text{Eq. (6-15)}} = 4.4968\,\frac{WL}{EA} - 1.3745(0.1)\,\frac{WL}{EA} = 4.3594\,\frac{WL}{EA}$$

and that the estimate according to Eq. (6-17) is

$$\delta_j^1\big|_{\text{Eq. (6-17)}} = 4.5198\,\frac{WL}{EA} + \left(\frac{1.3745}{1.1} - 1.3745\right)\frac{WL}{EA} = 4.3948\,\frac{WL}{EA}$$

Therefore we have from Eq. (6-19) that

$$4.3594\,\frac{WL}{EA} \leq \delta_j^{\text{actual}} \leq 4.3948\,\frac{WL}{EA} \tag{6-19d}$$

Equation (6-19d) informs us that the described modifications have brought the magnitude of the displacement to within, while still being near, its limit value of 4.4 WL/EA. It is important to note that the bounding technique has enabled us to come to this conclusion without ever having to conduct a re-analysis of the truss at any time during the modification process. In fact, it is only at this point that we need analyze the truss in order to verify the modifications that we have made to date. For the truss in Fig. 6-1 modified such that $a_5 = a_6 = 1.1A$, we find from analysis through Eqs. (6-1) to (6-3) that the displacement $\delta_j = u_3 = 4.3948\ WL/EA$, which satisfies the design condition that the displacement be not greater than $4.4\ WL/EA$.

That the actual value of $\delta_j = u_3$ for the modified truss is $4.3948\ WL/EA$, and that this is also the upper-bound value for Eq. (6-19d), illustrates that Eq. (6-17) may predict the exact value of the displacement δ_j in some instances. In fact, this will occur whenever the modifications to the truss are such that there is no change in the distribution of member axial forces. Such is the case for the truss in Fig. 6-1 when, as in the foregoing, the cross-section areas for members 5 and 6 are uniformly changed. In general, Eq. (6-17) provides a closer estimate of the displacement δ_j than does Eq. (6-15).

SUMMARY AND DISCUSSION

The sensitivity analysis techniques described in this chapter greatly improve the design process for structural frameworks under static loads. (In fact, while not considered herein, sensitivity analysis may be conducted for a broad range of structure types subject to a variety of static or dynamic loading.) Response sensitivities are easy to calculate, give valuable insight into structural behavior, and provide an effective means to carry out design modifications for a structure in a systematic and economical way. In essence, they allow optimal modifications to be made to a structure without the need for a tedious trial-and-error analysis process. The hands-on sensitivity-based design procedure described in this chapter is a simple version of a fully computer-automated design optimization capability described in Chapter 8, "How to Optimize Structural Steel Frameworks."

APPENDIX 6-A. NOTATION

The following symbols are used in this chapter.

A = cross-section area
a = cross-section area
$\mathbf{b}$ = virtual load vector
$\mathbf{C}$ = compatibility (topological) matrix
E = Young's modulus
I = cross-section moment of inertia
$\mathbf{K}$ = stiffness matrix
L = member length
M = bending moment
m = number of members
N = axial force

n = number of degrees of freedom
$\mathbf{P}$ = load vector
q = member deformation vector
$\mathbf{t}$ = virtual load vector
$\mathbf{u}$ = nodal displacement vector
W = load
x = reciprocal cross-section area
y = neutral-axis depth
δ = displacement
σ = stress

Chapter 7

HOW TO OPTIMIZE REINFORCED CONCRETE FRAMES

A procedure for the optimization of rectangular, three-dimensional, reinforced concrete frames is presented. The optimization of reinforced concrete frames is best accomplished in phases. In Phase I, the cross-sectional dimensions of the members are optimized. These dimensions are treated as continuous design variables, and any reliable nonlinear programming (NLP) algorithm such as those discussed in Appendix B may be used to solve the problem. In Phase II, these cross-sectional dimensions as well as the member axial forces and moment and shear envelopes are held fixed while the reinforcement is optimized member by member. Since the optimization in Phase II is performed individually for each member, the discrete exhaustive search methods described in Chapter 4 on "How to Optimize a Reinforced Concrete Column or Beam" should be used. This chapter is devoted primarily to the optimization problem in Phase I.

Even though the reinforcement is designed in Phase II, constraints and objectives in Phase I such as strength and cost are functions of both cross-sectional dimensions and reinforcement. Therefore in Phase I the optimal reinforcement design is approximated so that objective and constraint functions can be evaluated. A continuous smooth approximation is used so that efficient, gradient-based NLP algorithms can be used to solve the Phase I optimization problem.

The chapter begins by describing the formulation of the Phase I problem and the method of analysis. Particular attention is given to the assumptions and approximations made relative to the reinforcement design for beams and columns. Results from application of the procedure to a two-story frame are described in detail, and results

Richard J. Balling, email: ⟨rick@et.byu.edu⟩ (contact for correspondence), and Xiaoping Yao, Department of Civil and Environmental Engineering, Brigham Young University, Provo, Utah 84602.

from application of the procedure to a four-story frame are mentioned.

PROBLEM FORMULATION FOR PHASE I

The scope of the problem treated in this chapter is limited to the optimization of three-dimensional, rectangular, reinforced concrete frames consisting of rectangular columns and beams. The frame may also include slabs, but it is assumed that these have been designed beforehand. The constant parameters specified for the problem include:

Geometry: the number of stories and bays; story heights and bay widths; support and member connectivity conditions.
Loads: dead loads due to slabs, partitions, walls, nonstructural elements, and the like; live loads; snow and wind loads; seismic loads.
Material Properties: moduli of elasticity; yield strength of reinforcement; compressive strength of concrete; unit weights.
Unit Costs: weight-proportional and bar-proportional cost factors for reinforcement; volume-proportional cost factor for concrete.

The design variables for the Phase I optimization problem consist of the width b and depth h of each of the rectangular beams and columns in the frame. Members may be grouped so that all members in the group have the same b and h, and thus the entire group is represented by just two design variables. For beams that are cast monolithically with slabs, it is assumed that the slab has already been designed. Therefore the slab thickness t and the effective tee-beam flange width b_{eff} are constants, and the two design variables b and h represent the web width and total depth, respectively.

The objective in Phase I is to minimize the cost of the frame. This includes the concrete cost and the estimated reinforcement cost. Concrete cost is taken proportional to the total concrete volume consistent with standard practice. Reinforcement cost includes material and placement costs. The material cost of reinforcement is taken proportional to total reinforcement weight, and the placement cost is taken proportional to the total number of reinforcing bars.

Constraints for Phase I include deflection and drift constraints. Additional constraints whose meaning and evaluation are explained in the following sections include the following constraints for each column.

1. Column Strength Capacity
2. Maximum Reinforcement
3. Spacing

The following constraints are imposed at each critical moment location of each beam.

4. Flexural Strength Capacity
5. Maximum Reinforcement
6. Spacing

Critical negative moment locations in beams occur at supports, and critical positive moment locations occur where the moment is maximum in the interior of each span. Critical locations for shear in beams occur at a distance d from the face of each support where d is the distance from the extreme compressive concrete fiber to the centroid of the tension reinforcement. At each critical shear location in each beam, the following constraint is imposed.

7. Shear Strength Capacity

METHOD OF ANALYSIS FOR PHASE I

The NLP algorithm used in Phase I repeatedly calls for evaluations of the objective and constraint functions. A single evaluation of all such functions is referred to as an "analysis." The analysis for Phase I consists of two parts:

Frame Analysis

A linear elastic frame analysis based on the standard stiffness method is performed. The member weight and stiffness (EI, EA, GJ) are assumed to be proportional to gross cross-sectional properties that are determined directly from the design variables b and h. Output from the frame analysis consists of the deflection and drift constraints as well as the internal forces for each member. For columns, the internal forces consist of the factored axial force and moments (P_u, M_{uy}, M_{uz}) for each loading case. For beams, the internal forces consist of the factored moment envelope (M_u) and shear envelope (V_u) functions.

Individual Member Analysis

After executing the frame analysis, individual member analyses are performed for each member. Each individual member analysis receives its appropriate internal forces as input and evaluates the member cost as well as the constraints associated with the member. In order to make these evaluations, several assumptions must be made regarding the reinforcement. The assumptions and evaluations that take place in the individual member analyses are described in the following sections.

THE INDIVIDUAL MEMBER ANALYSIS FOR BEAMS

The concrete cost for a particular beam can be evaluated directly from the specified design variables b and h. The reinforcement cost and the constraints, however, require that one estimate the amount of reinforcement. To this end, it first becomes necessary to estimate the area of steel A_s at each critical moment location.

An assumption is made that in the optimal design, A_s is controlled at each critical moment location by either flexural strength or minimum reinforcement provisions of the ACI Specification (Building 1989). For a singly reinforced rectangular beam, if one assumes that flexural strength controls at a particular moment location, then one can solve for A_s from the equation

$$M_u = \phi M_n = \phi A_s f_y \left(d - \frac{A_s f_y}{2(0.85) f'_c b} \right) \tag{7-1}$$

where

M_u = factored moment at the critical moment location
ϕ = flexural strength reduction factor
M_n = flexural strength at the critical moment location
f_y = reinforcement yield strength
f'_c = concrete compressive strength

In Eq. (7-1) one must obtain a formula for d, the distance from the extreme concrete compression fiber to the centroid of the tension reinforcement. The following formula can be used.

$$d = h - \text{cover} - d_{\text{stir}} - \frac{n_{\text{layer}} d_b}{2} - \frac{(n_{\text{layer}} - 1)\ \text{space}}{2} \tag{7-2}$$

where

cover = required cover distance
space = required clear spacing between bars
d_{stir} = diameter of stirrups
d_b = diameter of main reinforcing bars
n_{layer} = number of layers of main reinforcing bars

At this point, one must assume values for d_{stir} and d_b. In the examples that follow in this chapter, it was assumed that d_{stir} = 0.952 cm (0.375 in.) which corresponds to a #3 bar, and d_b = 1.59 cm (0.625 in.) which corresponds to a #5 bar. One may also conservatively assume n_{layer} = 1, or approximate more rigorously:

$$n_{\text{layer}} = \frac{A_s(d_b + \text{space})}{\dfrac{\pi d_b^2}{4}\,(b - 2(\text{cover} + d_{\text{stir}}) - d_b)} \tag{7-3}$$

Substituting Eqs. (7-2) and (7-3) into Eq. (7-1) leads to a quadratic equation for A_s of the form:

$$\alpha A_s^2 + \beta A_s + \gamma = 0 \quad \text{with } \alpha > 0 \tag{7-4}$$

The coefficients α, β, and γ are in terms of the design variables b and h, the M_u received from frame analysis, and the constant parameters ϕ, f_y, f'_c, d_b, d_{stir}, cover, and space. A typical plot of A_s versus M_u with all other variables and parameters held fixed is shown in Fig. 7-1. The dashed parabolic curve represents the solution for A_s from the quadratic formula. For a given value of M_u there are two solutions for A_s. Naturally, the smaller value of A_s would be chosen. Note also that when $M_u > M_{u\max}$ there is no solution for A_s. Mathematically, $M_u > M_{u\max}$ represents the case where the discriminant in

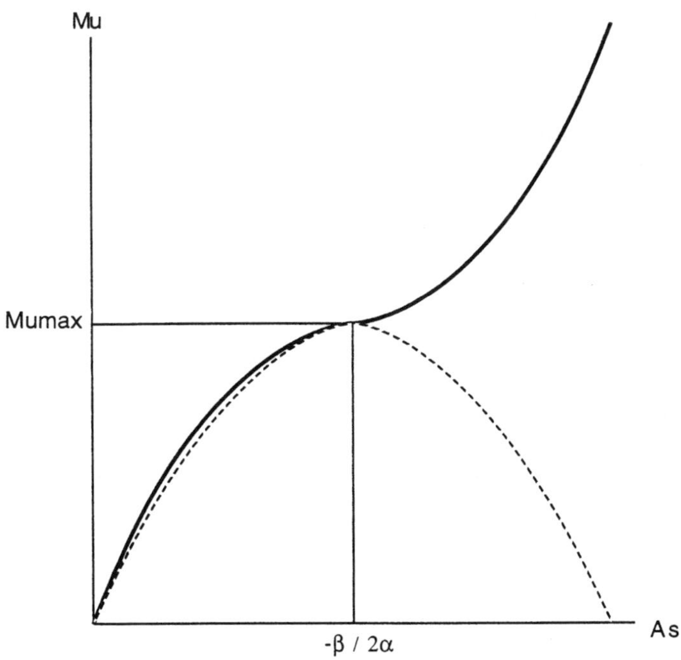

FIG. 7-1. M_u *Versus* A_s *for Beams.*

the quadratic formula is negative. Physically, M_{umax} represents the maximum factored moment that can be withstood at the section with unlimited reinforcement. If $M_u > M_{umax}$, the only thing that can be done to satisfy strength is to increase the b and h. Therefore a "flexural strength capacity" constraint must be included in the Phase I optimization problem that constrains $M_u \leq M_{umax}$ at each critical moment location. In other words, the discriminant in the quadratic formula must be constrained to be nonnegative:

$$\beta^2 - 4\alpha\gamma \geq 0 \tag{7-5}$$

During the course of Phase I optimization, the NLP algorithm may send values of b, h, and M_u to the individual member analysis that will lead to violation of the flexural strength capacity constraint. Nevertheless, a value for A_s is still needed in order to evaluate the remaining cost and constraint functions of this infeasible design. Therefore an artificial value of A_s is suggested for this situation (see solid line in Fig. 7-1):

$$\begin{aligned} &\text{if } \beta^2 - 4\alpha\gamma \geq 0, \quad A_s = \frac{-\beta - \sqrt{\beta^2 - 4\alpha\gamma}}{2\alpha} \\ &\text{if } \beta^2 - 4\alpha\gamma < 0, \quad A_s = \frac{-\beta + \sqrt{4\alpha\gamma - \beta^2}}{2\alpha} \end{aligned} \tag{7-6}$$

This concludes the discussion on the evaluation of A_s assuming that it is controlled by flexural strength. Now assume that A_s is controlled by minimum reinforcement. Then one can determine A_s from the following linear equation.

$$\rho = \frac{A_s}{bd} = \frac{1.38 \text{ MPa}}{f_y} = \frac{200 \text{ psi}}{f_y} \tag{7-7}$$

where one would substitute Eqs. (7-2) and (7-3) for d. A_s is calculated from Eq. (7-7) and compared to the A_s calculated from Eq. (7-6) and the greater of the two values is taken at each critical moment location.

With the A_s known at each critical moment location, the remaining costs and constraints can be evaluated. The "maximum reinforcement" constraint requires that at each critical moment location:

$$\rho = \frac{A_s}{bd} \leq 0.75\, \rho_b \tag{7-8}$$

where

ρ_b = reinforcement ratio for the balanced condition

The "spacing" constraint requires that b and h be large enough so the reinforcing bars fit within the section at each critical moment location:

$$\min[b - 2(\text{cover} + d_{\text{stir}}) - 2d_b - \text{space}, h - 2(\text{cover} + d_{\text{stir}}) - n_{\text{layer}}d_b - (n_{\text{layer}} - 1)\text{space}] \geq 0 \quad (7\text{-}9)$$

At each critical shear location, the "shear strength capacity" constraint ensures that the required shear strength provided by stirrups is less than the maximum allowed:

$$V_s = \frac{V_u}{\phi} - V_c \leq \sqrt{441 \text{ kPa } f_c'}\, bd = \sqrt{64 \text{ psi } f_c'}\, bd \quad (7\text{-}10)$$

where

V_u = factored shear at the critical shear location
ϕ = shear strength reduction factor
V_c = $\sqrt{27.6 \text{ kPa } f_c'}\, bd = \sqrt{4 \text{ psi } f_c'}$ = concrete shear strength

The reinforcement cost is evaluated as:

$$C_{\text{steel}}\gamma_{\text{steel}}\left(\frac{\pi d_b^2}{4}\sum_{i=1}^{n_b} L_{bi} + \frac{\pi d_{\text{stir}}^2 n_{\text{stir}} L_{\text{stir}}}{4}\right) + C_b n_b + C_{\text{stir}} N_{\text{stir}} \quad (7\text{-}11)$$

where

C_{steel} = material cost per weight of reinforcement
γ_{steel} = unit weight of steel
n_b = number of main reinforcing bars
L_{bi} = length of ith main reinforcing bar
n_{stir} = number of stirrups
L_{stir} = length of a single stirrup
C_b = placement cost per main reinforcing bar
C_{stir} = placement cost per stirrup

The number of main reinforcing bars at each critical moment location is easily determined from A_s and the assumed d_b.

$$n_b = \frac{4A_s}{\pi d_b^2} \quad (7\text{-}12)$$

The number of stirrups can be estimated by evaluating the following integral.

$$n_{\text{stir}} = \int_0^L \frac{dx}{s(x)} \quad (7\text{-}13)$$

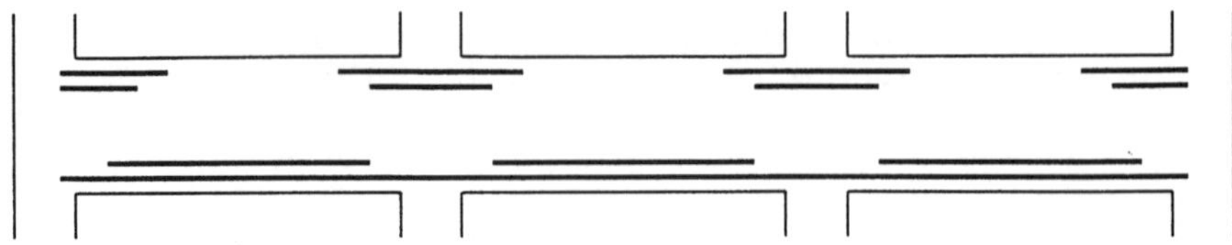

FIG. 7-2. Assumed Reinforcement Topology for Three-Span Beam.

where

L = length of the beam where stirrups are required
$s(x)$ = stirrup spacing function

The stirrup spacing function is controlled by either the minimum stirrup spacing requirement or the shear strength requirement at location x:

$$s(x) = \min\left[\frac{d}{2}, \frac{\pi d_{\text{stir}}^2 f_y d}{2V_s(x)}\right] \quad \text{where } V_s(x) = \frac{V_u(x)}{\phi} - V_c \tag{7-14}$$

Even though n_b and n_{stir} may not turn out to be integers from Eqs. (7-12) to (7-14), they are used directly in Eq. (7-11). The stirrup length L_{stir} can be determined directly from b and h. In order to determine the lengths L_{bi} of the main reinforcing bars, a reinforcement topology must be assumed such as the one shown in Fig. 7-2 which shows two groups of bars at each critical moment location. This topology was assumed in the examples that follow in this chapter where it was assumed that 33% of n_b is allocated to the longer bar group and 67% of n_b is allocated to the shorter bar group at each critical location. The minimum cutoff points can be determined from the factored moment envelopes in accordance with the reinforcement development provisions of the ACI Specification (ACI 1989).

THE INDIVIDUAL MEMBER ANALYSIS FOR COLUMNS

The concrete cost for a particular column can be evaluated directly from the specified design variables b and h. As was the case with beams, it is necessary to estimate the area of steel A_s in order to determine the reinforcement cost and the constraints.

An assumption is made that in the optimal design, A_s is controlled by either the column strength or minimum reinforcement provisions of the ACI Specification (ACI 1989). Assume for now that column strength under a particular loading case controls A_s:

$$P_u = \phi P_n \text{ at specified eccentricities } e_y = \frac{M_{uy}}{P_u} \text{ and } e_z = \frac{M_{uz}}{P_u} \tag{7-15}$$

Unlike the case of beams, column strength ϕP_n is not a simple formula in terms of A_s. In fact for a given column under biaxial bending, strength evaluation may require execution of a numerical procedure such as the Newton-Raphson procedure suggested by Sun and Lu (1992). Such a procedure should be used to evaluate the strength for three different A_ss:

$A_s = 0.01bh$
$A_s = 0.02bh$
$A_s = 0.08bh$

The numerical procedure for evaluating strength requires a knowledge of the number, size, and location of each bar in the cross-section. To this end, the number of bars is assumed to be four—one in each corner of the column. The diameter of the bars can then be calculated as:

$$d_b = \sqrt{\frac{A_s}{\pi}} \qquad (7\text{-}16)$$

After $P_u = \phi P_n$ corresponding to the preceding three values of A_s is determined, a quadratic curve can be fit through the three points as shown in Fig. 7-3. This curve is analogous to the $M_u - A_s$ curve for beams (see Fig. 7-1). The equation for this curve has the same form as Eq. (7-4) given earlier for beams. A "column strength capacity" constraint of the form of Eq. (7-5) is included in the Phase I optimization problem. The A_s is evaluated according to Eq. (7-6). This entire procedure for determining A_s is repeated for each loading case, and the maximum value is taken which also must be greater than the minimum reinforcement value:

$$A_s = 0.01bh \qquad (7\text{-}17)$$

After the A_s has thus been determined as the maximum from Eq. (7-17) and Eq. (7-6) for each loading case, the remaining costs and constraints can be evaluated. The "maximum reinforcement" constraint requires that

$$\rho = \frac{A_s}{bh} \leq 0.08 \qquad (7\text{-}18)$$

The "spacing" constraint requires that b and h be large enough so that four reinforcing bars fit within the section:

$$\min[b - 2(\text{cover} + d_{\text{tie}} + d_b) - \text{space},$$
$$h - 2(\text{cover} + d_{\text{tie}} + d_b) - \text{space}] \geq 0 \qquad (7\text{-}19)$$

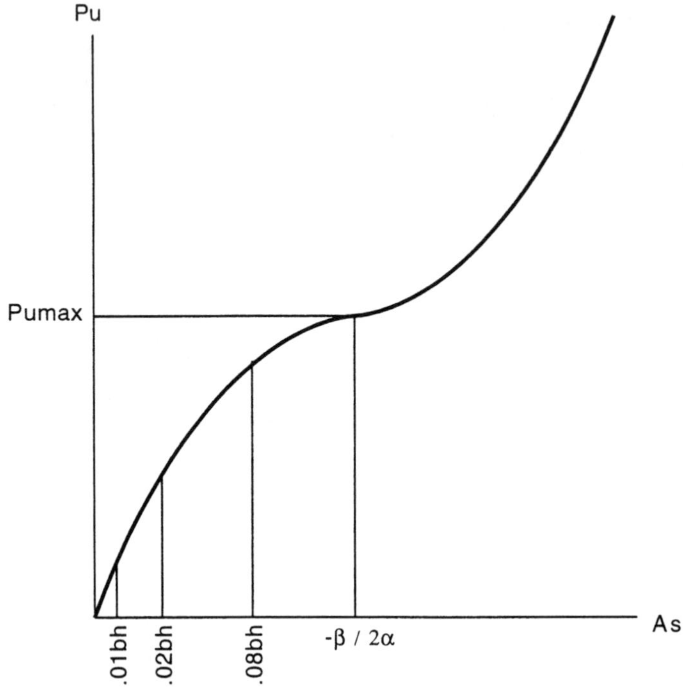

FIG. 7-3. P_u Versus A_s for Columns.

where

d_{tie} = diameter of ties = 0.952 cm (0.375 in.)

The reinforcement cost is evaluated as:

$$C_{steel}\gamma_{steel}\left(\pi d_b^2 L + \frac{\pi d_{tie}^2 n_{tie} L_{tie}}{4}\right) + 4C_b + C_{tie}n_{tie} \tag{7-20}$$

where

C_{steel} = material cost per weight of reinforcement
γ_{steel} = unit weight of steel
L = length of column
n_{tie} = number of ties
L_{tie} = length of a single tie
C_b = placement cost per main reinforcing bar
C_{tie} = placement cost per tie

The number of ties is computed as:

$$n_{\text{tie}} = \frac{L}{\min(16d_b, 48d_{\text{tie}}, b, h)} \tag{7-21}$$

Even though n_{tie} may not turn out to be an integer from Eq. (7-21), it is used directly in Eq. (7-20). The tie length L_{tie} can be determined directly from b and h.

EXAMPLES

The two-story frame shown in Fig. 7-4 was optimized according to the method described in this chapter. The members were grouped into the following.

1. bottom story beams parallel to x
2. top story beams parallel to x
3. bottom story exterior beams parallel to z
4. bottom story middle beam parallel to z
5. bottom story interior beams parallel to z
6. top story exterior beams parallel to z

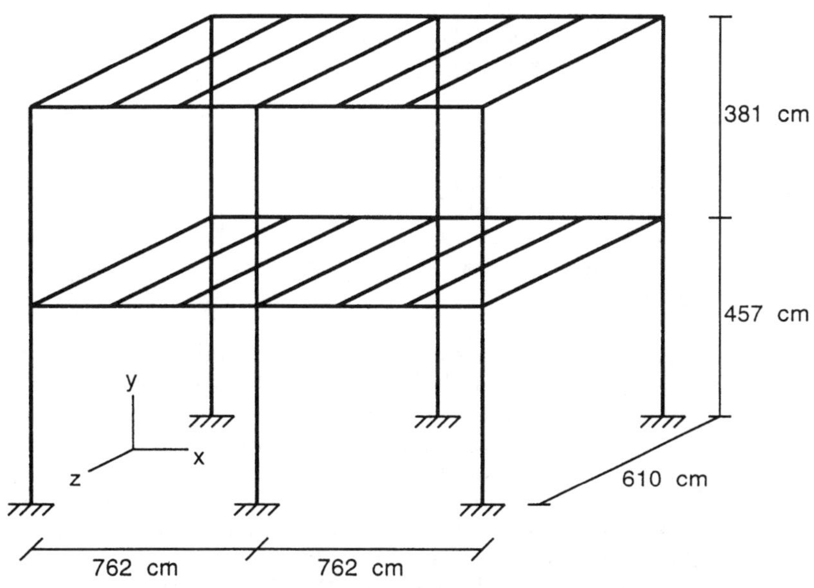

FIG. 7-4. Two-Story Frame.

7. top story middle beam parallel to z
8. top story interior beams parallel to z
9. exterior columns
10. interior columns

It is assumed that the interior beams are simply supported on the beams parallel to x. One-way slabs with thickness = 12.7 cm (5 in.) and weight = 2293 Pa (62.5 psf) rest on the interior beams. The slabs and beams are cast monolithically so that the beams are effectively tee-beams. The roof supports a roofing load of 575 Pa (12 psf) and a snow load of 1436 Pa (30 psf). The first story floor supports a partition load of 958 Pa (20 psf) and a live load of 3830 Pa (80 psf). The exterior wall is clay brick with thickness = 20.3 cm (8 in.) and weight = 3783 Pa (79 psf). The interior wall in the middle of the frame is steel and shear rock with weight = 2298 Pa (48 psf).

The frame is subjected to seismic loading in accordance with the procedure outlined in the Uniform Building Code (''Uniform'' 1994). The frame is in a seismic zone III and is designed as a shear wall system in which the frame is required to support 100% of the vertical load and 25% of the lateral load. Five loading cases were considered in which dead load, live load, snow load, and seismic load in each direction were critically combined.

Material properties were taken as:

$$f_y = 2.87 \text{ MPa } (60{,}000 \text{ psi})$$
$$f'_c = 192 \text{ KPa } (4000 \text{ psi})$$
$$\gamma_{\text{conc}} = 23.6 \text{ KN}/m^3 \text{ (150 pcf)}$$
$$\gamma_{\text{steel}} = 77.0 \text{ KN}/m^3 \text{ (490 pcf)}$$

Unit costs were taken as:

$$C_{\text{conc}} = \$131/m^3/(\$100/\text{yd}^3)$$
$$C_{\text{steel}} = \$0.056/N \; (\$0.25/\text{lb})$$
$$C_b = \$5/\text{bar}$$
$$C_{\text{stir}} = \$5/\text{stirrup}$$
$$C_{\text{tie}} = \$5/\text{tie}$$

Three different starting designs were generated: a ''conservative'' design (\$14,241); an ''infeasible'' design (\$11,536); and a ''competitive'' design (\$11,363). The OPTDESX software (Parkinson et al. 1992) was used to get basically the same optimum design (\$7,928) from the three starting designs in an average of about 12 minutes of CPU time on an HP 755-99 workstation. The Sequential Quadratic Programming (SQP) algorithm in the OPTDESX software package was used as the NLP algorithm (Powell 1978). Member dimensions were constrained between 15.2 cm (6 in.) and 76.2 cm (30 in.). The optimum member dimensions were:

	b	h
bottom story beams parallel to x	24.0 cm (9.45 in.)	76.2 cm (30.0 in.)
top story beams parallel to x	17.5 cm (6.89 in.)	72.2 cm (28.4 in.)
bottom story exterior beams parallel to z	15.3 cm (6.01 in.)	27.7 cm (10.9 in.)
bottom story middle beam parallal to z	17.0 cm (6.70 in.)	30.1 cm (11.8 in.)
bottom story interior beams parallel to z	18.9 cm (7.46 in.)	61.0 cm (24.0 in.)
top story exterior beams parallel to z	18.1 cm (7.12 in.)	73.6 cm (29.0 in.)
top story middle beam parallel to z	18.2 cm (7.16 in.)	72.4 cm (28.5 in.)
top story interior beams parallel to z	17.4 cm (6.84 in.)	49.7 cm (19.6 in.)
exterior columns	32.7 cm (12.9 in.)	32.6 cm (12.8 in.)
interior columns	36.9 cm (14.5 in.)	30.5 cm (12.0 in.)

These *b*s and *h*s along with the member axial forces and moment and shear envelopes were then submitted to Phase II where the reinforcement in the members was discretely designed according to the procedures outlined in Chapter 4 which required but a few minutes on the same computer. This led to an optimal design with a cost of $8,486. As an example, the final reinforcement design for bottom story beams parallel to x is shown in Fig. 7.5.

The four-story frame in Fig. 7-6 was also optimized. Beams parallel to x were grouped into 6 groups, beams parallel to z were grouped into 6 groups, and columns were grouped into 3 groups. Thus in Phase I there were a total of 30 design variables (b and h for each member group). Details of the optimization are not given here other than the fact that the entire process required about 2.6 hours of CPU time on an HP 755-99 workstation.

SUMMARY AND DISCUSSION

At this point, one may be disturbed by the fact that at the end of Phase II, the reinforcement design may be quite different from what was assumed in Phase I. For example, in Phase I the reinforcement topology in Fig. 7-2 was assumed for beams, it was assumed that #5 bars were used in beams, and it was assumed that the number of bars in columns was four. At the end of Phase II these assumptions may (and did in the case of the two-story frame) prove to be false. This fact along with the proper consideration of discreteness in Phase II explains the difference in cost for the two-story frame—$7,928 at the end of Phase I versus $8,486 at the end of Phase II. These discrepancies may cause one to question the optimum values for the *b*s and *h*s obtained in Phase I.

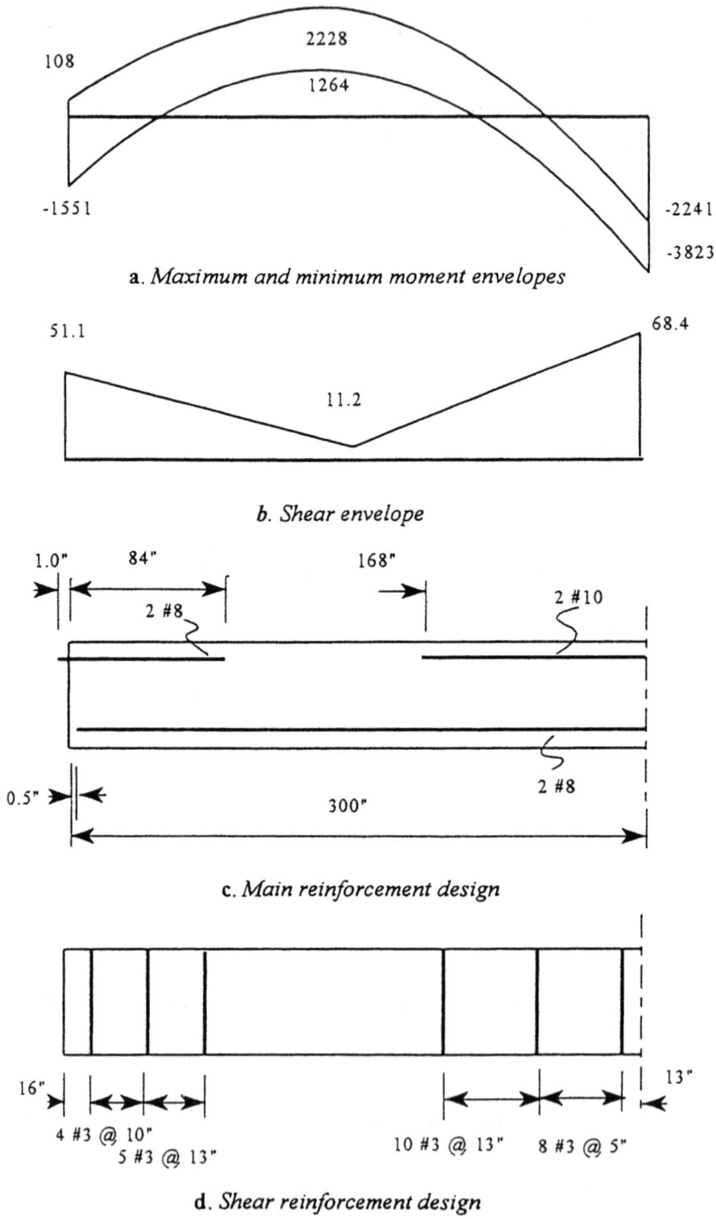

FIG. 7-5. Reinforcement Design for Bottom Story Beams Parallel to X.

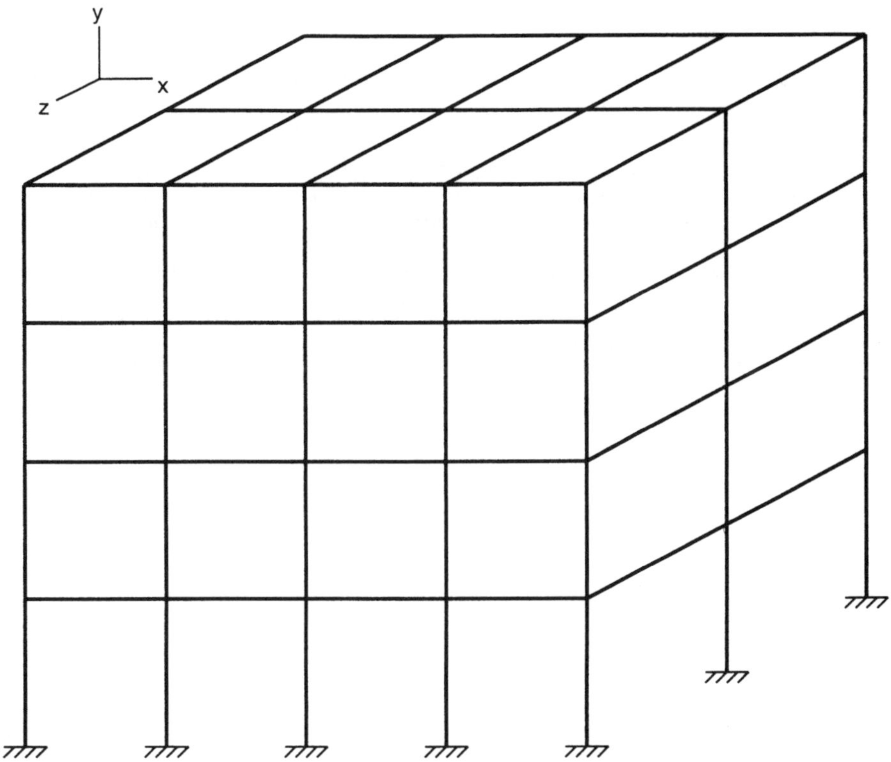

FIG. 7-6. Four-Story Frame.

Research has shown (Balling and Yao 1996) that after a very lengthy discrete optimization of the two-story frame where reinforcement topologies, numbers of bars, and diameters of bars were considered as design variables simultaneously with *b*s and *h*s, the optimum cost was $8,271, which is only 2.5% lower than the cost at the end of Phase II of the procedure described in this chapter. Furthermore, the average difference in the optimum values for *b*s and *h*s between the lengthy discrete optimization and the values determined in Phase I of the procedure described in this chapter was 14.5%. Some *b*s and *h*s were larger, but these were compensated by others that were smaller so that the optimum cost only dropped by 2.5%! This evidence supports the assumption made in this chapter that the details of reinforcement design do not significantly affect the optimum values of the *b*s and *h*s. Thus the reinforcement assumptions made in Phase I are not that critical to the final outcome.

REFERENCES

ACI (1989). Building code requirements for reinforced concrete, ACI 318-89. American Concrete Institute, Detroit, MI.

Balling, R. J., and Yao, X. (1997). "Optimization of reinforced concrete frames." *J. of Structural Engrg.*, ASCE, 123(3).

Parkinson, A. R., Balling, R. J., and Free, J. C. (1992). "OPTDESX: An X Window-based optimal design software system." AIAA Paper 92-4759, 4th AIAA/NASA/USAF/OAI Symposium on Multidisciplinary Analysis and Optimization, Cleveland, OH, September 21-23.

Powell, M. J. D. (1978). "Algorithms for nonlinear functions that use Lagrangian functions." *Mathematical Programming*, 14, 224.

Sun, B. - J., and Lu, Z. - T. (1992). "Design aides for reinforced concrete columns." *J. of Structural Engrg.*, ASCE, 118(11), 2986.

Uniform Building Code. (1994). International Conference of Building Officials, Whittier, CA.

Chapter 8

HOW TO OPTIMIZE STRUCTURAL STEEL FRAMEWORKS[1]

This chapter presents the details of an optimization strategy for the computer-automated design of least-weight structural steel frameworks, where members are sized using standard steel sections in conformity with design code requirements. The underlying features and functions of the corresponding computer software essential for application of the design strategy in professional practice are not dealt with herein, but may be briefly summarized as follows: capability to conduct first-order and second-order analysis of planar and space frameworks; account for the strength/stability of individual members and the stiffness of the structure as a whole; clause by clause verification of the design in conformance with all the relevant provisions of the governing steel design code, for example, American Institute of Steel Construction (1989, 1986) and Canadian Standards Association (1984); automatic sizing of members using commercially available standard steel sections; account for a broad range of section shapes (e.g., *W*, *T*, hollow-box, single and double angle, etc.); determination of local buckling classifications of member sections; calculation of effective length factors of members; calculation of the unbraced compression flange length of flexible members; account for fixed, pinned, spring, and roller supports, as well as for pinned and bolted connections; account for various gusset-plate thicknesses for back-to-back double-angle sections; account for strong-axis and weak-axis bending of member sections; capability to specify common section properties for groups of members, to impose limitations on

---

Donald E. Grierson, Civil Engineering, University of Waterloo, Ontario, Canada N2L 3G1, email: ⟨grierson@sail.uwaterloo.ca⟩.

[1]Excerpted in part, with permission, from the monograph on Advances in Design Optimization, edited by H. Adeli, Chapter 12, pages 451–491, published by Chapman and Hall, London, 1993.

section depths, and to fix the section properties of selected members; and account for a variety of nodal and member load types, as well as for effects due to temperature change and support settlement. A computer software system based on the optimization techniques described in this chapter, and which incorporates all of the foregoing features and functions (Grierson and Cameron 1994), is employed to design several structural steel frameworks, some from professional practice, to illustrate the scope and effectiveness of the design optimization strategy.

INTRODUCTION

The means by which a structural design evolves from an initial to a final form can be described as an optimization process wherein attempts are made to minimize costs or maximize benefits while satisfying design criteria reflecting performance and fabrication requirements. This chapter presents the details of a synthesis strategy to conduct computer-automated least-weight design of structural steel frameworks. (Although not considered herein, it is possible to design for objectives other than least weight, such as minimum cost or maximum compliance.) Members are automatically sized using commercial standard steel sections in full conformity with governing steel design code provisions for strength/stability and stiffness. Several design examples are presented to illustrate the scope and effectiveness of the optimization technique.

We note that design is herein defined as the detailed proportioning of the members of a structural steel framework to satisfy strength, stiffness, and fabrication criteria. This definition assumes that the structural layout and loadings have been finalized and it only remains to size the structural components. Within this context the design process may be reduced to the following sequence of tasks: assume an initial trial design; analyze the trial structure to determine behavioral responses; compare these responses to design code specifications; select a new trial design to either remedy any code violations or to improve on the economy of the previous design (herein, structural weight is taken as the measure of design economy); and re-analyze and continue the process until the design is deemed satisfactory. This trial-and-error iterative procedure often leads to a less than optimal final design after a tedious analysis-design process that owes its termination only to the skilled intuition and patience of the designer. The process becomes quite cumbersome because of the many facets of the design that must be coordinated. The initial design must be selected carefully because its quality has significant bearing on how quickly the design evolves to a satisfactory state. Design code verification is an exhaustive and time-consuming activity that requires selective application of critical code clauses at various stages of the synthesis process. The larger and more complicated the structure, the more difficult it becomes to have simultaneous concern for both design economy and design performance. On the other hand, contrary to the trial-and-error approach, computer-based optimization techniques can be effectively applied to overcome most of these difficulties.

DESIGN FORMULATION

The design problem is to seek a minimum weight structure while satisfying all strength and stiffness performance requirements simultaneously under static loading. For a structure having $i = 1, 2, \ldots, n$ members, and taking the cross-sectional area a_i as the sizing variable for each member i, the general form of the design problem may be stated as

Minimize

$$\sum_{i=1}^{n} w_i a_i \qquad \text{(8-1)}$$

Subject to

$$\delta_j \leq \underline{\delta}_j \leq \bar{\delta}_j \qquad (j = 1, 2, \ldots, d) \qquad \text{(8-2)}$$

$$\underline{\sigma}_k \leq \sigma_k \leq \bar{\sigma}_k \qquad (k = 1, 2, \ldots, s) \qquad \text{(8-3)}$$

$$a_i \in A_i \qquad (i = 1, 2, \ldots, n) \qquad \text{(8-4)}$$

Equation (8-1) defines the weight of the structure (w_i = weight coefficient for member i = material density × member length); Eqs. (8-2) define d constraints on displacements δ_j (under- and superscored quantities denote specified lower and upper bounds, resp.); Eqs. (8-3) define s constraints on stresses σ_k; Eqs. (8-4) require each cross-sectional area a_i to belong to the discrete set of areas $A_i = \{a_1, a_2, \ldots\}_i$ prevailing for the section shape specified for member i (e.g., W, T, hollow-box, double-angle, etc.).

The coefficients w_i in the objective function Eq. (8-1) remain constant throughout the synthesis process. This is also true for the displacement bounds δ_j and $\underline{\delta}_j$ in Eqs. (8-2). On the other hand, the allowable tensile and compressive stress bounds in Eqs. (8-3) must be updated over the design history because corresponding axial, flexural, shear, and torsional capacities are dependent on unbraced member lengths and the geometry of member sections, both of which vary as the synthesis proceeds. For example, the lengths of unbraced compression flanges depend on the bending moment distribution, which changes with modifications in the stiffness distribution for the structure. Or, the width-thickness ratios of section plate elements change as different standard sections are selected for a member at different stages of the synthesis process. The tensile stress bound $\bar{\sigma}_k$ accounts for the type of member connection (bolted, etc.), whereas the compressive stress bound $\underline{\sigma}_k$ depends on the stress state (axial, flexural, combined, etc.) and accounts for both local section buckling and overall member buckling.

In their present form, the displacement and stress constraints Eqs. (8-2) and (8-3) are implicit nonlinear functions of the cross-sectional areas a_i. To facilitate computer solution, it is necessary to express these constraints as explicit functions of the sizing variables a_i. In fact, while introducing some approximations into the formulation, it is computationally advantageous to express these constraints as linear functions of the design variables. This suggests, then, to approximate each displacement and stress by its first-order Taylor series expansion. Moreover, since displacement and stresses in elastic structures under static loads generally vary inversely with member sizes a_i, it follows that better quality explicit linear constraint approximations result if the Taylor series expansions are expressed in terms of the reciprocal-sizing variables

$$x_i = \frac{1}{a_i} \qquad (i = 1, 2, \ldots, n) \tag{8-5}$$

The first-order Taylor series approximation of each displacement δ_j in terms of reciprocal sizing variables x_i is

$$\delta_j = \delta_j^0 + \sum_{i=1}^{n} \left(\frac{\partial \delta_j}{\partial x_i}\right)^0 (x_i - x_i^0) \tag{8-6}$$

whereas that for each stress σ_k is

$$\sigma_k = \sigma_k^0 + \sum_{i=1}^{n} \left(\frac{\partial \sigma_k}{\partial x_i}\right)^0 (x_i - x_i^0) \tag{8-7}$$

where the superscript (0) indicates known quantities for the current design stage (e.g., the initial "trial" design). The quantities $(\partial\delta_j/\partial x_i)^0$ and $(\partial\sigma_k/\partial x_i)^0$ are displacement and stress gradients referenced to the current design, and the x_i are the sizing variables for the next optimization. It is readily shown that

$$\delta_j^0 = \sum_{i=1}^{n} \left(\frac{\partial \delta_j}{\partial x_i}\right)^0 x_i^0 \tag{8-8}$$

$$\sigma_k^0 = \sum_{i=1}^{n} \left(\frac{\partial \sigma_k}{\partial x_i}\right)^0 x_i^0 \tag{8-9}$$

Therefore, from Eqs. (8-1) to (8-9), the statement of the design optimization problem becomes:

Minimize

$$\sum_{i=1}^{n} w_i / x_i \tag{8-10}$$

Subject to

$$\underline{\delta}_j \le \sum_{i=1}^{n} \left(\frac{\partial \delta_j}{\partial x_i}\right)^0 x_i \le \bar{\delta}_j \qquad (j = 1, 2, \ldots, d) \tag{8-11}$$

$$\underline{\sigma}_k \le \sum_{i=1}^{n} \left(\frac{\partial \sigma_k}{\partial x_i}\right)^0 x_i \le \bar{\sigma}_k \qquad (k = 1, 2, \ldots, s) \tag{8-12}$$

$$x_i \in X_i \qquad (i = 1, 2, \ldots, n) \tag{8-13}$$

where the elements of each discrete set $X_i = \{x_1, x_2, \ldots\}$ in Eq. (8-13) are the reciprocals of the cross-sectional areas comprising the corresponding discrete set A_i in Eqs. (8-4); the reciprocal areas x_i correspond to the database of standard steel sections having the shape specified for member i.

The displacement and stress gradients in Eqs. (8-11) and (8-12) may be evaluated using a number of different sensitivity analysis techniques. For example, a finite difference technique based on a central difference scheme may be employed to approximate the gradients as

$$\frac{\partial \delta_j}{\partial x_i} = \frac{\delta_j(x_i + \Delta x_i) - \delta_j(x_i - \Delta x_i)}{2\Delta x_i} \tag{8-14}$$

$$\frac{\partial \sigma_k}{\partial x_i} = \frac{\sigma_k(x_i + \Delta x_i) - \sigma_k(x_i - \Delta x_i)}{2\Delta x_i} \tag{8-15}$$

Despite their apparent simplicity, however, Eqs. (8-14) and (8-15) are computationally expensive to evaluate because they involve $2n$ analyses of the structure, where n is the number of variables x_i undergoing specified small perturbations Δx_i. Alternatively, a semi-analytical technique may be employed to establish the displacement and stress gradients. Displacement gradients are first considered in the following.

Each individual nodal displacement δ_j of concern to the design is related to the overall vector of nodal displacements $\mathbf{u}$ for the structure as

$$\delta_j = \mathbf{b}_j^T \mathbf{u} \tag{8-16}$$

where $\mathbf{b}_j$ is a Boolean vector that identifies δ_j from among all nodal displacements (e.g., if $\mathbf{b}_j^T = [1, 0, \ldots, 0]$, then $\delta_j = u_1$) and, depending on the requirements of the design, the displacements $\mathbf{u}$ are found from either first-order or second-order analysis of the current structure by solving the equilibrium conditions

$$\mathbf{Ku} = \mathbf{P} \tag{8-17}$$

for each vector $\mathbf{P}$ of applied loads, where $\mathbf{K}$ is the structure stiffness matrix, to get

$$\mathbf{u} = \mathbf{K}^{-1}\mathbf{P} \tag{8-18}$$

Now, differentiate Eq. (8-16) with respect to each reciprocal-sizing variable x_i to get

$$\frac{\partial \delta_j}{\partial x_i} = \mathbf{b}_j^T \frac{\partial \mathbf{u}}{\partial x_i} + \mathbf{u}^T \frac{\partial \mathbf{b}_j}{\partial x_i} = \mathbf{b}_j^T \frac{\partial \mathbf{u}}{\partial x_i} \tag{8-19}$$

which recognizes that $\partial \mathbf{b}_j / \partial x_i = 0$. Next, differentiate Eq. (8-17) w.r.t. x_i to get

$$\mathbf{K} \frac{\partial \mathbf{u}}{\partial x_j} + \frac{\partial \mathbf{K}}{\partial x_i} \mathbf{u} = \frac{\partial \mathbf{P}}{\partial x_i} = \mathbf{0} \tag{8-20}$$

which recognizes for the design of steel building frameworks that $\partial \mathbf{P} / \partial x_i = \mathbf{0}$ since applied loads $\mathbf{P}$ are generally taken independent of changes in member sizes x_i. Solve Eq. (8-20) to find

$$\frac{\partial \mathbf{u}}{\partial x_i} = -\mathbf{K}^{-1} \frac{\partial \mathbf{K}}{\partial x_i} \mathbf{u} \tag{8-21}$$

The stiffness gradient $\partial \mathbf{K} / \partial x_i$ in Eq. (8-21) can be calculated using a finite difference approximation. Alternatively, an analytical approach can be used to find $\partial \mathbf{K} / \partial x_i$ for steel building frameworks since for each member i it is generally possible to express the member stiffness matrix $\mathbf{K}_i$ as a sole function of the reciprocal-sizing variable x_i, either exactly or instantaneous-approximately. For example, consider a structural steel member subject to a three-dimensional force field involving axial force, biaxial shear forces, torsional moment, and biaxial bending moments. The geometrical properties of the member cross-section are: A_x, A_y, A_z = axial and shear areas; I_x, I_y, I_z = torsional and bending moments of inertia. For commercial standard steel sections, each of the properties A_y, A_z, I_x, I_y, and I_z may be expressed in terms of the axial area A_x as follows (Chan 1993; Chan and Grierson 1993; Grierson 1993).

$$1/A_Y = C_{AY}/(A_X + C'_{AY})$$
$$1/A_Z = C_{AZ}/(A_X + C'_{AZ})$$
$$1/I_X = C_{IX}/(A_X + C'_{IX})$$
$$1/I_Y = C_{IY}/(A_X + C'_{IY})$$
$$1/I_Z = C_{IZ}/(A_X + C'_{IZ})$$

where the coefficients C and C' are determined by linear regression analysis and have different values depending on the type and size of the section. These relationships permit the stiffness matrix for each structural steel member i to be instantaneously approximated in terms of the reciprocal of its cross-section area $1/a_i$ $(= 1/A_X)$ alone, such that the structure stiffness matrix can be expressed as

$$K = \sum_{i=1}^{n} K_i = \sum_{i=1}^{n} \frac{1}{x_i} K_i^* \tag{8-22}$$

where $\mathbf{K}_i^*$ is a constant matrix. Evidently, Eq. (8-22) is exact for pin-jointed truss structures that experience only axial forces but approximate for structures that also experience flexure, torsion, and shear. In the latter case, the matrix $\mathbf{K}_i^*$ is updated from stage to stage of the design process to account for the changing relationships between the section area a_i and the other section parameters concerned with flexural, torsional, and shear behavior.

Now, differentiate Eq. (8-22) w.r.t. x_i to get the stiffness gradient as

$$\frac{\partial K}{\partial x_i} = \frac{\partial}{\partial x_i}\left(\sum_{i=1}^{n} \frac{1}{x_i} K_i^*\right) = \frac{\partial}{\partial x_i}\left(\frac{1}{x_i} K_i^*\right) = -\frac{K_i^*}{x_i^2} = -\frac{K_i}{x_i} \tag{8-23}$$

which recognizes that $\partial \mathbf{K}_r/\partial x_i = 0$ when $r \neq i$. It is now possible to calculate the displacement gradient $\partial \delta_j/\partial x_i$ through Eq. (8-19) using a pseudo-load or virtual-load method. The pseudo-load method proceeds by observing that the gradient vector $\partial \mathbf{u}/\partial x_i$ in Eq. (8-21) is actually the solution of the system of equilibrium equations for the structure subject to the vector of pseudo-loads $-(\partial \mathbf{K}/\partial x_i)\mathbf{u}$. The resulting vector $\partial \mathbf{u}/\partial x_i$ is then substituted into Eq. (8-19) along with the vector $\mathbf{b}_j$ to find the displacement gradient $\partial \delta_j/\partial x_i$. For $\partial \delta_j/\partial x_i$ to be found for all n sizing variables, this method involves analyzing the structure through Eq. (8-21) for n different pseudo-load cases. Alternatively, the virtual-load method substitutes a priori for $\partial \mathbf{u}/\partial x_i$

from Eq. (8-21) into Eq. (8-19) to get

$$\frac{\partial \delta_j}{\partial x_i} = -b_j^T K^{-1} \frac{\partial K}{\partial x_i} u \tag{8-24}$$

which becomes from Eq. (8-23)

$$\frac{\partial \delta_j}{\partial x_i} = \frac{1}{x_i} (b_j^T K^{-1} K_i u) \tag{8-25}$$

It is then observed that if a unit virtual load is associated with the real displacement δ_j in Eq. (8-16), by the principle of virtual-work equivalence the vector $\mathbf{b}_j$ may be viewed as a vector of virtual loads associated with the vector of real displacements $\mathbf{u}$ in Eq. (8-16). The vector of virtual nodal displacements $\mathbf{u}_j$ corresponding to the virtual loads $\mathbf{b}_j$ is found as

$$\mathbf{u}_j = \mathbf{K}^{-1}\mathbf{b}_j \tag{8-26}$$

and then substituted into Eq. (8-25) to find the displacement gradient as

$$\frac{\partial \delta_j}{\partial x_i} = \frac{1}{x_i} (u_j^T K_i u) \tag{8-27}$$

Contrary to the pseudo-load method, the virtual-load method only involves a single analysis of the structure for the virtual loads $\mathbf{b}_j$ to find the gradient $\partial \delta_j / \partial x_i$ for all n sizing variables. On the other hand, if the gradients of m different displacements are to be found, then the virtual-load method will involve m analyses of the structure whereas the pseudo-load method will still involve n analyses regardless of the number of displacements of concern to the design of the structure. Either method is quite efficient as each analysis of the structure to establish response gradients is really nothing more than a back-substitution operation for different pseudo or virtual loads since the inverse or factorized stiffness matrix $\mathbf{K}^{-1}$ is already available from the baseline analysis Eq. (8-18). The virtual-load method Eq. (8-27) is adopted to calculate displacement gradients for the design examples presented later in this chapter.

Stress gradients are similarly calculated analytically using the virtual-load method described in the foregoing for displacement gradients. Here, each individual member stress σ_k of concern to the design is related to the overall vector of nodal displacements $\mathbf{u}$ for the structure as

$$\sigma_k = \mathbf{t}_k^T\mathbf{u} \tag{8-28}$$

where the vector $\mathbf{t}_k$ is row k of the global-axis stress matrix for the member associated with the stress σ_k. The vector $\mathbf{t}_k$ is invariant for truss structures comprised of axial members alone. For flexural structures, however, the vector $\mathbf{t}_k$ is a function of the neutral-axis position for the member section associated with the stress σ_k, and therefore it varies as the section design changes over the synthesis history. Upon viewing $\mathbf{t}_k$ as a vector of virtual loads applied to the structure to cause virtual nodal displacements

$$\mathbf{u}_k = \mathbf{K}^{-1}\mathbf{t}_k \tag{8-29}$$

the gradient of the stress σ_k with respect to change in each reciprocal-sizing variable x_i is found as

$$\frac{\partial \sigma_k}{\partial x_i} = \frac{1}{x_i}\,(\mathbf{u}_k^T \mathbf{K}_i \mathbf{u}) \tag{8-30}$$

Equation (8-30) is based on the assumption that the vector $\mathbf{t}_k$ is invariant with changes in x_i for the current design stage, for both truss and flexural structures. The error inherent in this assumption for flexural structures becomes negligible as the synthesis process converges to an optimum where no changes take place in the sizing variable x_i for a number of successive design iterations.

The calculation of displacement and stress gradients using the virtual-load method is illustrated in Chapter 6, "Sensitivity Analysis for Structural Design."

DESIGN PROCESS

By virtue of the approximate nature of the performance constraints Eqs. (8-11) and (8-12), the design synthesis is conducted through an iterative process that involves solving the optimization problem Eqs. (8-10) to (8-13) during each design cycle. Attention is focused here on presenting the details of the iterative synthesis process that specifically pertain to the design of practical steel frameworks using standard steel sections in conformity with the provisions of a governing steel code, for example, American Institute of Steel Construction (1989, 1986) and Canadian Standards Association (1984).

The synthesis process begins by selecting an initial "trial" design of the structure. A simple way to do this is to identify the standard section shape for each member (e.g., *W*, *T*, etc.) and then select the largest section from the corresponding database of available sections (i.e., as commercially supplied by steel mills). The advantage of this approach is that it is easy to apply to establish an initial design. Also, as the largest available sections are being used, this approach clearly signals that the design is probably ill-posed if the initial design is found to be infeasible for the performance requirements of the governing steel design code. A disadvantage of this approach is that the initial design is generally not very well proportioned. However, a reasonably proportioned design is generally found after one cycle of the synthesis process by solving the weight optimization problem for continuous-valued sizing variables by temporarily replacing the discrete sizing constraints Eq. (8-13) with the simple bounding conditions

$$\underline{x}_i \leq x_i \leq \bar{x}_i \qquad (i = 1, 2, \ldots, n) \tag{8-31}$$

where $\underline{x}_i = 1/\bar{a}_i$ and $\bar{x}_i = 1/\underline{a}_i$, in which $\bar{a}_i$ and $\underline{a}_i$ are the largest and smallest section areas available for member i, respectively. The foregoing approach to establish an initial design is adopted for the design examples presented later in this chapter.

Having a reasonably well-proportioned initial design, structural analysis is conducted to determine the corresponding displacements and stresses for each load case. Either first-order or second-order analysis is performed depending on the structure and the requirements of the governing steel design code. If first-order analysis is the basis for the design calculations, then effective length factors greater than unity are adopted for flexural members; if second-order analysis is employed, then all member effective length factors are typically set equal to unity.

Having the stresses and displacements from structural analysis, the database of standard steel sections is searched, member by member, to find the least-weight sections that satisfy the stress-related strength/stability requirements of the governing steel design code while simultaneously ensuring that all displacement-related stiffness constraints are satisfied. This searching activity, hereafter referred to as member-by-member optimization, has two distinct aspects. The concern for strength/stability involves selecting standard sections taking into account axial tension and compression, bending, shear, and torsion effects, and combinations of these effects, for individual members considered in isolation from the structure. This aspect of the search is somewhat more straightforward than when displacements are of concern where member sections must be economically selected while at the same time ensuring that the stiffness of the assembled structure is satisfactory. To this end, an effective technique adopted herein is based on calculating displacement gradients so as to establish the relative influence that individual members have on displacements, which then allows member sections to be selected while having concern for both displacements and economy (Cameron 1989; Cameron et al. 1992).

Having the member sections resulting from the member-by-member optimization procedure, structural analysis is again conducted and the strength/stability and stiffness requirements for the design are checked for the resulting stresses and displacements. If no design violations are detected, the synthesis process continues on to the next stage involving formal optimization. Otherwise, the design is infeasible and member-by-member optimization is conducted again.

The formal optimization stage of each cycle of the synthesis process involves formulating and solving the optimization problem posed by Eqs. (8-10) to (8-13). For computational efficiency, it is desirable to temporarily delete relatively inactive displacement and stress constraints from the formulation. For steel frameworks, this generally results in a significant reduction in the size of the active constraint set and thus either allows for faster computation for a certain size structure, or for the design of a larger structure than would be otherwise possible on a computer with fixed memory capacity. Deleted constraints are continuously monitored and added to the constraint set if and when they become active for a subsequent iteration of the synthesis process. Moreover, to overcome possible design history oscillations that impede convergence of the synthesis process, it is recommended to retain a constraint in the active set even if it subsequently becomes passive (Grierson and Lee 1984).

For the current standard section sizes for the members of the structure, sensitivity analysis is conducted to establish the gradients $(\partial \delta_j / \partial x_i)^0$ and $(\partial \sigma_k / \partial x_i)^0$ of the active displacement and stress constraints in Eqs. (8-11) and (8-12). It is important to note that the computational effort required to find these gradients is nominal since the inverse or factorized form of the structure stiffness matrix is available for the current cycle of the synthesis process.

To complete the formulation of the optimization problem Eqs. (8-10) to (8-13) for the current design stage, it remains to update each set of discrete sizing variables X_i in Eq. (8-13) from the corresponding database of standard steel sections to reflect only those sections having adequate strength/stability and stiffness properties for the current set of active stress and displacement constraints. In fact, the total number of sections reflected in each set X_i is limited to restrict the extent of movement in the design space during solution of the optimization problem, thereby preserving the integrity of the first-order displacement and stress gradients in Eqs. (8-11) and (8-12). The updated subset of standard sections for each member (or fabrication group of members) is arranged to bracket the current design section, thereby allowing for the selection of either a smaller or larger section as the solution of the optimization defined by Eqs. (8-10) to (8-13).

The solution of the design problem defined by Eqs. (8-10) to (8-13), referred to as the formal optimization stage of each design cycle, can be found using one of a number of different algorithms. An optimality criteria (OC) method involves temporarily replacing the discrete sizing constraints Eq. (8-13) with the simple bounding conditions Eq. (8-31) and deriving an optimality criterion based on the stationary conditions for the Lagrangian function involving Eqs. (8-

10) to (8-12). The Lagrange variables, which identify the active performance constraints for the design, are then determined along with the sizing variables x_i through a recursive algorithm that converges after a number of cycles to a nonstandard design (i.e., a set of x_i values that does not correspond to standard steel sections). A pseudo-discrete algorithm based on the optimality criterion is then applied to progressively assign a standard section to each of the members. The details of this OC method are presented in Appendix C along with an example application to a simple framework.

Another optimization algorithm for solving the design problem Eqs. (8-10) to (8-13) is a generalized optimality criteria technique that is often referred to in the literature as the Dual method (Schmit and Fleury 1979; Fleury 1979). Unlike the conventional OC method, the Dual method directly accounts for discrete sizing variables and the corresponding constraints Eq. (8-13). The details of the Dual method, which is used for the design examples presented later in this chapter, are presented in Appendix D along with an example application to a simple framework.

EXAMPLE APPLICATIONS

The least-weight designs of three different steel frameworks are presented in the following. The designs are conducted using a software system based on the optimization procedure described earlier (Grierson and Cameron 1994). The first design example has been selected more for its illustrative value than for its practicality. The remaining two design examples are presented briefly and are representative of professional design practice.

Example 1. 3-Bay by 3-Bay by 2-Story Frame

The 3-bay by 3-bay by 2-story steel framework in Fig. 8-1 is to be designed in accordance with the strength/stability and stiffness provisions of the American Allowable Stress Design steel code AISC-89 (American Institute of Steel Construction 1989). The bay spacings and story heights in the X, Z, and Y-axis directions are 20, 16.4, and 10 feet, respectively.

First-order effects alone are to be accounted for, and sidesway is to be permitted. The in-plane and out-of-plane effective length factors K_x and K_y for the continuously connected X-direction girders and the Y-direction columns are to be calculated on the basis of their end stiffnesses (Johnston 1976). The effective length factors for the pinned-end Z-direction girders and the diagonal bracing members are specified to be $K_x = K_y = 1.0$. The maximum allowable slenderness ratios for each member are specified to be $KL/r = 200$ in compression and $KL/r = 300$ in tension (where L = member length and r = section radius of gyration).

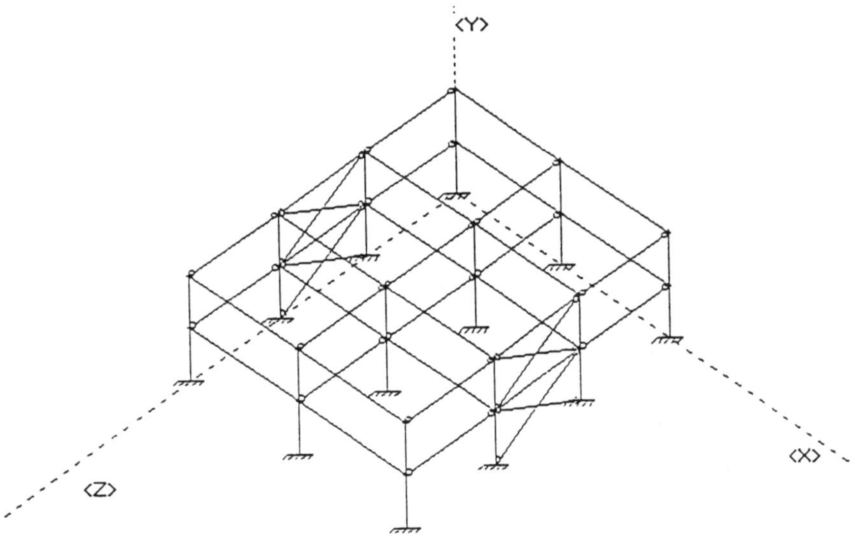

FABRICATION GROUPS

GX1-1 = First Story X-Girders
GX2-2 = Second Story X-Girders
GZ1-1 = First Story Z-Girders
GZ2-2 = Second Story Z-Girders
IC1-2 = Interior Columns
EC1-2 = Exterior Columns
BRACE = Diagonal Braces

FIG. 8-1. Example 1: Steel Building Framework.

Member sections are to be selected from available AISC (1989) standard steel sections. All column and girder members are to have wide-flange (*W*) shapes, and all diagonal bracing members are to have square hollow section (SHS) shapes. For these specified section types, the Young's modulus for the steel = 29,000 ksi, and the shear modulus = 11,200 ksi, the yield stress = 36 ksi and 50 ksi for the *W* and SHS sections, respectively, and the ultimate stress = 58 ksi and 62 ksi for the *W* and SHS sections, respectively.

To satisfy practical fabrication requirements, the members of the structure are to be grouped as follows (see Fig. 8-1).

1. The eight continuously connected girder members located in the *X*-axis direction at story level 1 (i.e., the second floor) are all to have the same *W*-shape section (the depth and size to be determined by the design). This fabrication group is designated as $GX1 - 1$.
2. The eight continuously connected girder members located in the *X*-axis direction at story level 2 (i.e., the roof) are all to have the

same W-shape section. This fabrication group is designated as $GX2 - 2$.

3. The nine pinned-end girder members located in the Z-axis direction at story level 1 are all to have the W-shape section $W10 \times 33$ (i.e., fixed for the design). This fabrication group is designated as $GZ1 - 1$.
4. The nine pinned-end girder members located in the Z-axis direction at story level 2 are all to have the W-shape section $W10 \times 33$. This fabrication group is designated as $GZ2 - 2$.
5. The four interior column members in the Y-axis direction (i.e., two interior columns for each of the two story levels) are all to have the same W-shape section with a nominal depth of 12 in. (the actual section size to be determined by the design). This fabrication group is designated as $IC1 - 2$.
6. The 20 exterior column members in the Y-axis direction (i.e., ten exterior columns for each of the two story levels) are all to have the same W-shape section with a nominal depth between 10 and 14 in. (the appropriate depth and size to be determined by the design). This fabrication group is designated as $EC1 - 2$.
7. The eight pinned-end bracing members located in the two exterior $Z - Y$ faces of the structure are all to have the same SHS-shape section (the size to be determined by the design). This fabrication group is designated as BRACE.

Member cross-sections are to be oriented as follows (see Fig. 8-1).

1. Section webs for X-direction girders are to lie in a vertical $X - Y$ plane such that strong axis bending takes place about the Z-axis.
2. Section webs for Y-direction columns are to lie in a vertical $X - Y$ plane such that strong-axis bending takes place about the Z-axis.
3. Section webs for Z-direction girders are to lie in a vertical $Y - Z$ plane such that strong-axis bending takes place about the X-axis.

The service loading on the structure is (see Fig. 8-1):

1. Wind loading acts horizontally in the positive X-axis direction of magnitude 11.0 kips at each exterior node of the $Z - Y$ face of the structure for which the X-coordinate = 0. This loading is designated as WIND:X.
2. Wind loading acts horizontally in the positive Z-axis direction of magnitude 9.0 kips at each exterior node of the $X - Y$ face of the structure for which the Z-coordinate = 0. This loading is designated as WIND:Z.

3. Gravity uniformly distributed loading acts vertically in the negative Y-axis direction of magnitude 1.4 kips/ft on each of the X-axis girders in the second floor and the roof. This loading is designated as GRAVITY:X.
4. Gravity point loading acts vertically in the negative Y-axis direction of magnitude 14.5 kips at midspan of each of the Z-axis girders in the second floor and the roof. This loading is designated as GRAVITY:Z.

Two load combinations are considered for the design of the structure, as follows

$$\text{GRAVITY}:X + \text{GRAVITY}:Z + \text{WIND}:Z$$
$$\text{GRAVITY}:X + \text{GRAVITY}:Z + \text{WIND}:X$$

Both load combinations define service loading schemes for which the stress-related provisions of the AISC-89 (1989) design code are to be satisfied. The second load combination also defines the service loading scheme for which the lateral sway in the X-axis direction is limited to 0.8 in. at the roof level (i.e., $h/300$).

The least-weight design of the framework is conducted using the synthesis process involving formal and member-by-member optimization described earlier in this chapter. The iterative synthesis history and final design for the steel framework are shown in Fig. 8-2. The AISC standard steel section shapes and designations given in Fig. 8-2 apply for the seven fabrication groups defined in Fig. 8-1. The initial structure weight given in Fig. 8-2 refers to an arbitrarily selected conservative design (i.e., largest section sizes satisfying fabrication requirements). As shown in Fig. 8-2, the optimization process converged after five iterations to a least-weight final design satisfying the imposed X-axis sway limitation and all relevant strength/stability requirements of the AISC-89 (1989; Cameron and Grierson 1991) steel design code.

The design optimization involved 7 sizing variables, 4 displacement constraints, and 364 stress constraints, and required approximately 10 minutes execution time on a 486-based microcomputer. If formal optimization is excluded and member-by-member optimization is alone applied, a very similar final design is found after only approximately 1 minute of computer time. However, this result is peculiar to this example as the use of formal optimization generally yields lower-weight designs compared to those produced by member-by-member optimization alone. In fact, as demonstrated in the following example, the two optimization strategies generally work in tandem to produce a design that is an improvement over any one of them applied alone.

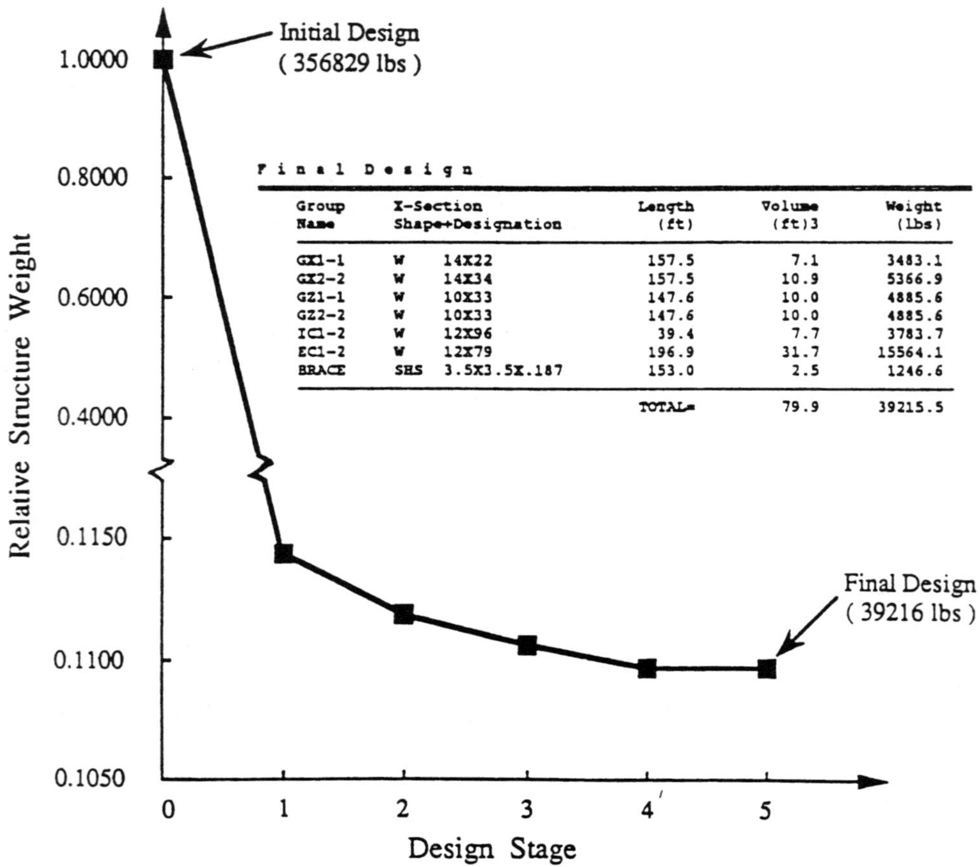

Final Design

Group Name	X-Section Shape+Designation		Length (ft)	Volume (ft)3	Weight (lbs)
GX1-1	W	14X22	157.5	7.1	3483.1
GX2-2	W	14X34	157.5	10.9	5366.9
GZ1-1	W	10X33	147.6	10.0	4885.6
GZ2-2	W	10X33	147.6	10.0	4885.6
IC1-2	W	12X96	39.4	7.7	3783.7
EC1-2	W	12X79	196.9	31.7	15564.1
BRACE	SHS	3.5X3.5X.187	153.0	2.5	1246.6
			TOTAL=	79.9	39215.5

FIG. 8-2. Example 1: Synthesis History and Final Design.

Example 2. Planar Frame

The planar steel framework with trussed roof and pinned supports in Fig. 8-3 is part of a mill crane building and is to be designed in accordance with user-specified displacement constraints and the strength/stability provisions of the Canadian Limit-States-Design steel design code CAN3-S16.1-M84 (Canadian Standards Association 1984). Member sections are to be selected from the Canadian database of standard steel sections.

The crane framework is subject to 14 design load cases where, as indicated in Fig. 8-3, each load case is a particular combination of dead, live, wind, and crane loading, coupled with a thermal effect caused by a 50°C temperature change for the bottom chord members of the roof truss (due to elevated temperatures within the building enclosure). Load cases 1 to 6 = 1.25 dead + 1.5 live + temperature + 1.5 crane load at nodes 27 to 32; load cases 7 to 12 = 1.25 dead + 1.5 wind + temperature + 1.5 crane load at nodes 27 to 32; load case 13 = live + crane load at node 32; and load case 14 = wind + crane load at node 32. Vertical displacement at node 32 is limited to 50 mm for load case 13, and horizontal displacement at node 26 is limited to 20 mm for load case 14.

Out-of-plane bracing is applied at all 56 nodes of the framework, including at the intermediate column nodes 51, 52, and 53 (see Fig. 8-3). The framework has 103 members, consisting of 9 column members and 94 roof truss members. For the roof truss, the vertical and diagonal members are pin-connected, whereas the members comprising the top and bottom chords are continuous-connected over each of the two spans and pin-connected only at the exterior and interior columns.

The column members and the top and bottom chord members in the roof truss are specified to have wide-flange (*W*) section shapes oriented such that bending takes place about the major axis and having compression flange bracing only at members' ends. The vertical and diagonal members in the roof truss are specified to have double-angle section shapes with long legs back to back (⅃L) separated by a 10 mm thick gusset plate.

To satisfy fabrication requirements, six groups of members are identified for which all members in each group are specified to have common section properties. The six fabrication groups are: exterior column members (COLext); interior column members (COLint); top chord truss members (CHtop); bottom chord truss members (CHbot); vertical truss members (Vert); diagonal truss members (Diag).

The steel material properties for the design are: Young's modulus = 200,000 MPa; shear modulus = 77,000 MPa; yield stress = 300 MPa; ultimate stress = 450 MPa; and coefficient of thermal expansion = 0.117×10^{-4} °C^{-1}.

Second-order (*P*-delta) analysis is adopted as the underlying basis for the design. The in-plane effective length factor is specified to be

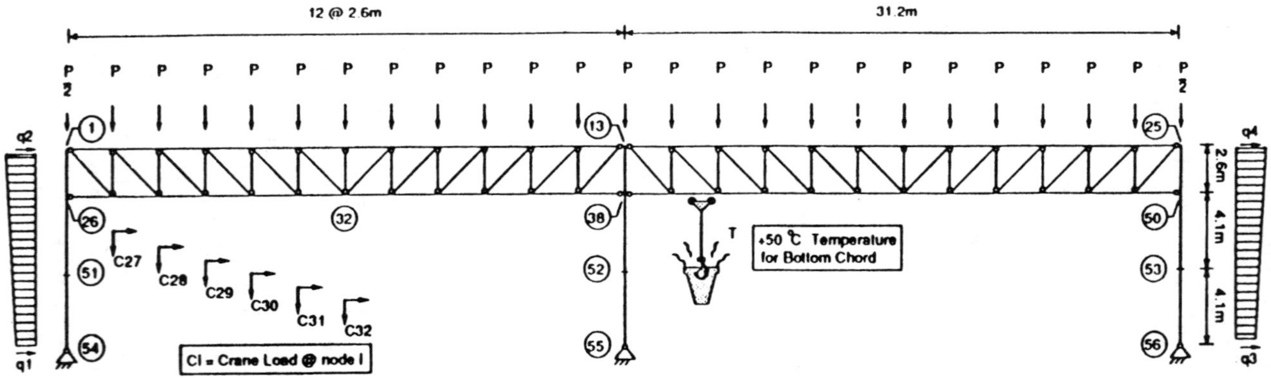

LOAD CASE	P = Dead + Live	Crane	Wind = q1, q2	Wind = q3, q4	T (°C)
1	1.25D + 1.5L	1.5C27			+50
2	1.25D + 1.5L	1.5C28			+50
3	1.25D + 1.5L	1.5C29			+50
4	1.25D + 1.5L	1.5C30			+50
5	1.25D + 1.5L	1.5C31			+50
6	1.25D + 1.5L	1.5C32	1.5w1, 1.5w2	1.5w3, 1.5w4	+50
7	1.25D	1.5C27	1.5w1, 1.5w2	1.5w3, 1.5w4	+50
8	1.25D	1.5C28	1.5w1, 1.5w2	1.5w3, 1.5w4	+50
9	1.25D	1.5C29	1.5w1, 1.5w2	1.5w3, 1.5w4	+50
10	1.25D	1.5C30	1.5w1, 1.5w2	1.5w3, 1.5w4	+50
11	1.25D	1.5C31	1.5w1, 1.5w2	1.5w3, 1.5w4	+50
12	1.25D	1.5C32			+50
13	L	C32			
14		C32	w1, w2	w3, w4	

Where: D = 40.0 kN, L = 30.0 kN, Crane = 80.0 kN (down) & 15.0 kN (right)

w1 = 1.5 kN/m, w2 = 2.5 kN/m, w3 = 1.05 kN/m, w4 = 1.75 kN/m

(1 m = 3.28 feet, 1 kN = 0.225 kips, 1 kN/m = 68.5 plf)

FIG. 8-3. Example 2: Mill Crane Building Framework.

K_x = 1.0 for each roof truss member and top column member, and K_x = 2.0 is specified for each lower column member. The out-of-plane effective length factor is specified to be K_y = 1.0 for each member. The maximum allowable slenderness ratios for each member are specified to be KL/r = 200 in compression and KL/r = 300 in tension.

The least-weight design of the framework is conducted using the previously described optimization procedure. Table 8-1 presents the design history results from the initial design cycle 1 to the final design cycle 3 for the six member fabrication groups for the framework (Grierson and Cameron 1989). The structure weight is given for each design stage along with the response ratios for the most critical strength and displacement constraints for the corresponding design. (A response ratio = actual/allowable response and therefore the maximum allowable response ratio for feasibility is unity.) The framework design found at the beginning of design cycle 1 by formal optimization weighs 19,665 kg but the response ratio is 1.292 for a strength constraint, which implies 29.2% infeasibility. (Recall that the approximate nature of the constraints for the formal optimization may result in designs that are infeasible relative to the actual constraint functions.) The member-by-member optimization restores design feasibility but, in the process, increases the structure weight to 19,725 kg. The resulting design is nearly active for strength response (ratio = 0.978) but somewhat inactive for displacement response (ratio = 0.638), suggesting that it may be possible to redistribute member stiffnesses to achieve further weight reduction without violating the displacement constraint. For the feasible design from cycle 1 as its starting basis, the formal optimization for cycle 2 determines a design that weighs 17,110 kg but which is 13.8% infeasible for a displacement constraint. The member-by-member optimization restores design feasibility but, in the process, increases the structure weight to 17,500 kg. The resulting design is nearly active for both strength response (ratio = 0.996) and displacement response (ratio = 0.910), suggesting that the synthesis process is converging to a least-weight structure. For the feasible design from cycle 2 as its starting basis, the formal optimization for cycle 3 determines a design that weighs 16,795 kg but which is 4.8% infeasible for a strength constraint. The member-by-member optimization determines a heavier design weighing 17,190 kg, but now the critical strength constraint is identically satisfied (ratio = 1.000) and the critical displacement constraint is nearly active (ratio = 0.969). The tight degree of satisfaction of both the strength and displacement constraints indicates that the synthesis process has converged to a least-weight feasible design of the framework, as given in the last column of Table 8-1. The iterative design process involved 2,886 strength and displacement constraints and required approximately 5 minutes execution time on a 486-based microcomputer.

As an illustration that formal optimization and member-by-member optimization together produce the lightest design, the design in the first column in Table 8-1 was taken as the starting basis for

TABLE 8-1 Example 2: Synthesis History and Final Design (Last Column)

Design Cycle	1 Section Designation		2 Section Designation		3 Section Designation	
	Formal Optimization	Member-by-Member Optimization	Formal Optimization	Member-by-Member Optimization	Formal Optimization	Member-by-Member Optimization
Member group:						
COLext	W610 × 174	W610 × 174	W610 × 113	W610 × 140	W610 × 125	W610 × 125
COLint	W610 × 195	W610 × 195	W530 × 123	W610 × 125	W610 × 125	W610 × 125
CHtop	W360 × 57	W250 × 49	W250 × 49	W250 × 49	W250 × 49	W250 × 49
CHbot	W410 × 85	W310 × 97	W250 × 89	W310 × 86	W250 × 80	W310 × 86
Vert	][100 × 90 × 10	][125 × 90 × 10	][125 × 90 × 10	][125 × 90 × 10	][125 × 90 × 10	][125 × 90 × 10
Diag	][150 × 100 × 10	][100 × 75 × 13	][125 × 90 × 10	][125 × 90 × 10	][125 × 90 × 10	][125 × 90 × 10
Structure weight (kg)	19,665	19,725	17,110	17,500	16,795	17,190
Critical strength response ratio	1.292	0.978	0.994	0.996	1.048	1.000
Critical displacement response ratio	0.632	0.638	1.138	0.910	0.971	0.969

Note: $Wd \times m \equiv$ wide-flange section; depth d (mm) and mass m (kg m^{-1})
][$l_1 \times l_2 \times t \equiv$ double-angle section with long legs back to back; long leg length l_1 (mm), short leg length l_2 (mm), and thickness t (mm).

member-by-member optimization applied alone and the iterative synthesis process converged after three cycles to a feasible design weighing 18,070 kg. This design is 18,070 − 17,190 = 880 kg or 5.1% heavier than the final design in Table 8-1 found using formal and member-by-member optimization in tandem.

Example 3. Truss Footbridge

The Warren pony truss structure shown in Fig. 8-4 is a pedestrian footbridge with a 5% camber that is to be designed in accordance with the provisions of the CAN3-S16.1-M84 steel design code (Canadian Standards Association 1984) and the highway bridge code for Ontario, Canada (Ontario Ministry of Transportation and Communications 1983). The bridge has a span of 9.14 m (30 ft) and a width of 1.83 m (6 ft). The member joints are all moment-resisting rigid connections.

To satisfy practical construction requirements, the 47 members of the bridge structure are collected into six fabrication groups: top chord members (T_CHORD); bottom chord members (B_CHORD); diagonal bracing members in bridge side walls (DIAG); diagonal bracing members in bridge floor (BRACE); parallel floor beams (BEAM); and vertical end-members in side walls (VERT).

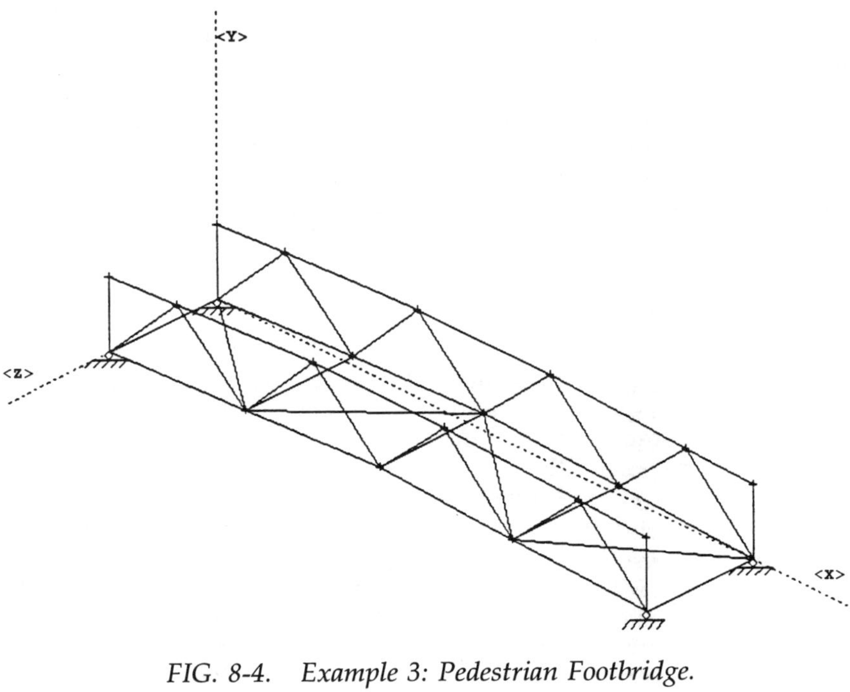

FIG. 8-4. Example 3: Pedestrian Footbridge.

The in-plane and out-of-plane effective length factors for the T_CHORD and VERT members are taken to be $K_x = 2.0$ and $K_y = 1.0$, respectively, and those for the B_CHORD and BEAM members are taken to be $K_x = K_y = 1.0$. The effective length factors for the DIAG members are taken to be $K_x = 1.2$ and $K_y = 1.0$, and those for the BRACE members are taken to be $K_x = K_y = 0.2$. The maximum allowable slenderness ratios for each member are specified to be $KL/r = 200$ in compression and $KL/r = 300$ in tension.

The members of the bridge are to have rectangular hollow section (RHS) shapes, with the exception that the floor bracing members (BRACE) are to have square hollow section (SHS) shapes. Member sections are to be selected from the Canadian database of standard RHS and SHS steel sections. Young's modulus = 200,000 MPa, shear modulus = 77,000 MPa, yield stress = 350 MPa, and ultimate stress = 450 MPa. Member sections are to be oriented to undergo strong-axis bending, with the exception that weak-axis bending is to be experienced by the sections of the beam members (BEAM) in the bridge floor.

The bridge is subject to gravity dead and live loading in the Y-axis direction (see Fig. 8-4), wind loading in the Y-axis and Z-axis directions, rail loading on the top chord members in the Z-axis direction, and temperature loading on all members. The bridge is to be designed for eight different factored combinations of these loads.

The pedestrian footbridge is designed using the optimization procedure. The iterative synthesis history and final design for the structure are shown in Fig. 8-5. The optimization process converged after six iterations to a least-weight design satisfying all relevant strength/stability requirements of the governing design code (displacement constraints were not imposed for the design), and required approximately 1 minute execution time on a 486-based microcomputer. In fact, a near optimal design only 2% heavier than the final design is found after completion of the first design stage in approximately 0.2 minutes execution time.

SUMMARY AND DISCUSSION

The practical design of structural steel frameworks is effectively carried out using the optimization techniques described in this chapter. Member-by-member and formal optimization employed in tandem permits the simultaneous consideration of both strength and stiffness aspects of the design, thereby eliminating much of the guesswork from the synthesis process. The use of optimization techniques results in reduced designer time and more economical designs as compared to the conventional trial-and-error techniques. In fact, optimal structural steel design typically represents a 3–10% saving in steel weight compared to conventional design, resulting in significant monetary savings for large scale structures.

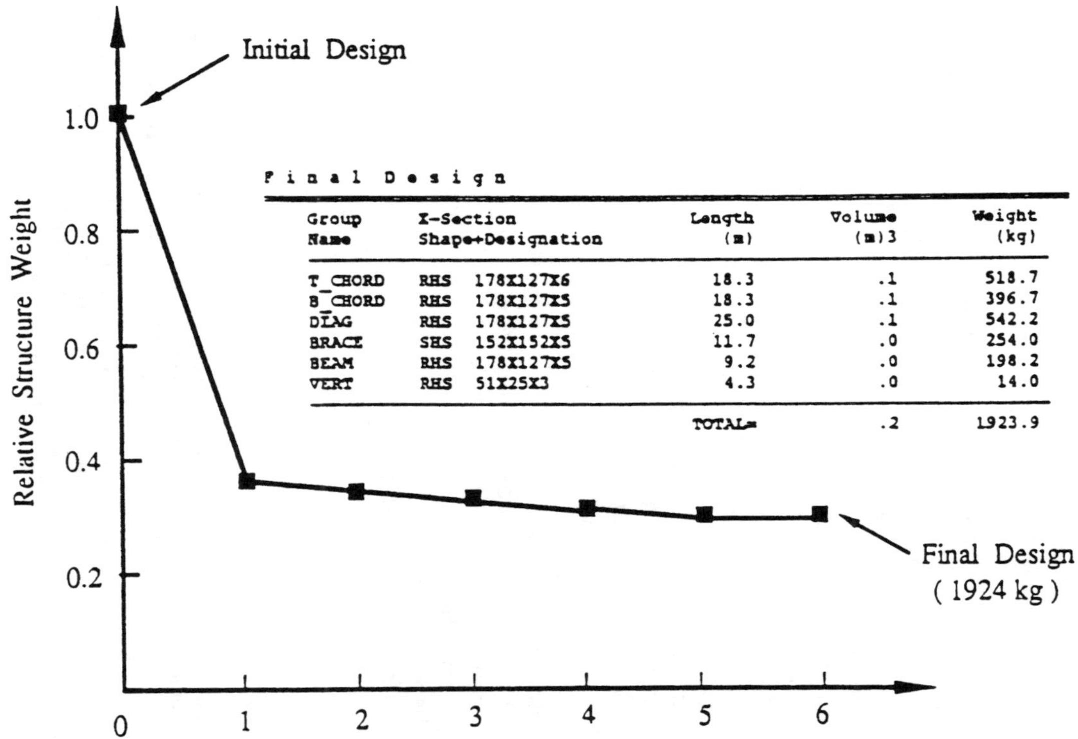

Final Design

Group Name	X-Section Shape+Designation	Length (m)	Volume (m)3	Weight (kg)
T_CHORD	RHS 178X127X6	18.3	.1	518.7
B_CHORD	RHS 178X127X5	18.3	.1	396.7
DIAG	RHS 178X127X5	25.0	.1	542.2
BRACE	SHS 152X152X5	11.7	.0	254.0
BEAM	RHS 178X127X5	9.2	.0	198.2
VERT	RHS 51X25X3	4.3	.0	14.0
		TOTAL=	.2	1923.9

FIG. 8-5. Example 3: Synthesis History and Final Design.

APPENDIX 8-A. NOTATION

The following symbols are used in this chapter.

A = cross-section area; set of cross-section areas
a = cross-section area
$\mathbf{b}$ = virtual load vector
C = coefficient
d = number of displacement conditions
I = cross-section moment of inertia
$\mathbf{K}$ = stiffness matrix
K = effective length factor
L = member length
n = number of members
OC = optimality criteria
$\mathbf{P}$ = load vector
r = cross-section radius of gyration
RHS = rectangular hollow section
s = number of stress conditions
SHS = square hollow section
$\mathbf{t}$ = virtual load vector
$\mathbf{u}$ = nodal displacement vector
w = weight coefficient
W = wide-flange shape
X = set of reciprocal cross-section areas
x = reciprocal cross-section area
Δ = perturbation
δ = displacement
]⌊ = double-angle section
σ = stress

REFERENCES

American Institute of Steel Construction (1989). *Specifications for structural steel buildings: Allowable stress design and plastic design.*

American Institute of Steel Construction (1986). *Manual of steel construction: load and resistance factor design.*

Cameron, G. E. (1989). *A knowledge-based expert system for structural steel design.* Ph.D. Thesis, Civil Engineering, University of Waterloo, Canada.

Cameron, G. E., Chan, C. - M., Xu, L., and Grierson, D. E. (1992). "Alternate methods for the optimal design of slender steel frameworks." *Computers and Structures,* 44(4), 735–741.

Cameron, G. E., Xu, L., and Grierson, D. E. (1991). "Discrete optimal design of 3D frameworks." *ASCE Structure Congress/10th Electronic Computation Conference,* Indianapolis, IN.

Canadian Standards Association (1984). *CAN3-S16.1M84 Steel structures for buildings (limit states design).*

Chan, C. - M. (1993). *An automatic resizing technique for the design of tall steel building frameworks.* Ph.D. Thesis, University of Waterloo, Canada.

Chan, C. - M. and Grierson, D. E. (1993). "An efficient resizing technique for the design of tall steel buildings subject to multiple drift constraints." *Int. J. of Structural Design of Tall Buildings*, 2, 17–32.

Fleury, C. (1979). "Structural weight optimization by dual methods of convex programming." *Int. J. of Numerical Methods in Engineering*, 14, 1761–1783.

Grierson, D. E. (1993). *Practical optimization of structural steel frameworks*. Chapter 12 in the monograph on Advances in Design Optimization, Edited by H. Adeli, Chapman and Hall, London.

Grierson, D. E. and Cameron, G. E. (1994). *SODA: Structural optimization design and analysis*. User Manual, Acronym Software Incorporated, Waterloo, Ontario, Canada.

Grierson, D. E. and Cameron, G. E. (1989). "Microcomputer-based optimization of steel structures in professional practice." *Microcomputers in Civil Engineering*, 4(4), 289–296.

Grierson, D. E. and Lee, W. - H. (1984). "Optimal synthesis of steel frameworks using standard sections." *J. of Structural Mechanics*, 12(3), 335–370.

Johnston, B. G., Editor (1976). *Guide to stability design criteria for metal structures*, Third Edition, John Wiley and Sons, Inc., New York.

Ontario Ministry of Transportation and Communications (1983). *Ontario Highway Bridge Design Code.*

Schmit, L. A., Jr. and Fleury, C. (1979). "An improved analysis / synthesis capability based on dual methods—ACCESS 3." *Proceedings of AIAA/ASME/AHS 20th Structures, Structural Dynamics and Materials Conference*, St. Louis, MO.

Chapter 9

HOW TO OPTIMIZE TALL STEEL BUILDING FRAMEWORKS

Due to the complex nature of a modern tall building consisting of thousands of structural members, the traditional trial-and-error design method is generally highly iterative and time consuming. This chapter presents an optimal sizing technique for the design of practical tall steel building frameworks. A computer-based technique is developed to formalize the numeric tasks of the analysis-design cycle and produce minimum cost design of tall steel frameworks of given topology under static gravity and lateral loadings. Most economical member sizes are automatically selected from databases of commercial standard steel sections, and service lateral drift, ultimate member strength, and discrete sizing constraints are simultaneously satisfied in accordance with building code and constructability requirements.

The design optimization approach is based on an Optimality Criteria method, which is well suited for large-scale tall building frameworks. By exploiting the fact that member force distributions are relatively insensitive to changes in member sizes for building frameworks, the design optimization technique generally converges rapidly in a few cycles. A full-scale 50-story building example is presented to illustrate the applicability, efficiency, and practicality of the automatic optimal sizing technique.

INTRODUCTION

Modern tall steel buildings are complex large-scale structures. Their design is generally a complicated, laborious, and time-consuming

Chun-Man Chan, M. ASCE, Assistant Professor, Department of Civil and Structural Engineering, Hong Kong University of Science and Technology, Clear Water Bay, Kowloon, Hong Kong, email: ⟨cecmchan@usthk.ust.hk⟩.

task. The design process involves the coordinated application of a preliminary conceptual design procedure, and then repetitive application of structural analysis, design evaluation, and member resizing procedures. Once the structural layout and loadings for a building framework are established, most of the engineering effort is expended in an iterative analysis and design process. The design of tall buildings is primarily controlled by lateral stiffness criteria rather than member strength requirements (Council 1979). Modifications of member sizes to satisfy the stiffness constraints are rather difficult because all members in the structure have to be considered simultaneously under multiple loading conditions. In fact, for a large-scale tall building with thousands of structural members, a consideration of all design performance constraints including multiple stiffness and strength requirements is incomprehensible using the traditional trial-and-error design method. Oftentimes in practice, an optimally sized tall building structure is not sought due to time and budget consideration and the absence of a practical and efficient optimization tool.

Research on structural optimization was motivated in large part by the aerospace industry in the 1960s and 1970s. Very little practical optimization work has been carried out for the civil engineering building industry. An early attempt for optimal steel building framework design was reported by Brown and Ang (1966), who developed a program based on a mathematical programming approach using the Gradient Projection method. Although the work produced a structural synthesis program using the 1964 AISC specification, the efficiency of the computer program was very poor even for two-dimensional frameworks. To improve efficiency, Tabak and Wright (1981) developed a program using the Optimality Criteria (OC) approach. However, their program was developed without accounting for design code specifications in the optimization procedure. Other researchers such as Cheng and Truman (1983), Khan (1984), and Sadek (1992) also adopted the OC approach for the design of steel frameworks. However, none of these studies had explicit concern for the lateral stiffness problem associated with practical tall steel frameworks and designs were finalized using continuous member sizes instead of discrete standard steel sections. The first commercial structural optimization software for practical building design is called SODA (Grierson and Cameron 1990), which is capable of handling both stiffness and member strength constraints in accordance with steel design standards. Although this software is remarkably useful, its capacity is still limited to relatively small skeleton steel frameworks.

In addition to university-based researchers, several design professionals, Velivasakis and DeScenza (1983), Baker (1989), Charney (1991), and Gilsanz and Carlson (1991) have also recently developed software for sizing members of tall steel building frameworks to satisfy lateral stiffness criteria. Their methods are essentially equivalent to the early OC method developed by Venkayya et al. (1969). Based on the concept of uniform strain energy density, an optimal design can be achieved indirectly by requiring every structural member of a framework to contribute the same amount of work per unit volume of material to resist the occurrence of a particular displacement of concern to the design. Although this technique is quite efficient, it is useful only for single displacement constraint problems and is valid only for statically determinate truss structures.

This chapter presents a practical optimization technique for the design of tall steel building frameworks. Specifically, a computer-based method is developed to formalize the numeric-intensive tasks of the analysis-design cycle and produce minimum weight designs for tall steel building frameworks subject to multiple interstory drift, strength, and sizing constraints in accordance with building code and fabrication requirements. The design optimization problem is first formulated for a general asymmetrical three-dimensional framework and then the details of the resizing technique are developed. Using the principle of virtual work and section properties' regression relationships, the original implicit design problem is expressed in an explicit continuous form that is readily solved numerically. A rigorously derived OC algorithm is developed to solve the explicit design problem until convergence of member sizes and satisfaction of constraints occur. To achieve a final optimal design using standard steel sections, a pseudo-discrete OC technique is applied to progressively assign standard steel sections to the members of the structure while maintaining the least change in structure weight. Finally, a full-scale 50-story three-dimensional practical framework example is presented to illustrate the applicability, efficiency, and practicality of the automatic optimal design method.

DESIGN PROBLEM FORMULATION

Consider a general three-dimensional steel building framework having $i = 1, 2, \ldots, N$ members (or member fabrication groups), $j = 1, 2, \ldots, M$ stories, $k = 1, 2, \ldots, S$ column lines, under $l = 1, 2, \ldots, L$ lateral loading conditions. The problem of finding the minimum structural weight of the framework can be posed as the following discrete optimization problem.

Minimize

$$W(A_i) = \sum_{i=1}^{N} w_i A_i \tag{9-1a}$$

Subject to

$$d_{kjl} = (\delta_{kjl} - \delta_{kj-1l})/h_j \leq d_j^U \qquad (k = 1, 2, \ldots, S);$$
$$(j = 1, 2, \ldots, M);\ (l = 1, 2, \ldots, L) \tag{9-1b}$$

$$\sigma_p \leq \sigma_p^U \qquad (p = 1, 2, \ldots, P) \tag{9-1c}$$

$$A_i \in \mathbf{A}_i \qquad (i = 1, 2, \ldots, N) \tag{9-1d}$$

Equation (9-1a) defines the structure weight, where A_i is the axial cross-sectional area for member i and w_i is the corresponding weight coefficient (material density $\times$ member length); Eq. (9-1b) defines the multiple interstory drift constraints for the structure, where δ_{kjl} and δ_{kj-1l} are the lateral deflections on a column line k at two adjacent story levels (j and $j - 1$) under lateral loading condition l, h_j is the corresponding story height, and d_j^U is the allowable jth story drift limit; Eq. (9-1c) defines P member strength constraints, where σ_p represents a stress state for a member and σ_p^U is the corresponding allowable member strength; Eq. (9-1d) requires each cross-sectional A_i to belong to the set of areas $\mathbf{A}_i = \{A_1, A_2, \ldots\}_i$ prevailing for the standard steel section profile (e.g., W14-shape) specified for member i.

To facilitate computer solution of the design optimization problem Eqs. (9-1), and as discussed later, it is expedient to initially treat the sizing variables A_i as continuous variables that are restricted as

$$A_i^L \leq A_i \leq A_i^U \tag{9-1e}$$

where A_i^L and A_i^U are the corresponding lower and upper size bounds specified for member i. Moreover, it is also necessary to express the implicit drift and strength constraints Eqs. (9-1b) and (9-1c) as explicit functions of the sizing variables A_i, as described in the following.

Drift Constraints

Under the action of lateral loadings, asymmetrical building frameworks not only translate laterally but also rotate with torsional twisting. Unlike symmetric buildings where every point on a floor plane translates the same amount and a drift value can thus be expressed at a master node for each lateral loading condition, asymmetric buildings have different displacements at different points on the floor plane because of the additional torsional effect. Therefore control of drift must be applied on each individual column line of an asymmetric framework.

With the assumption that floor diaphragms are rigid in their planes, the primary lateral translation δ_{kjl} of column line k at the jth floor under the lth lateral loading condition can be related to the corresponding master floor node displacements through a master-slave transformation as follows.

$$\delta_{kjl} = [1 \ \ \Delta_{kjl}] \begin{bmatrix} \delta_{jl} \\ \theta_{jl} \end{bmatrix} = T_{kjl} \begin{bmatrix} \delta_{jl} \\ \theta_{jl} \end{bmatrix} \tag{9-2}$$

where δ_{jl} and θ_{jl} are the lateral translation and rotation at the jth floor master node; T_{kjl} denotes the transformation matrix relating the kth slave column node to the jth floor master node under the lth lateral loading condition. As illustrated in Fig. 9-1, if $l = 1$ represents the X-direction loading, then $\delta_{jl} = \delta_j^X$ and $\Delta_{kjl} = -\Delta Y_{kj}$; and similarly if $l = 2$ represents the Y-direction loading, then $\delta_{j2} = \delta_j^Y$ and $\Delta_{kj2} = \Delta X_{kj}$. The transformation matrix T_{kjl} is readily established since the geometric layout and the master floor nodes of the building framework are predefined.

To formulate explicit drift constraints for the slave column lines, one needs to first express explicitly the displacements at the master floor nodes. By the principle of virtual work, the lateral displacement at the jth floor master node under the lth lateral loading condition can be expressed as

$$\delta_{jl} = \sum_{i=1}^{N} \int_0^{L_i} \left(\frac{F_{Xl} f_X}{EA} + \frac{F_{Yl} f_Y}{GA_Y} + \frac{F_{Zl} f_Z}{GA_Z} + \frac{M_{Xl} m_X}{GI_X} + \frac{M_{Yl} m_Y}{EI_Y} + \frac{M_{Zl} m_Z}{EI_Z} \right)_i dx \tag{9-3}$$

where L_i is the length of member i; E, G are the axial and shear elastic material moduli; A, A_Y, A_Z are the axial and shear areas for the cross-section; I_X, I_Y, I_Z are the torsional and flexural moments of inertia for the cross-section; F_{Xl}, F_{Yl}, F_{Zl}, M_{Xl}, M_{Yl}, M_{Zl} are the member forces and moments due to the actual lth lateral loading condition; f_X, f_Y, f_Z, m_X, m_Y, m_Z are the member forces and moments due to a unit virtual load applied to the framework at the location of and in the sense of displacement δ_{jl}.

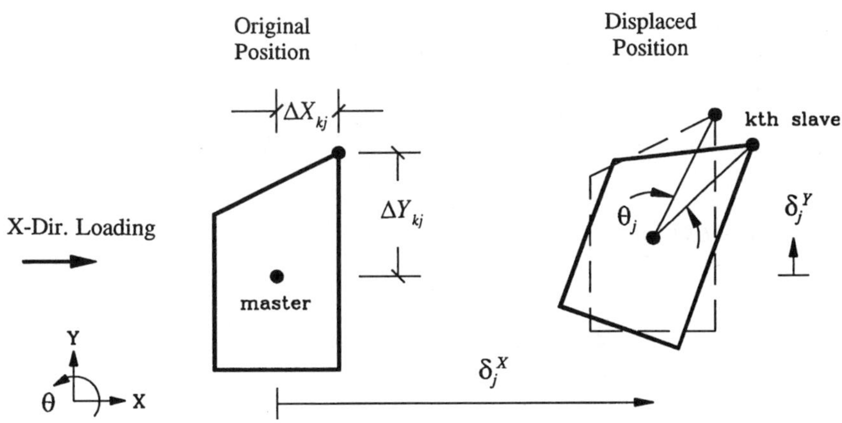

FIG. 9-1. Relationship Between Master and Slave Nodes on Rigid Floor Plane.

Now, for commercial standard steel sections the cross-section properties A_Y, A_Z, I_X, I_Y, and I_Z may all be instantaneously expressed in terms of the axial area A as follows (Chan 1993).

$$1/A_Y = C_{AY}(1/A) + C'_{AY} \tag{9-4a}$$

$$1/A_Z = C_{AZ}(1/A) + C'_{AZ} \tag{9-4b}$$

$$1/I_X = C_{IX}(1/A) + C'_{IX} \tag{9-4c}$$

$$1/I_Y = C_{IY}(1/A) + C'_{IY} \tag{9-4d}$$

$$1/I_Z = C_{IZ}(1/A) + C'_{IZ} \tag{9-4e}$$

where the coefficients C and C' are determined by linear regression analysis and have different values depending on the type and size of the section [e.g., see Tables 9-1 and 9-2 for W14 and W24 sections from the AISC-LRFD design manual (Manual 1986)]. Given a particular section type and depth for each member, the displacement δ_{jl} from Eq. (9-3) can be concisely expressed solely in terms of the member cross-section areas A_i through Eqs. (9-4) as

$$\delta_{jl} = \sum_{i=1}^{N} \left(\frac{C_{ijl}}{A_i} + C'_{ijl} \right) \tag{9-5}$$

where, from Eqs. (9-3) and (9-4), the coefficients C_{ijl} and C'_{ijl} are given by

$$C_{ijl} = \int_0^{L_i} \left(\frac{F_{XI} f_X + M_{YI} m_Y C_{IY} + M_{ZI} m_Z C_{IZ}}{E} + \frac{F_{YI} f_Y C_{AY} + F_{ZI} f_Z C_{AZ} + M_{XI} m_X C_{IX}}{G} \right)_i dx \tag{9-6a}$$

$$C'_{ijl} = \int_0^{L_i} \left(\frac{M_{YI} m_Y C'_{IY} + M_{ZI} m_Z C'_{IZ}}{E} + \frac{F_{YI} f_Y C'_{AY} + F_{ZI} f_Z C'_{AZ} + M_{XI} m_X C'_{IX}}{G} \right)_i dx \tag{9-6b}$$

Applying the L actual lateral loadings, together with two virtual lateral loadings and one virtual torque corresponding to the three degrees of freedom per floor at the master node, an explicit lateral

TABLE 9.1 Relationships between Cross-Section Area A and Shear Areas A_Y and A_Z for AISC W14 and W24 Sections

Sections	$\frac{1}{A_Y} = \frac{C_{AY}}{A} + C'_{AY}$		Max.	$\frac{1}{A_Z} = \frac{C_{AZ}}{A} + C'_{AZ}$		Max.
(No. of Sections)	C_{AY}	C'_{AY}	% error	C_{AZ}	C'_{AZ}	% error
W14X22–26 (2)	1.4354911	0.0952506	0.00	2.5618151	−0.0962252	0.00
W14X30–38 (3)	1.6353798	0.0843246	1.26	2.0862633	−0.0437900	1.16
W14X43–53 (3)	3.0048877	0.0010805	0.42	1.5728685	−0.0068556	0.04
W14X61–82 (4)	3.8144416	−0.0199141	1.14	1.3836832	0.0000917	0.49
W14X90–132 (5)	4.7186962	−0.0158902	1.10	1.2944072	−0.0004526	0.47
W14X145–176 (3)	4.9439546	−0.0162764	0.02	1.2641882	−0.0000348	0.15
W14X193–257 (4)	4.6603358	−0.0097951	0.70	1.2745534	−0.0003102	0.19
W14X283–426 (6)	4.4247100	−0.0068696	0.28	1.2777818	−0.0003280	0.41
W14X455–730 (6)	4.1089656	−0.0046189	0.25	1.2627338	−0.0001827	0.14
W24X55–62 (2)	1.3929148	0.0214267	0.00	3.0904030	−0.0494249	0.00
W24X68–84 (3)	1.4291581	0.0306909	0.55	2.5179705	−0.0301700	0.55
W24X94–103 (2)	1.8572561	0.0128252	0.00	2.0489677	−0.0109334	0.00
W24X104–131 (3)	2.3277754	0.0071301	0.20	1.7554814	−0.0050792	0.01
W24X146–192 (4)	2.4884181	0.0044622	0.44	1.6587269	−0.0030770	0.23
W24X207–306 (5)	2.8964781	−0.0030112	0.30	1.5187585	−0.0005579	0.20
W24X335–492 (5)	2.8209222	−0.0024542	0.68	1.5138177	−0.0005307	0.60

TABLE 9-2 Relationships between Cross-Section Area A and Moment of Inertias I_Z, I_Y, and I_X for AISC W14 and W24 Sections

Sections	$\frac{1}{I_Z} = \frac{C_{IZ}}{A} + C'_{IZ}$		Max. %	$\frac{1}{I_Y} = \frac{C_{IY}}{A} + C'_{IY}$		Max. %	$\frac{1}{I_X} = \frac{C_{IX}}{A} + C'_{IX}$		Max. %
(No. of Sections)	C_{IZ}	C'_{IZ}	error	C_{IY}	C'_{IY}	error	C_{IX}	C'_{IX}	error
W14X22–26 (2)	0.0392400	−0.0010211	0.00	1.2736421	−0.0533897	0.00	82.5200066	−7.9530423	0.00
W14X30–38 (3)	0.0353453	−0.0005698	0.79	0.5711482	−0.0137538	1.02	57.3168240	−3.8983881	4.30
W14X43–53 (3)	0.0319730	−0.0002026	0.15	0.3140174	−0.0028040	0.06	28.3914777	−1.3119968	2.43
W14X61-82 (4)	0.0298816	−0.0001096	0.38	0.1810829	−0.0007904	0.71	17.5919341	−0.5400378	3.65
W14X90–132 (5)	0.0290363	−0.0000971	0.48	0.0780857	−0.0001925	0.47	13.2299043	−0.2647232	6.62
W14X145–176 (3)	0.0285349	−0.0000839	0.15	0.0689388	−0.0001381	0.09	6.7408450	−0.0927814	2.05
W14X193–257 (4)	0.0280568	−0.0000771	0.13	0.0680315	−0.0001246	0.19	3.6149842	−0.0354938	2.74
W14X283–426 (6)	0.0271200	−0.0000652	0.35	0.0674557	−0.0001159	0.16	1.5839994	−0.0098414	6.73
W14X455–730 (6)	0.0244159	−0.0000440	0.66	0.0634434	−0.0000838	0.32	0.6200119	−0.0022544	9.57
W24X55–62 (2)	0.0140903	−0.0001290	0.00	0.7929359	−0.0145824	0.00	38.7216771	−1.5427694	0.00
W24X68–84 (3)	0.0134348	−0.0001225	0.23	0.3894727	−0.0052045	0.50	28.2704125	−0.8784636	2.73
W24X94–103 (2)	0.0119560	−0.0000613	0.00	0.2488717	0.0001898	0.00	15.9685482	−0.3863678	0.00
W24X104–131 (3)	0.0110073	−0.0000373	0.08	0.1371777	−0.0006215	0.02	15.7421923	−0.3050556	2.46
W24X146–192 (4)	0.0106682	−0.0000300	0.16	0.1206501	−0.0002557	0.72	7.5894514	−0.1037394	3.84
W24X207–306 (5)	0.0099410	−0.0000172	0.20	0.1200412	−0.0002494	0.18	3.1506929	−0.0271871	6.39
W24X335–492 (5)	0.0097758	−0.0000156	0.41	0.1149098	−0.0002011	0.91	1.2726781	−0.0067761	6.36

translation for the kth slave column node can be written in terms of sizing variables A_i using Eqs. (9-2) and (9-5), as follows.

$$\delta_{kjl} = T_{kjl} \begin{bmatrix} \sum_{i=1}^{N} \left(\frac{C_{ijl}}{A_i} + C'_{ijl} \right)_{\delta} \\ \sum_{i=1}^{N} \left(\frac{C_{ijl}}{A_i} + C'_{ijl} \right)_{\theta} \end{bmatrix} \quad (j = 1, 2, \ldots, M); \quad (l = 1, 2, \ldots, L) \tag{9-7}$$

Assuming that the slave column node locations with respect to the corresponding master node do not change from one floor to another (i.e., $\Delta_{kjl} = \Delta_{kj-1l}$ and consequently $T_{kjl} = T_{kj-1l}$), the explicit interstory drift constraints for all $k = 1, 2, \ldots, S$ critical column lines can be expressed as

$$d_{kjl} = \frac{\delta_{kjl} - \delta_{kj-1l}}{h_j} = \sum_{i=1}^{N} \left(\frac{e_{ikjl}}{A_i} + e'_{ikjl} \right) \leq d_j^U \quad (k = 1, 2, \ldots, S); \quad (j = 1, 2, \ldots, M); \quad (l = 1, 2, \ldots, L) \tag{9-8}$$

where

$$e_{ikjl} = T_{kjl} \begin{bmatrix} \left(\frac{C_{ijl} - C_{ij-1l}}{h_j} \right)_{\delta} \\ \left(\frac{C_{ijl} - C_{ij-1l}}{h_j} \right)_{\theta} \end{bmatrix}; \quad e'_{ikjl} = T_{kjl} \begin{bmatrix} \left(\frac{C'_{ijl} - C'_{ij-1l}}{h_j} \right)_{\delta} \\ \left(\frac{C'_{ijl} - C'_{ij-1l}}{h_j} \right)_{\theta} \end{bmatrix} \tag{9-9a,b}$$

Strength Constraints

Although the serviceability drift requirement is of paramount importance for the design of tall buildings, the adequacy of the structural safety of these frameworks cannot be overlooked. One approach to account for the member strength requirements is to express them explicitly in terms of design variables in the same manner as the drift constraints. Though direct, this approach has a major obstacle in its implementation. As each member has at least several strength constraints for different load combinations, a practical tall steel building having several thousand members will result in an excessive number of strength constraints and will require enormous computer effort for solution.

As the design of a tall building framework is generally controlled by its lateral stiffness, member strength requirements can be treated as secondary constraints because most of them are usually far from being active. This implies that member strength requirements can be treated separately from the lateral drift constraints. Moreover, unlike the lateral drift requirements, which are system-level constraints, member strength requirements are local-level constraints. Recognizing this fact, and that only a relatively small number of them are critical to a tall building design, suggests the following alternative approach to account for member strength constraints in a member-by-member manner. Immediately after a framework is analyzed under different gravity and lateral loading conditions, a strength design process is applied to determine the minimum size of each member from a given set of standard sections to satisfy the strength criteria in accordance with the governing design standard. These strength sizes are then taken as the lower size bounds on the design variables for the current design cycle involving displacement constraints. In this way, the design optimization of a tall steel building framework involves explicit accounting for only lateral drift constraints whereas the strength constraints are implicitly accounted for through the lower bounds on sizing variables.

Explicit Design Problem

For convenience of future discussion, the subscripts (k, j, l) in Eqs. (9-8) are changed to a single subscript s to represent the sth constraint in a collective set of D drift constraints, where the total number of drift constraints D is equal to the product of $S \times M \times L$ for an M-story framework having concern for drift control of S slave column lines under L lateral loading conditions. As such, from Eqs. (9-1a), (9-1e), and (9-8), the design optimization problem involving drift and sizing constraints can be stated concisely in terms of continuous-valued sizing variables A_i as follows.

Minimize

$$W(A_i) = \sum_{i=1}^{N} w_i A_i \tag{9-10a}$$

Subject to

$$d_s(A_i) = \sum_{i=1}^{N} \left(\frac{e_{is}}{A_i} + e'_{is} \right) \leq d_s^U \qquad (s = 1, 2, \ldots, D) \tag{9-10b}$$

$$A_i^L \leq A_i \leq A_i^U \qquad (i = 1, 2, \ldots, N) \tag{9-10c}$$

in which local-level member strength requirements are accounted for by adopting strength-based sizes as the minimum member sizes A_i^L in Eqs. (9-10c).

AUTOMATIC RESIZING TECHNIQUE

Upon formulating the design optimization problem for a tall steel building framework, the next task is to apply a suitable method to solve the problem. A rigorously derived Optimality Criteria (OC) method, which has been shown to be computationally very efficient for large-scale structures (Berke and Khot 1974, 1988; Khot et al. 1979; Venkayya et al. 1973), is employed. The OC method is based upon the satisfaction of a set of necessary conditions for the optimal design. It essentially consists of complementary phases: the first phase concerns the derivation of a set of necessary conditions that the optimal design must satisfy, and the second phase involves the application of a recursive algorithm to satisfy the optimality conditions and thus indirectly achieve the optimal design. This OC method has gained tremendous popularity due to its high efficiency. It is particularly suitable for large-scale structures since it only has a weak dependence on the size of the structure. Moreover, the convergence of the OC method does not depend on the starting design, but rather on the structural behavior. The OC technique used to solve the explicit design optimization problem Eqs. (9-10) is herein presented. A simple tutorial example of the OC technique can be found in Appendix C of this Guide.

Optimality Criteria

Temporarily omitting the sizing constraints Eqs. (9-10c), from Eqs. (9-10a) and (9-10b), the design optimization problem can be reformulated as the minimization of the following unconstrained Lagrangian function (Wismer and Chattergy 1978).

$$L(A_i, \lambda_s) = \sum_{i=1}^{N} w_i A_i + \sum_{s=1}^{D} \lambda_s \left[\sum_{i=1}^{N} \left(\frac{e_{is}}{A_i} + e'_{is} \right) - d_s^U \right] \qquad (9\text{-}11)$$

where the Lagrange multipliers are such that $\lambda_s > 0$ if constraint s is active (i.e., $d_s = d_s^U$) or $\lambda_s = 0$ if constraint s is inactive (i.e., $d_s < d_s^U$). Assume temporarily that all drift constraints are active.

Differentiate Eq. (9-11) with respect to the sizing variables A_i and rearrange terms to obtain the stationary conditions

$$\sum_{s=1}^{D} \lambda_s \frac{e_{is}}{w_i A_i^2} = 1 \qquad (i = 1, 2, \ldots, N) \qquad (9\text{-}12)$$

Equation (9-12) denotes the necessary conditions to be satisfied for an optimal structure. As the Lagrangian multiplier λ_s can be interpreted as a measure of the sensitivity of the optimal design to the sth constraint, Eq. (9-12) specifies that the weighted sum of the virtual strain energy densities for each member i must be the same and indeed equal to unity at optimum (Chan 1993; Chan and Grierson 1993).

Recursive Algorithm

To solve the optimization problem, a recursive algorithm is applied to drive an initial design to the final design satisfying Eq. (9-12) at the optimum. To this end, multiply both sides of Eq. (9-12) by A_i^{η}, take the ηth root and then apply a first-order binomial expansion to obtain the linear recursive relations

$$A_i^{\nu+1} = A_i^{\nu}\left[1 + \frac{1}{\eta}\left(\sum_{s=1}^{D}\frac{\lambda_s e_{is}}{w_i A_i^2} - 1\right)_{\nu}\right] \qquad (i = 1, 2, \ldots, N) \qquad (9\text{-}13)$$

where ν and $\nu + 1$ indicate successive iterations, and η is a step-size or relaxation parameter that controls the convergence rate of the recursive process; as η becomes larger, the value of A_i^{ν} is modified to a lesser extent, and vice versa.

Before the recursive relations Eq. (9-13) can be utilized to find A_i, the unknown Lagrange multipliers λ_s must first be determined. To establish the current λ_s^{ν} values, consider the change $(d_t^{\nu+1} - d_t^{\nu})$ in the tth constraint due to the changes $(A_i^{\nu+1} - A_i^{\nu})$ in the design variables; that is,

$$(d_t^{\nu+1} - d_t^{\nu}) = \sum_{i=1}^{N}\left(\frac{\partial d_t}{\partial A_i}\right)_{\nu}(A_i^{\nu+1} - A_i^{\nu}) \qquad (9\text{-}14)$$

where, from Eqs. (9-10b) for drift constraint t,

$$\left(\frac{\partial d_t}{\partial A_i}\right)_{\nu} = -\left(\frac{e_{it}}{A_i^2}\right)_{\nu} \qquad (9\text{-}15)$$

and, from Eq. (9-13) for the sizing variables,

$$A_i^{\nu+1} - A_i^{\nu} = \frac{A_i^{\nu}}{\eta}\left(\sum_{s=1}^{D}\frac{\lambda_s e_{is}}{w_i A_i^2} - 1\right)_{\nu} \qquad (i = 1, 2, \ldots, N) \qquad (9\text{-}16)$$

Suppose that drift constraint t becomes active after the $(\nu + 1)$th iteration so that $d_i^{\nu+1} = d_t^{U}$, and then substitute Eqs. (9-15) and (9-16) into Eq. (9-14) and rearrange terms to obtain, for all D drift constraints,

$$\sum_{s=1}^{D}\lambda_s^{\nu}\left(\sum_{i=1}^{N}\frac{e_{is}e_{it}}{w_i A_i^3}\right)_{\nu} = \sum_{i=1}^{N}\frac{e_{it}}{A_i^{\nu}} - \eta(d_t^{U} - d_t^{\nu}) \qquad (t = 1, 2, \ldots, D) \qquad (9\text{-}17)$$

Having the current set of design variables A_i^{ν}, the corresponding values for the Lagrange multipliers λ_s^{ν} are readily determined by solving the set of simultaneous equations (9-17). Having the current values of the λ_s^{ν}, the new set of design areas $A_i^{\nu+1}$ is found through the recursive relations Eq. (9-13).

During the evaluation of the OC iterations using Eqs. (9-13) and (9-17), some members may be found to exceed their limiting sizes as defined in Eqs. (9-10c). Once a sizing variable reaches a limiting value, it is not allowed to change any further and is deemed to be inactive. To account for this sizing restriction, the set of simultaneous equations (9-17) is modified to include only active sizing variables that are free to participate in the design optimization as follows.

$$\sum_{s=1}^{D} \lambda_s^{\nu} \left(\sum_{i=1}^{NA} \frac{e_{is} e_{it}}{w_i A_i^3} \right)_{\nu} = \sum_{i=1}^{NA} \frac{e_{it}}{A_i^{\nu}} - \eta(d_t^U - d_t^{\nu}) \qquad (t = 1, 2, \ldots, D) \qquad (9\text{-}18)$$

where NA is the current number of active sizing variables.

Thus far, all drift constraints have been assumed to be active, that is, to be at their drift limits at the optimum. Although a framework may tend to have most of its drift constraints active or nearly active at the optimum so as to result in a minimum weight design, it may not necessarily always be the case. As indicated previously, the sign for the Lagrange multiplier λ_s determines whether the corresponding drift constraint is active or not, and the following iterative technique for solving Eq. (9-18) allows for a dynamic account of this sign restriction. Specifically, the Gauss-Seidel iterative technique (Hornbeck 1975) is applied to solve Eq. (9-18) using the following recursive formulae

$$\lambda_t^{\tau+1} = \frac{1}{a_{tt}} \left(b_t - \sum_{s=1}^{t-1} a_{ts} \lambda_s^{\tau+1} - \sum_{s=t+1}^{D} a_{ts} \lambda_s^{\tau} \right)$$

$$(t = 1, 2, \ldots, D) \qquad (9\text{-}19)$$

where τ and $\tau + 1$ are two successive iterations of the Gauss-Seidel process, and

$$a_{tt} = \left(\sum_{i=1}^{NA} \frac{e_{it}^2}{w_i A_i^3} \right)_{\nu} \qquad (9\text{-}20a)$$

$$a_{ts} = \left(\sum_{i=1}^{NA} \frac{e_{is} e_{it}}{w_i A_i^3} \right)_{\nu} \qquad (9\text{-}20b)$$

$$b_t = \sum_{i=1}^{NA} \frac{e_{it}}{A_i^{\nu}} - \eta(d_t^U - d_t^{\nu}) \qquad (9\text{-}20c)$$

Any $\lambda_t^{\tau+1}$ value found to be negative at any stage of applying Eq. (9-19) is temporarily set to zero to reflect the fact that the corresponding drift constraint is presently inactive. Upon converging to $\lambda_t^{\tau+1} = \lambda_t^{\tau}$ through Eq. (9-19), the Gauss-Seidel iterative technique has determined the solution $\lambda_s^{\nu} = \lambda_s^{\tau+1}$ $(s = 1, 2, \ldots, D)$ of Eq. (9-18).

Equations (9-13) and (9-18) form the basis for an automatic OC resizing method to solve the continuous optimization problem posed by Eqs. (9-10). Upon convergence of the values for A_i and λ_s (i.e., $A_i^{\nu+1} = A_i^{\nu}$ and $\lambda_s^{\nu} = \lambda_s^{\nu-1}$), a continuous-valued optimal design solution is obtained.

Initial Design Selection

From the recursive resizing relations Eq. (9-16), it can be observed that the rate of convergence of the OC process depends on the initial set of member sizes A_i^0 adopted at the beginning of the design process. In the following, an initial preprocessor is developed to proportion the initial design after the first analysis to establish a reasonable starting design for the iterative OC process. A single drift constraint design problem is first explored. Then, a simple "closed form" solution for the single constraint problem is derived and exploited as an initial preprocessor for the iterative OC process.

If a single drift constraint alone is of concern, that is, $s = 1$, the stationary conditions Eqs. (9-12) become

$$\lambda_s \frac{e_{is}}{w_i A_i^2} = 1 \qquad (i = 1, 2, \ldots, N) \tag{9-21}$$

Rearranging Eq. (9-21), each member section area A_i can be obtained as

$$A_i = \sqrt{\lambda_s} \cdot \sqrt{\frac{e_{is}}{w_i}} \qquad (i = 1, 2, \ldots, N) \tag{9-22}$$

To determine the as yet unknown Lagrange multiplier λ_s in Eq. (9-22), substitute for A_i from Eq. (9-22) into Eq. (9-10b) and set $d_s = d_s^U$ to get

$$d_s^U = \sum_{i=1}^{N} \left(\frac{e_{is}}{\sqrt{\lambda_s}} \cdot \sqrt{\frac{w_i}{e_{is}}} + e_{is}' \right) \tag{9-23}$$

which can be further simplified to

$$\sqrt{\lambda_s} = \frac{\sum_{i=1}^{N} \sqrt{w_i e_{is}}}{d_s^U - \sum_{i=1}^{N} e_{is}'} \tag{9-24}$$

Finally, substitute the calculated square root value of λ_s from Eq. (9-24) into Eq. (9-22) to obtain the optimal member areas A_i as

$$A_i = \left(\frac{\sum_{i=1}^{N} \sqrt{w_i e_{is}}}{d_s^U - \sum_{i=1}^{N} e'_{is}} \right) \cdot \sqrt{\frac{e_{is}}{w_i}} \qquad (i = 1, 2, \ldots, N) \qquad \text{(9-25)}$$

At the beginning of the design process for the multiple drift constraint problem Eqs. (9-10), an initial set of member sizes must be adopted to commence the design. Although the OC method does not impose any restriction on the initial member sizes to be selected, it is recommended that the maximum available section sizes be used as the initial design. An advantage of selecting a maximum-size initial design is that a clear signal is given that the design problem is probably ill-posed if the corresponding member sizes are found to be infeasible for any of the drift constraints. If no drift constraint is found to be violated after the first analysis, the next task is to establish a reasonable starting point for the OC design process.

To establish a reasonably proportioned starting design for the OC process, the top-story building drift alone is initially considered as being representative of the overall sway behavior of a framework. Using Eq. (9-25), member sizes are then calculated to satisfy the corresponding top-story drift constraint. If the design involves a three-dimensional building framework, two single-drift constraint problems are considered concerning the top-story master node lateral displacements corresponding to the two lateral loadings acting in the two orthogonal directions for the building. In such a circumstance, envelope member sizes determined from the two single-drift constraint problems are taken as the starting design for the iterative OC process. In the case of a statically indeterminate framework, a second analysis of the structure using the member sizes determined from Eq. (9-25) is carried out before the OC resizing process begins. This is because substantial changes in member sizes are usually encountered upon applying Eq. (9-25) at this stage, and it is therefore necessary to allow for member force redistribution and to update the virtual strain energy coefficients e_{is} and the correction coefficients e'_{is} in Eq. (9-10b).

Experience (Chan 1993) has shown that the preceding initial preprocessor is a very effective tool to initially proportion a building framework. Generally speaking, applying the preprocessor before activating the OC resizing process can reduce the number of OC iterations for the first design cycle by a factor of about three.

Discrete Section Selection

For the design of practical building frameworks, it is unacceptable to conclude the design using the continuous-valued member sizes obtained from the described OC resizing procedure. A common approach to selecting discrete section sizes for the members of a structure is to simply round up all the continuous member sizes to their next available discrete section sizes. However, although such a simple round-up technique may be direct, it produces a more conservative design than necessary. An effective pseudo-discrete section selection strategy is developed in the following to achieve a smooth progressive transition from the continuous-variable design to the optimal final design using discrete standard steel sections (Chan 1992).

Recognizing that the most economical solution is when the discrete section sizes are exactly the same as the continuous sizes, the least possible disturbance from the continuous-variable results is most desirable in the transition from continuous to discrete section selection. In order to realize the minimum such deviation, an effective strategy is to select the member(s) involving the least change in material weight when the continuous size is increased to the next available discrete size, that is, the member with the least value of $w_i x(A_{i(\text{discrete})} - A_{i(\text{continuous})})$. Once the currently most economical member (or member group) is identified and assigned a standard section, it is thereafter treated as being passive to the optimization process. To account for the passive members, the relations Eq. (9-18) for the Lagrangian multipliers λ_s are progressively modified as

$$\sum_{s=1}^{D} \lambda_s^{\nu} \left(\sum_{i=1}^{N-\xi} \frac{e_{is} e_{it}}{w_i A_i^3} \right)_{\nu} = \sum_{i=1}^{N-\xi} \frac{e_{it}}{A_i^{\nu}} - \eta(d_t^U - d_t^{\nu}) \qquad (t = 1, 2, \ldots, D) \qquad (9\text{-}26)$$

where ξ is the number of passive members currently assigned standard sections (initially, ξ is equal to the number of members reaching their limiting sizes at the convergence of the continuous-valued design solution). Since the assignment of a discrete section to a member results in a more conservative design than the current optimal continuous solution, the other, as yet, unsized (active) members can then be further optimized using the OC procedure Eqs. (9-13) and (9-26) until all active λ_s and A_i converge again. In this way, members are progressively assigned standard sizes while maintaining design economy and constraint feasibility. When all members have been assigned standard sections (i.e., $\xi = N$), a final design with optimal standard sections has been found.

With the view to decrease the required computational effort for large-scale structures, it can be specified that several members (or member groups) having the least disturbance to the continuous solution be assigned their next available standard sections in a batch manner. For example, if a batch size of 10% of the initial set of active members at the completion of the continuous solution stage is chosen, and the same batch size is selected each time thereafter, the final optimal discrete design will be achieved in 10 section selection cycles.

Indeed, such a 10% batch size has been found to reduce substantially the computation time for the pseudo-discrete section selection process without appreciably affecting the minimum weight result (Chan 1993).

Although the pseudo-discrete OC technique is intuitive in nature and there is no guarantee that it produces the ultimate optimal discrete solution, it is very practical for large-scale tall building frameworks using discrete standard steel sections. Its effectiveness is primarily due to the fact that whenever a member group is identified and assigned a standard section, the influence of using a larger size than necessary is accounted for by further optimizing the other unsized members. By so doing, not only is the minimum change of objective function realized but the feasibility of the solution is also maintained.

Overall Design Procedure

A design cycle is defined as a process that involves an application of structural analysis together with a subsequent continuous OC optimization. For the case of a statically determinate structure, the result of the first design cycle gives the optimal solution. However, if the structure is statically indeterminate, the explicit design formulation Eqs. (9-10) is only approximate because any changes in the member sizes will cause a redistribution of internal forces and, hence, a change in the OC recursive relations. Consequently, after each design cycle the structure is re-analyzed for the updated sizes and the continuous optimization OC process is applied again. This iterative process involving multiple design cycles continues until the weight of the structure converges to a minimum-weight structure defined by continuous-valued member sizes. Then a pseudo-discrete section selection process is applied to transform to an optimal design involving discrete-valued member sizes corresponding to commercially available standard steel sections.

The overall design optimization procedure using the OC resizing technique is outlined step by step in the following, and presented graphically in the two flow charts given in Figs. 9-2 and 9-3.

Step 1.

Assume initial member sizes for the building structure. Normally, all the members can initially have their maximum available sizes A_i^U.

Step 2.

Analyze the structure and then apply the initial preprocessor to first satisfy the overall building drift at the top of the building so as to establish a reasonable starting design A_i^0; set the iteration index $\nu = 0$ and commence the OC optimization process.

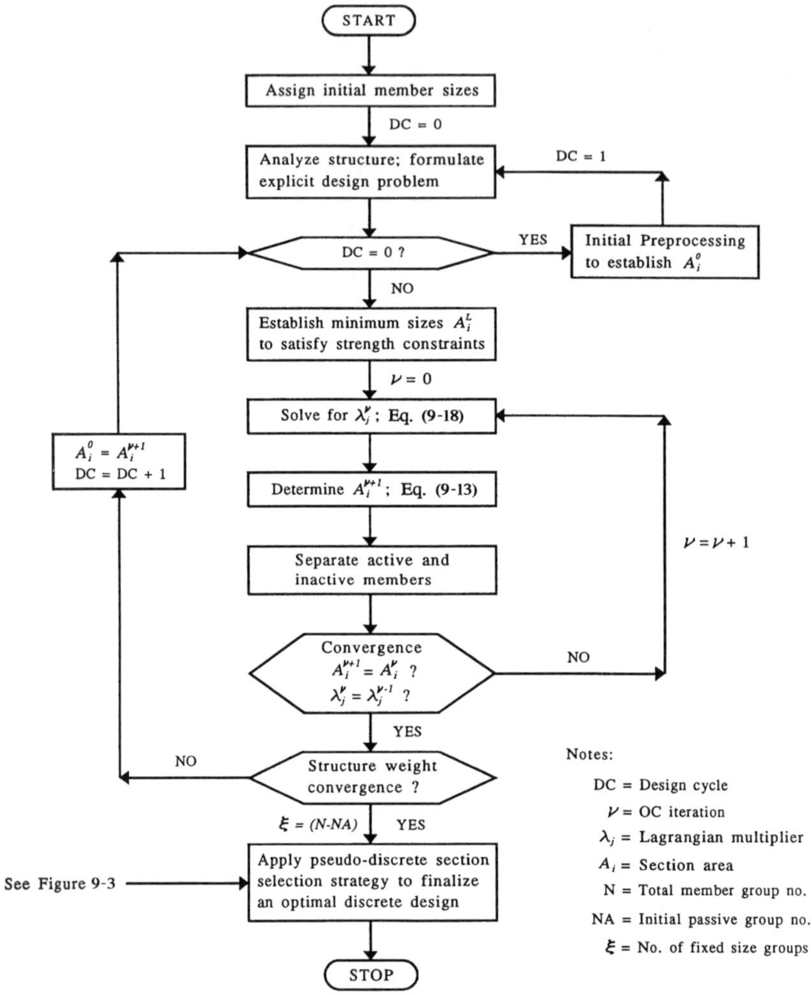

FIG. 9-2. Flow Chart of Overall Design Optimization Procedure.

Step 3.

Analyze the structure, design the members to satisfy strength requirements, and formulate the explicit design optimization problem Eqs. (9-10).

Step 4.

For the current A_i^ν, use the Gauss-Seidel iterative technique to solve Eqs. (9-18) for the corresponding set of Lagrangian multipliers λ_s^ν. Any Lagrangian multiplier value found to be negative at any stage of the recursive solution of Eqs. (9-18) is temporarily set to zero.

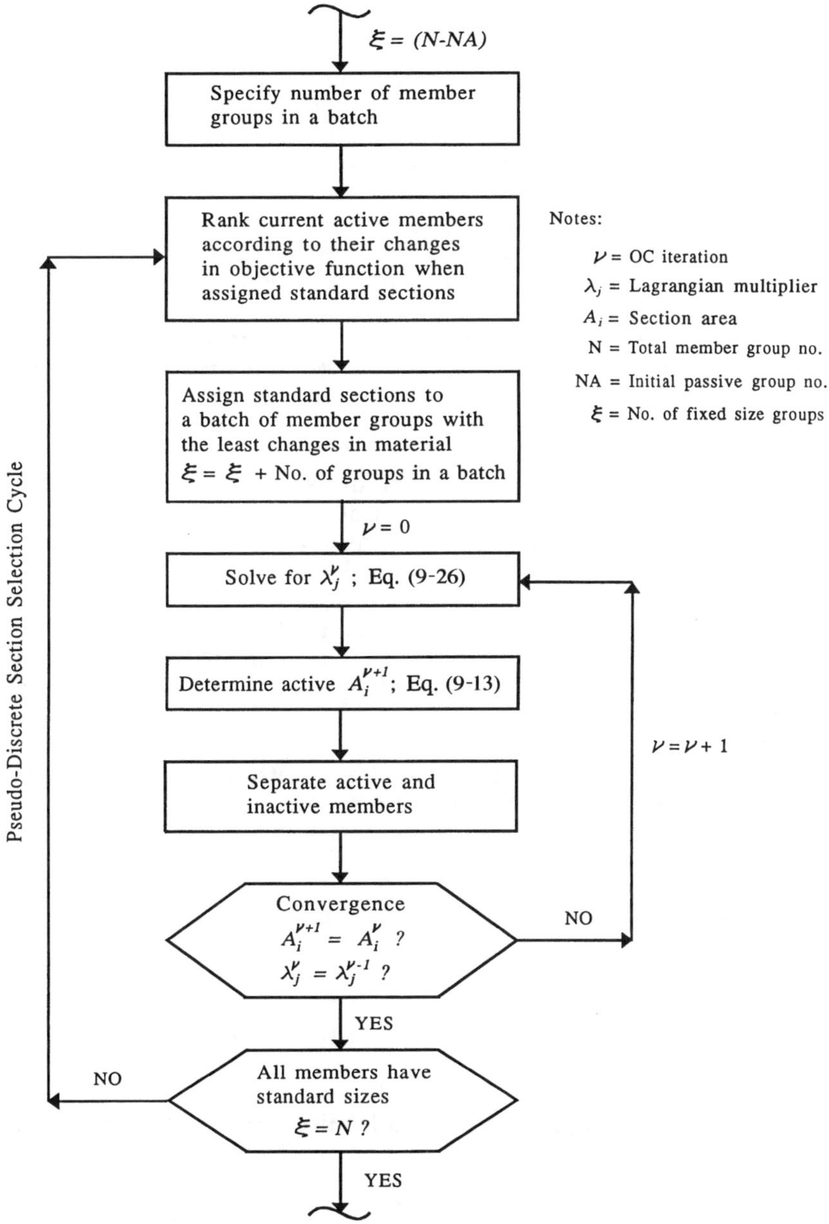

FIG. 9-3. Flow Chart of Pseudo-Discrete Section Selection Procedure.

Step 5.
For the current λ_s^ν and A_i^ν, find the new set of $A_i^{\nu+1}$ from Eqs. (9-13).

Step 6.
For the current $A_i^{\nu+1}$, any member group found to exceed its size limits [i.e., A_i^L or A_i^U in Eq. (9-10c)] is set to the corresponding size limit and deemed to be inactive; the number NA of active members used in Eq. (9-18) is then determined.

Step 7.
Check convergence of the OC recursive process. If all $A_i^{\nu+1} = A_i^\nu$ and $\lambda_s^\nu = \lambda_s^{\nu-1}$, go to Step 8; otherwise, update Eq. (9-18) for the current $A_i^{\nu+1}$, set $\nu = \nu + 1$ and return to Step 4.

Step 8.
Check convergence of the structure weight. If the structure weight from Eq. (9-10a) is the same for two successive design cycles, go to Step 9; otherwise, set $\nu = 0$ and return to Step 3 for next design cycle.

Step 9.
At the convergence of the continuous OC process, set initial $\xi = (N - NA)$, that is, the current number of inactive members, and specify the number of members (say NB) to be assigned standard sections in a batch.

Step 10.
For the current active $A_i^{\nu+1}$ from the continuous OC process, rank active members according to the increment in structure weight they undergo when they are assigned to their next available discrete section sizes.

Step 11.
Identify the batch active members with the least changes in material weight and assign these members their next available discrete standard sections. Set $\xi = \xi + NB$ and $\nu = 0$, and proceed to Step 12.

Step 12.
For the current A_i^ν, solve Eq. (9-26) for λ_s^ν using the Gauss-Seidel iterative technique.

Step 13.
For the current λ_s^ν and current A_i^ν, find the new set of active $A_i^{\nu+1}$ from Eq. (9-13), and then identify the new sets of active and inactive member groups.

Step 14.
Check convergence of the pseudo-discrete OC recursive process. If all active $A_i^{\nu+1} = A_i^\nu$ and $\lambda_s^\nu = \lambda_s^{\nu-1}$, go to Step 15; otherwise, set $\nu = \nu + 1$ and return to Step 12.

Step 15.
Terminate with the final optimal building structure if all member groups have assigned discrete standard steel sections; otherwise, return to Step 10 for the next pseudo-discrete section selection cycle.

DESIGN EXAMPLE

The 50-story 7-bay by 10-bay trussed framework with 5,400 members shown in Fig. 9-4 is now considered to verify the feasibility of the proposed design optimization technique for large-scale three-dimensional tall building frameworks. For a bay width of 15 ft and a story height of 12 ft, the framework has a height-to-width aspect ratio of 5.7 in the X-direction and 4.0 in the Y-direction. Fig. 9-4 shows the three-dimensional computer model of the framework. Details of the framework are shown in its plan and elevation views in Figs. 9-5

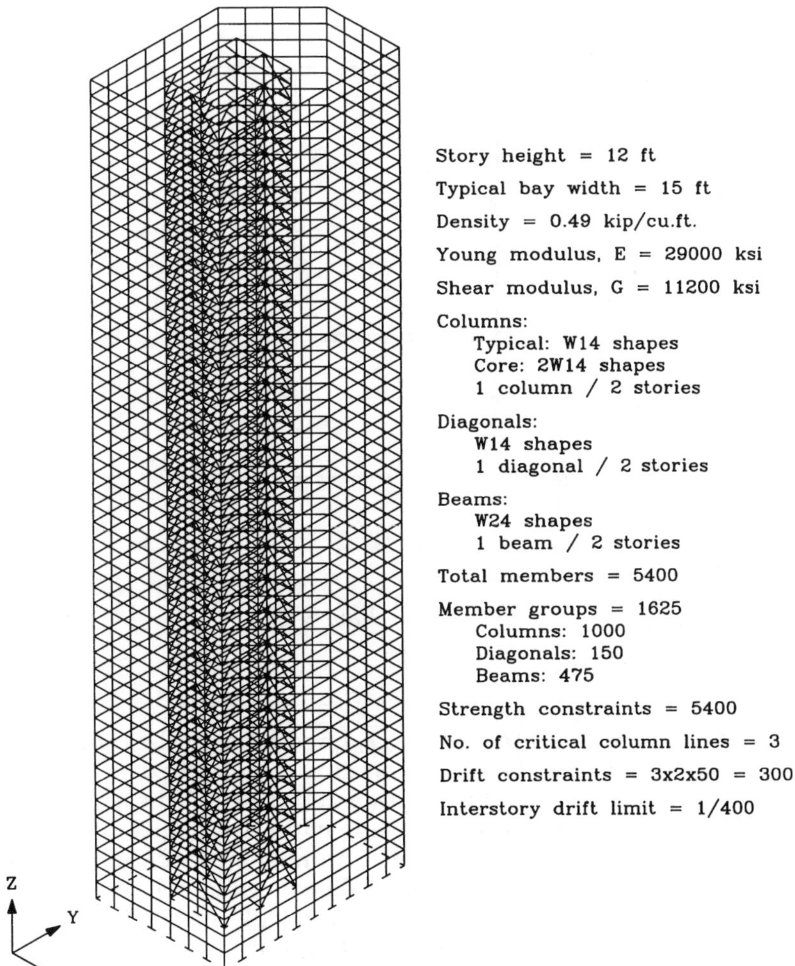

FIG. 9-4. 3D Model of 50-Story Framework.

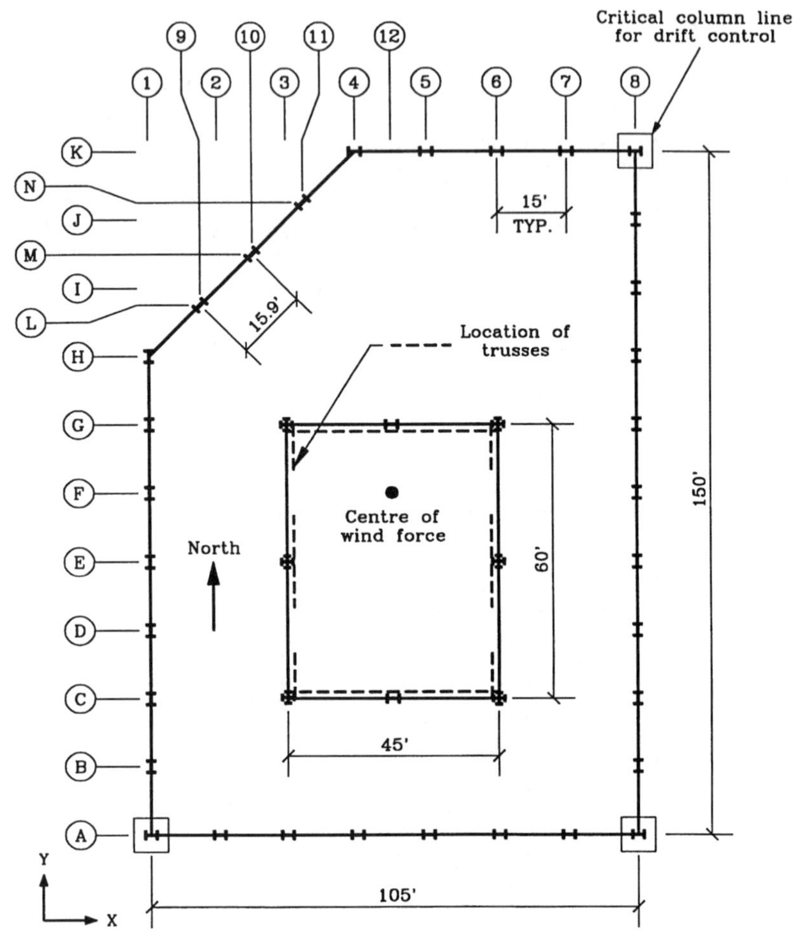

FIG. 9-5. Plan View.

to 9-7. As illustrated in these figures, one corner of the building is cut off at a 45° angle and the core is eccentrically located in the *Y*-direction (i.e., shifted towards the south exterior face). These features create a built-in asymmetry that causes natural twisting of the framework under lateral loadings. The framework consists of exterior moment frames and a braced core. All beams and columns are rigidly connected and the diagonal braces are simply connected. Two-story *K*-bracing modules are used on both the south and north faces of the core, as shown in Fig. 9-6. Single-story knee-bracing is used in the west and east faces of the core to ensure accessibility to the elevators in the core, as shown in Fig. 9-7.

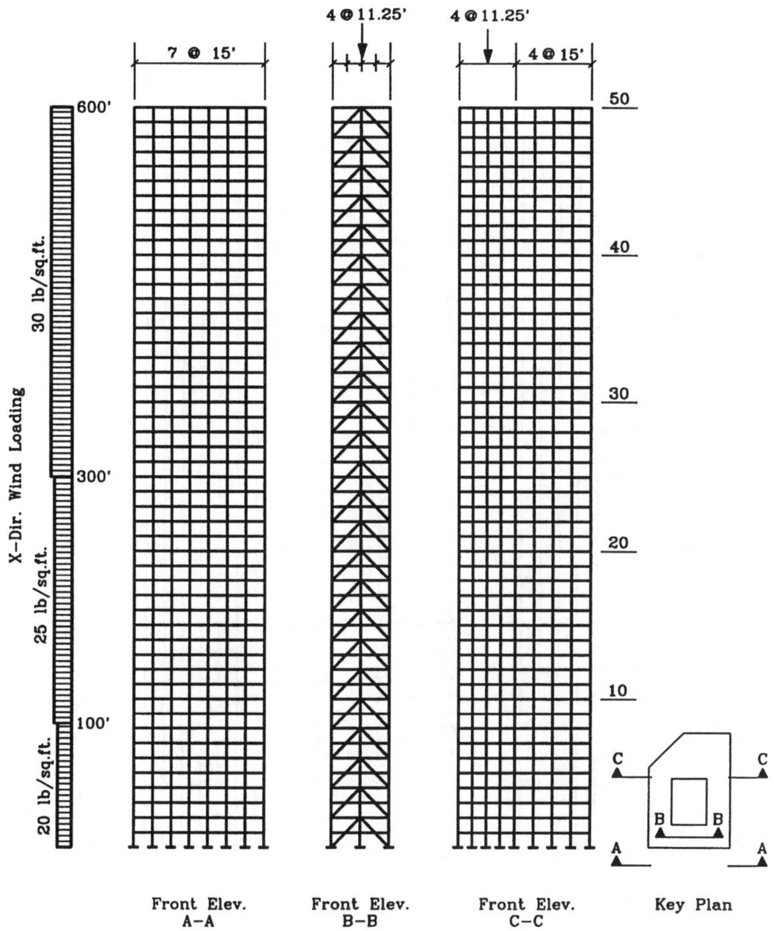

FIG. 9-6. Front Elevations.

Four basic load cases are considered for this design example. Namely, gravity dead load (D), specified live load (L), specified wind load in the principle X-direction (W_X), and specified wind load in the orthogonal Y-direction (W_Y). The two individual gravity loading conditions concerning dead and live loads are applied as uniformly distributed loads on the beams of each floor, as shown in Fig. 9-8. The lateral wind loadings shown in Figs. 9-6 and 9-7 are applied to the framework as horizontal point loads at the center of force for each floor in the two orthogonal X and Y directions. From among the four load cases, eight different load combinations are identified for the strength design of each structural member in accordance with the

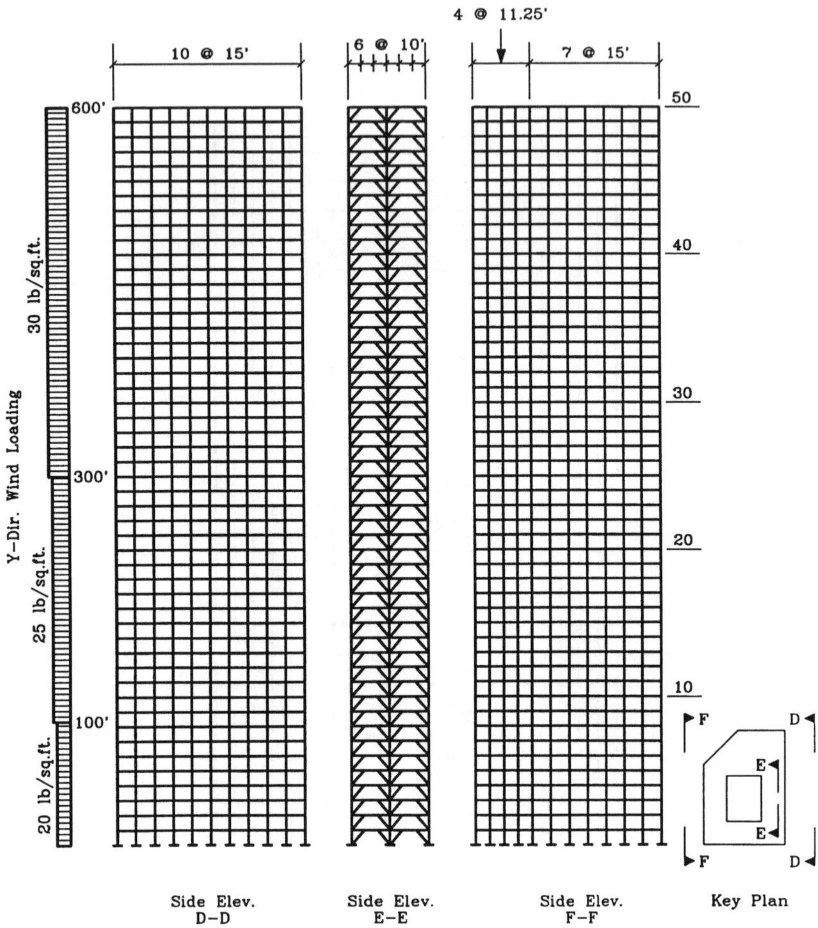

FIG. 9-7. Side Elevations.

American AISC LRFD design standard (Manual 1986), as shown in the following.

$$1.4D \tag{9-27a}$$

$$1.2D + 1.6L \tag{9-27b}$$

$$1.2D + 0.5L + 1.3W_X \tag{9-27c}$$

$$1.2D + 0.5L - 1.3W_X \tag{9-27d}$$

$$1.2D + 0.5L + 1.3W_Y \tag{9-27e}$$

$$1.2D + 0.5L - 1.3W_Y \tag{9-27f}$$

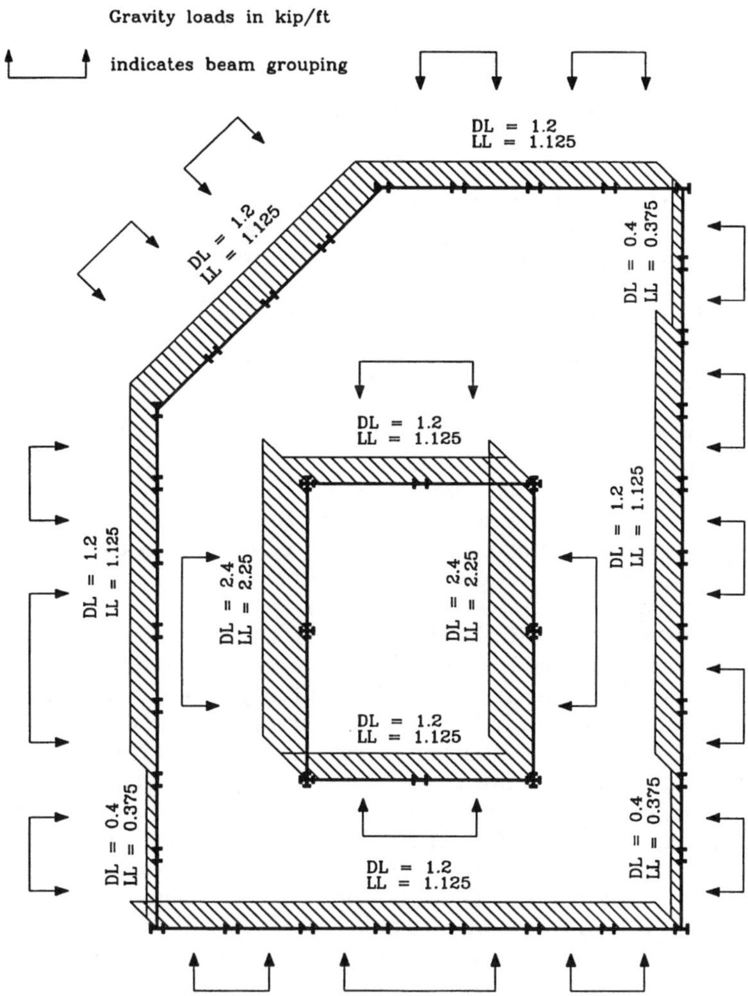

FIG. 9-8. Uniform Distributed Gravity Beam Loads and Beam Grouping.

$$0.9|D| - 1.3|W_X| \qquad \text{(9-27g)}$$

$$0.9|D| - 1.3|W_Y| \qquad \text{(9-27h)}$$

Load combinations Eqs. (9-27a) and (9-27b) involve only vertical gravity loads. Various combinations of total gravity loads and lateral wind loads are described in combinations Eqs. (9-27c) to (9-27f). Combinations Eqs. (9-27g) and (9-27h) account for any uplift or reversal effect of the lateral wind loads.

As twisting may occur for this asymmetric framework, interstory drifts that account for the effect of building torsion at the most distant column lines are controlled within a typical allowable drift ratio of 1/400. The column lines A1, A8, and K8 at the building corners (see Fig. 9-5) are the most critical column lines for the stiffness optimization. The adverse effects of member shear deformations and second-order $P - \Delta$ behavior are accounted for in the analysis of the lateral load responses. Since the $P - \Delta$ analysis employed in this study uses an equivalent story-shear concept that requires the presence of vertical gravity loads, a typical story weight of 2,300 kips is employed for this framework for the calculation of the second-order $P - \Delta$ effects. American AISC standard sections are used to size the members as follows: beams are W24 shapes; diagonals are W14 shapes; and columns are also W14 shapes except that the cruciform columns in the core (see Fig. 9-8) use pairs of two W14 shapes oriented perpendicular to each other. To satisfy practical construction requirements, beams that are grouped together on each floor as shown in Fig. 9-8 are specified to have the same section over two adjacent stories, whereas columns in each line are grouped together as having a common section over two adjacent stories, as are diagonals in each span.

To commence the optimization process, the framework is first analyzed with the members taken to have their maximum available section sizes as the initial design. The initial preprocessor is then applied to establish a reasonably proportioned starting design satisfying the two top-story drift constraints corresponding to the two orthogonal W_X and W_Y wind loadings. Upon conducting a second reanalysis after the initial preprocessing, the automatic OC resizing method is then implemented to achieve the optimal continuous-variable design. To finalize the design using discrete standard sections, the pseudo-discrete OC resizing technique is applied upon convergence of the continuous solution. During this discrete section selection stage, a batch of the currently most economical member groups, equal in number to 10% of the member groups, is identified and the corresponding groups are assigned their next available discrete standard sections, such that the discrete optimization is completed after 10 batches have been selected and so assigned.

Rapid and steady convergence to the optimal design is found for this large-scale framework. The history of the design process for the framework is shown in Fig. 9-9. For a weight convergence criterion that terminates the design process when the structure weights for two successive design cycles are within 0.5%, the iterative continuous-variable design process is terminated after the fifth design cycle, and the pseudo-discrete OC algorithm finalizes the design to a structure weight of 11,316 kips, which corresponds to 15.4 lbs of steel per square foot of floor area. The optimal design process required 12.5 hours of CPU time on a SUN SPARC 1 station (with 32-bit CPU running at 12.5 MIPS). With the complexity and scale of this 50-story

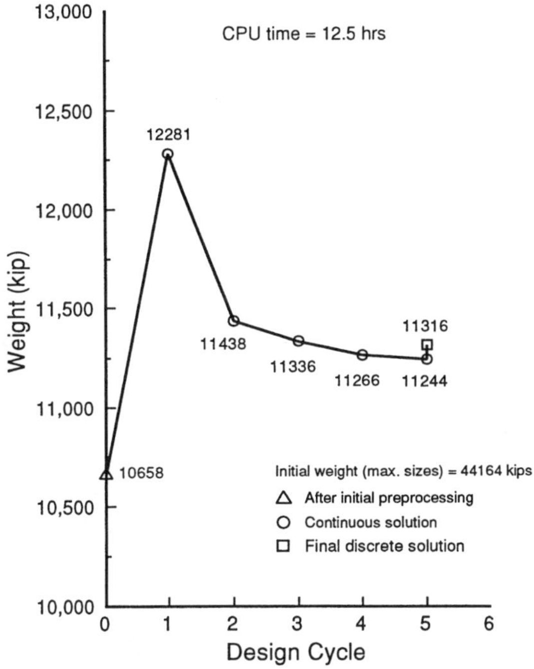

FIG. 9-9. Design History for 50-Story 3D Framework.

three-dimensional framework, such a relatively short computation time on such a low-end computer clearly demonstrates the efficiency and practicality of the automatic OC resizing technique for practical tall steel building frameworks. In fact, a rather stringent weight convergence criterion (0.5%) is used for this example; if a 1% criterion is used, the final design is achieved after the third design cycle in approximately 7.5 hours of CPU time. As shown in Fig. 9-9, the first design cycle result is within 10% of the final result. Such a quick convergence can be explained by the fact that the internal member force distribution is quite insensitive to changes in member sizes for building frameworks. Hence, such behavior of building frameworks results in good quality approximations of the explicit drift constraints with consequent rapid convergence to the optimal design.

Fig. 9-10 shows the deflected profiles of the framework after the initial preprocessing and after completion of the pseudo-discrete OC resizing technique in both the X and Y directions. Although the two top-story drift constraints in the two orthogonal directions are satisfied during the initial preprocessing, re-analysis determines some violation in this regard and, as well, significant torsional twisting of the framework as shown by the dashed lines in Fig. 9-10. Note as

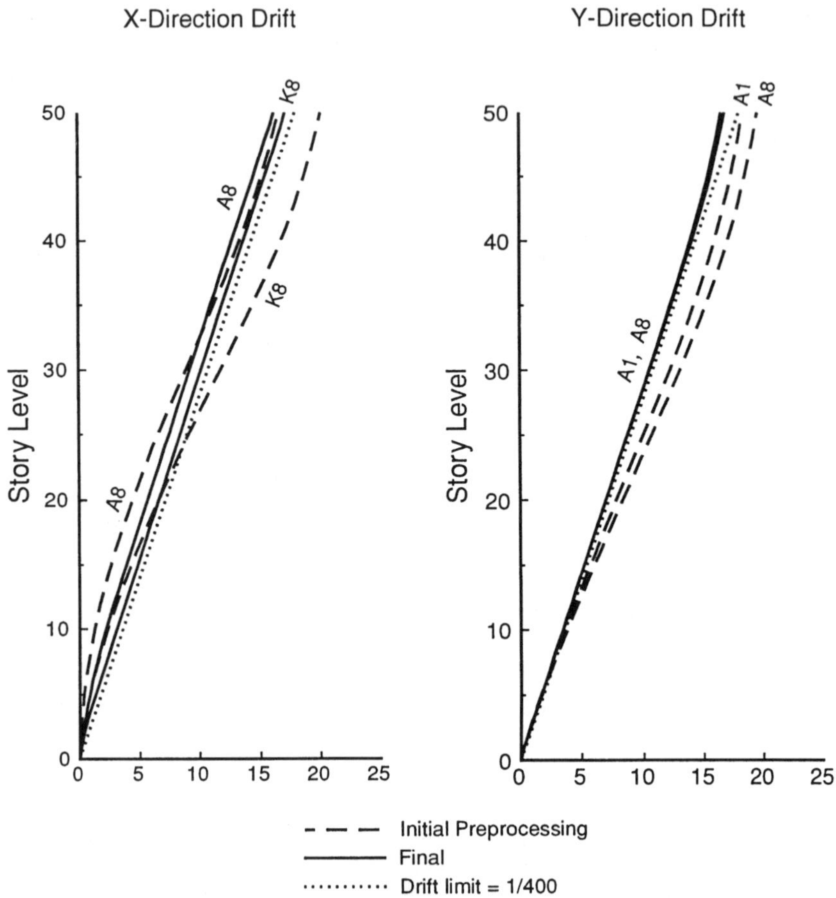

FIG. 9-10. Deflected Profiles for 50-Story 3D Framework.

well that the interstory drift constraints at the central height of the framework are also substantially violated after the initial preprocessing stage. However, it is found that little or no twisting occurs for the optimal design as shown by the solid lines in Fig. 9-10. For the *Y*-direction drift, both the south-west column line A1 and the south-east column line A8 are found to deflect almost identically under the wind loading applied in the *Y*-direction. For the *X*-direction drift, a slight difference in lateral deflection of about 1 in. is found between the south-east column line A8 and the north-east column line K8 in the upper portion of the building, indicating a twisting rotation of about 1 inch in 150 ft or 1/1800 radian. Such a result seems to indicate that for the optimal design of an asymmetric framework, the lateral load resisting system will be sized by the OC procedure such as to distribute its stiffness so that little or no building torsion occurs.

A careful look at the distribution of the final discrete section sizes indicates that a majority of the member sizes is found to be controlled by lateral drift requirements and to be within upper and lower size bounds. However, the bottom two columns of the interior column lines G3 and G6 located at the corners of the north face of the core are overstressed 18.7 and 8.4%, respectively, even when the maximum available cruciform section (a pair of W14X730s) is used. To rectify the problem, one easy solution is to adopt a higher strength of 50 ksi in place of the 36 ksi steel previously used for the overstressed sections, such that a percentage increase in strength of 38% is achieved.

Since the framework is composed of shear trusses at the core and a rigid frame at the exterior, a shear truss-frame interaction behavior is observed. The trussed core behaves likes a flexural cantilever beam that more effectively resists the applied lateral loads near the bottom half of the framework. Larger section sizes are generally found for the columns, beams, and diagonals of the interior core towards the base of the framework. On the other hand, the exterior rigid frame behaves like a shear beam that more effectively resists the lateral loading towards the upper half of the framework. Generally, larger column and beam sizes occur at about the two-third height of the exterior frame. Such a result of the distribution of optimal sizes seems to match closely engineering intuition concerning the structural behavior of the framework.

SUMMARY

The automatic resizing technique developed in this chapter is both practical and suitable for tall steel building design. The efficiency of the iterative resizing technique presented herein is mainly influenced by the number of constraints and is only weakly dependent on the number of variables. As such, it provides an effective strategy for the optimal design of tall building structures involving many sizing variables and comparatively few drift constraints. Moreover, the resizing method will generally converge quite rapidly because the member force distributions for such structures are somewhat insensitive to changes in member sizes. Not only is a least-weight design achieved while satisfying a comprehensive set of design constraints, but there is also a considerable saving in designer time and cost. As well, the optimal design results are shown to adhere closely to engineering intuition concerning the structural behavior of such frameworks. An interesting finding is that the optimal design of an asymmetric 3D building framework corresponds to a state where there is little or almost no building torsion.

APPENDIX 9-A. NOTATION

The following symbols are used in this chapter.

A = cross-sectional area of a member
$\mathbf{A}$ = a set of cross-section areas of the available discrete standard sections
a, b = coefficient terms of the set of linear simultaneous equations for Lagrange multipliers
C, C' = regression coefficients; virtual strain energy coefficients
D = dead gravity load; total number of interstory drift constraints
d = interstory drift
E = modulus of elasticity of steel (29000 ksi or 200000 MPa)
e, e' = virtual strain energy coefficients
F, f = actual and virtual member forces
G = elastic shear modulus of steel (11200 ksi or 77000 MPa)
h = story height
I = moment of inertia
L = length of member; number of lateral loading conditions; Lagrangian function; live load
M, m = actual and virtual internal member moments
N = total number of member groups for a building framework
NA = total number of active member groups at one stage of the continuous optimization process
NB = number of members selected in a batch to be assigned standard steel sections
S = total number of slave column lines
T = transformation matrix to relate lateral displacement to the master node
W = structure weight of a building framework; wind loading
Δ = geometric distance of a floor node and the corresponding master node
δ = lateral translation
ε = virtual strain energy density of a member
η = step-size or relaxation parameter
λ = Lagrange multiplier
θ = rigid floor rotation
ξ = number of member groups with assigned discrete standard steel sections

Subscripts and Superscripts

i = index of a member or member group
j = index of a story

k = index of a slave column line
l = index of a lateral loading condition
L = lower bound index
s, t = index of a drift constraint
U = upper bound index
X, Y, Z = local X, Y, and Z axes of a member cross-section
ν = iteration number of the recursive optimality criteria algorithm
τ = iteration number of Gauss-Seidel solution process

APPENDIX 9-B. CONVERSION TO SI UNITS

1 in. = 25.4 mm
1 ft = 0.3048 m
1 lb = 0.454 kgf
1 kip = 4.448 kN
1 lb/ft^2 = 4.88 kgf/m^2
1 ksi = 6.895 MPa

REFERENCES

Baker, W. (1989). "Sizing techniques for lateral bracing." *Steel Structures*, J. Iffland, ed. ASCE, New York, 298–307.

Berke, L. and Khot, N. S. (1974). "Use of optimality criteria methods for large scale systems." AGARD-LS-70, 1.1–1.29.

Berke, L. and Khot, N. S. (1988). "Performance characteristic of optimality criteria methods." *Proceedings of IUTAM Symposium on Structural Optimization*, G. I. N. Rozvany and B. L. Karihaloo, eds., Melbourne, Australia, 39–46.

Brown, D. M. and Ang, H. S. (1966). "Structural optimization by nonlinear programming." *J. of Structural Division*, ASCE, 92(ST6), 319–340.

Chan, C. - M. (1992). "An optimality criteria algorithm for tall steel building design using commercial standard sections." *Structural Optimization*, 5, 26–29.

Chan, C. - M. (1993). *An automatic resizing technique for the design of tall steel building frameworks*. Ph.D. Thesis, University of Waterloo, Canada.

Chan, C. - M. and Grierson, G. E. (1993). "An efficient resizing technique for the design of tall steel buildings subject to multiple drift constraints." *The Structural Design of Tall Buildings*, Vol. 2, No. 1, John Wiley & Sons, Ltd., West Sussex, England, 17–32.

Charney, F. A. (1991). "The use of displacement participation factors in the optimization of wind sensitive buildings." *Proceedings of the Ninth Structures Congress*, ASCE, Indianapolis, Indiana, USA, 545–547.

Cheng, F. Y. and Truman, K. Z. (1983). "Optimization algorithm of 3-D building systems for static and seismic loading." *Modeling and Simulation in Engineering*, W. F. Ames, ed., North-Holland Pub. Co., 315–326.

Council of Tall Building & Urban Habitat. (1979). *Structural design of tall steel buildings. Monograph on the planning and design of tall buildings*, Vol. SB, ASCE, New York.

Gilsanz, R. and Carlson, A. (1991). "Optimization in building design." *Proceedings of the Second International Conference on Computer Aided Optimum Design of Structures 91*, Hernandez and Brebbia, eds., Computational Mechanics Publications, Boston, MA, 541–551.

Grierson, D. E. and Cameron, G. E. (1990). *"SODA—structural optimization design and analysis."* Release 3.1, User Manual, Waterloo Engineering Software, Waterloo, Canada.

Hornbeck, R. W. (1975). *Numerical method.* Prentice-Hall Inc., Englewood Cliffs, NJ.

Khan, M. R. (1984). "Optimality criterion techniques applied to frames having general cross-sectional relationships." *AIAA J.*, 22(5), 669–676.

Khot, N. S., Berke, L., and Venkayya, V. B. (1979). "Comparison of optimality criteria algorithms for minimum weight design of structures." *AIAA J.*, 17(2), 182–190.

Manual of steel construction—load and resistance factor design. (1986). American Institute of Steel Construction, Inc., Chicago, IL, First Edition.

Sadek, E. A. (1992). "Optimization of structures having cross-sectional relationships using an optimality criterion method." *Computers and Structures*, 43(5), 959–969.

Tabak, E. I. and Wright, P. M. (1981). "Optimality criteria method for building frames." *J. of Structural Division*, ASCE, 107(ST7), 1327–1342.

Velivasakis, E. and DeScenza, R. (1983). "Design optimization of lateral load resisting frameworks." *Proceedings of the Eighth Conference on Electronic Computation*, ASCE, New York, 130–143.

Venkayya, V. B., Khot, N. S., and Berke, L. (1973). "Application of optimality criteria approaches to automated design of large practical structures." *Second Symposium Structural Optimization, AGARD-CP-123*, Milan, Italy.

Venkayya, V. B., Khot, N. S., and Reddy, V. S. (1969). "Energy distribution in an optimum structural design." Technical Report AFFDL-TR-68-156, Wright-Patterson Air Force Base, Ohio.

Wismer, D. A. and Chattergy, R. (1978). *Introduction to nonlinear optimization.* North-Holland Pub. Co., New York.

Chapter 10

HOW TO OPTIMIZE FRAMES USING PLASTIC DESIGN CONCEPT

The objective of this chapter is to acquaint practicing engineers with optimal plastic design of frames. It is hoped that the information presented stimulates the use of optimal plastic design methods for frames.

INTRODUCTION

In the 1960s, advances in the theory of plastic structures led to the development of limit design methods for minimum weight structures. Some of the fundamental contributions in this area are attributed to Foulkes, Heyman, and others. The idealization of structural behavior as rigid-plastic allows the problem of finding the minimum weight structure to be formulated as a linear programming problem, which can then be subjected to efficient and systematic solution methods.

The objective of this chapter is to discuss and review the basic concepts and assumptions underlying the optimal plastic design of structures. First, the concept of plastic design is explained. Then, the formulation of the optimization problem is discussed. Finally, the limitations and strengths of the method are discussed.

In the following sections, the necessary background information is first reviewed and then the optimal plastic design of frames is formally introduced.

Shahram Pezeshk, Member ASCE, Associate Professor of Civil Engineering, The University of Memphis, Memphis, TN 38152; email: ⟨s-pezeshk@memphis.edu⟩.

BACKGROUND INFORMATION

Limit Design

Modern design of structures is based on both the elastic and plastic analyses. An elastic analysis of a structure is required to study the performance, especially with regard to serviceability, under the loading for which the structure is designed. The plastic analysis cannot replace the elastic analysis but supplements it by giving useful information about the collapse load and the mode of collapse. The main objective of plastic analysis is to determine the collapse load of a structure when resisting capacities of its members are known. Design of structures based on the plastic approach, referred to as "limit design," is increasingly used and accepted by various codes of practice.

Plastic Hinge Concept

For steel with an idealized stress-strain curve of Fig. 10-1(a), the bending moment-rotation relationship at any section of a member has the form of Fig. 10-1(b). It is seen from the figure that when the fully plastic moment M_p is reached, rotation takes place with no increase in the bending moment, and the section performs as a hinge rotating with constant moment M_p. Such a hinge is called a "plastic hinge." For structures composed of members with constant cross-section, plastic hinges will develop first at points with peak moments.

Collapse Mechanism

When a sufficient number of hinges develop, the structure becomes an unstable system referred to as a "collapse mechanism." The magnitude of the applied load causing the formation of a collapse mechanism is termed the "ultimate load."

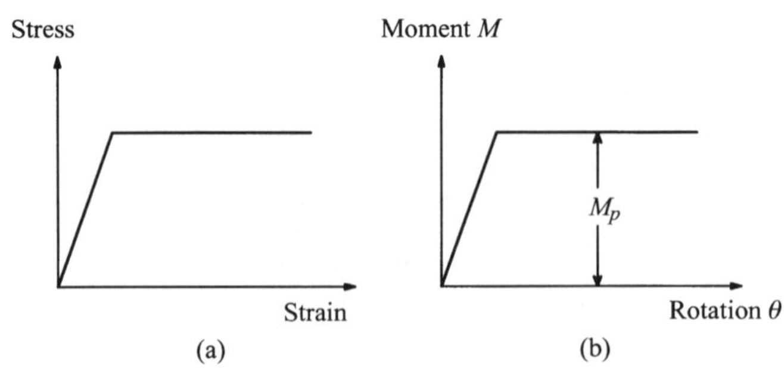

FIG. 10-1. (a) Idealized Stress Strain Curve; (b) Moment-Rotation Relationship.

Energy Consideration

The principle of conservation of energy can be applied to study whether a collapse mechanism satisfies a specified safety factor criterion. In view of the actual behavior of structures, it is reasonable to assume that the displacements are small before the loads attain their ultimate value, so that changes in geometry before a collapse mechanism forms are negligible. Applying a virtual displacement, δ (= $h\theta$), to the collapse mechanism of Fig. 10-2, with each plastic hinge undergoing a rotation θ from energy considerations the external work E imparted by the external load P_u, moving along displacement δ must be equal to the internal work U absorbed by the plastic hinges. However, to design a structure that will satisfy the safety factor criterion, it is necessary that its energy absorption capacity U be equal to, or greater than, the external energy E imparted by the loads. Therefore for a possible collapse mechanism i we have

$$U_i \geq E_i \tag{10-1}$$

The total energy dissipated by plastic hinges U_i is given in the general form as (Neal 1985)

$$U_i = \sum_{j=1}^{J} M_{pj}\theta_{ij} \tag{10-2}$$

where

J = number of plastic hinges to develop the collapse mechanism i
θ_{ij} = rotation for plastic hinge j
M_{pj} = plastic moment for plastic hinge j

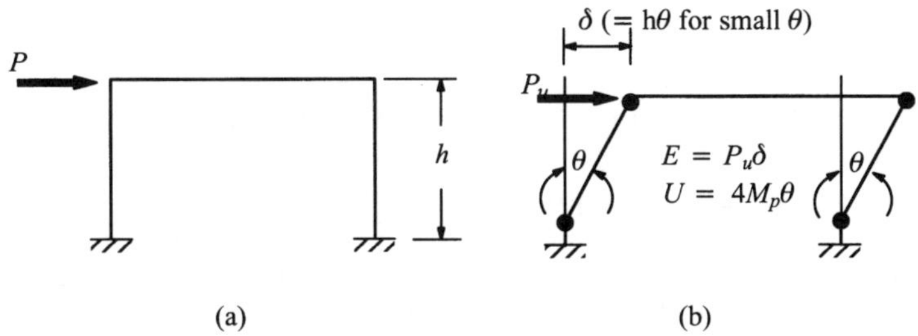

FIG. 10-2. (a) Frame Under Service Load P, (b) Frame Under Ultimate Load P_u (Collapse Mechanism).

The preliminary concepts explained so far are used in the application of graphical and linear programming to the minimum weight design of a steel frame.

ASSUMPTIONS FOR UTILIZING PLASTIC STRENGTH IN DESIGN

It is assumed in the following that the dimensions of the frame and the factored values of the characteristic loads are prescribed. Furthermore, each frame considered is assumed to be composed of uniform prismatic members. The normal assumptions for the calculation of plastic collapse loads are made, and the plastic moments of the members are taken to be unaffected by axial and shear force. In addition, the effect of elastic flexibility, combined with the deterioration of stiffness due to plasticity is not considered. It is also assumed that premature failure of individual members is avoided in order to allow plastic hinges to form at the critical sections and to achieve moment redistribution. The latter requires a certain amount of rotational capacity to be available at the hinges. The most important considerations for members subjected to bending moment are local buckling and lateral torsional buckling, and design requirements are necessary to ensure not only the formation of plastic hinges but also sufficient rotational capacity. The local buckling requirement is usually given as the maximum permissible width-to-thickness ratios of the cross-sectional elements and the lateral buckling requirement is defined in terms of the unbraced length.

The effect of elastic and plastic deformations of the structure, causing finite changes of geometry and thus modifying the equilibrium conditions, is ignored. The finite change of geometry is called the $P - \Delta$ effect because the changes in moment may be thought of as due to a typical load P changing its line of action by an amount Δ, where Δ is the deflection of a node of the structure perpendicular to the direction of P (Horne and Morris 1974).

In addition to the preceding factors, it is necessary to choose a method of linearizing the objective function as explained in the next section.

OPTIMAL PLASTIC DESIGN FORMULATION

The major step in design is to assign suitable sections to all the members of the frame so that the applied load system or systems can be carried at the specified load factors. The starting point is to obtain an optimum design according to rigid plastic theory, but the resulting design has to be modified for a number of factors not allowed for in the basic design process. The factors not allowed for are those that introduce nonlinearity.

This procedure involves at the outset the arbitrary selection of the relative values of the plastic moments of the members. A different selection of these relative values would result in a different design. For a given frame and loading, there are a large number of possible designs, and it is relevant to consider which design can achieve the minimum cost.

When an optimum design is being considered, the main purpose is to minimize a function that expresses the total weight or cost of a frame. Though such a function represents an exact solution of the problem, it cannot be used in that form. This is because the constraints provided by the moment equilibrium consideration relate the plastic moment capacities of the individual members and not their weights. However, by assuming a linear relation between the plastic moment capacity M_p and the weight W_i of member i, we have (Neal 1985; Horne and Morris 1974; Rubinstein and Karagozian 1966)

$$W_i = (a_i + b_i M_{pi})L_i \tag{10-3}$$

The total weight function for a structure with n members is

$$W = \sum_{i=1}^{n} W_i \tag{10-4}$$

Therefore

$$W = \sum_{i=1}^{n} a_i L_i + \sum_{i=1}^{n} b_i M_{pi} L_i \tag{10-5}$$

where a_i and b_i are constants, M_{pi} is the plastic moment of member i, and L_i is the length of member i. Since the first term and b_i are constant and do not influence optimization, the "effective" weight function W becomes (without change of symbolism):

$$W = \sum_{i=1}^{n} M_{pi} L_i = \sum_{j=1}^{g} M_{pj} L_j \tag{10-6}$$

where M_{pj} is the group plastic moment (a scalar quantity), L_j is the total length of all members in group j, and g represents the total number of groups in a structure. The concept of grouping can be adopted so that members in any particular group are automatically required to have the same moment capacity and hence the same properties, thereby permitting the important practical advantage of repetitive sections (Horne and Morris 1974).

APPLICATION

The following sections first describe the formulation for the optimal plastic design of a simple portal frame by defining the objective function and constraints. Then the optimization problem is solved using both a graphical and a linear programming approach.

Problem Description and Design Solution Using a Graphical Approach

The concept of minimum weight design using plastic design concepts is best explained through an illustrative example problem. Therefore a discussion is presented on optimal design of a rectangular frame with factored loading pattern and dimensions as shown in Fig. 10-3. For simplicity, convenience, and for all practical purposes, it will be required that the two columns be identical ($g = 2$). The objective is to select column and beam sections that will minimize the weight of the frame. The beam is to have a plastic moment of M_b, and the columns are each to have a plastic moment of M_c. The weight function as defined in Equation (10-6) is

$$W = \sum_{j=1}^{g} M_{pj} L_j = 16\lambda M_c + 20\lambda M_b \qquad (10\text{-}7)$$

where λ is a constant representing the weight of a linear foot of a member with a plastic moment capacity of 1 ft kip.

For this type of frame and loading there are two independent mechanisms, the beam and the sway mechanisms. These can be combined to form a third mechanism. Since it is not known at the outset whether M_b is less than or greater than M_c, each of these mechanisms can take two forms, with any hinge occurring at a joint between the beam and a column appearing either in the beam or the column. The resulting six possible mechanisms of collapse are shown in Fig. 10-3 together with the corresponding work equations.

Using the expressions for U and E derived in Fig. 10-3, the inequality given in Eq. (10-1) takes the form of

$$\begin{aligned}
&M_c \geq 6 && \text{Fig. 10-3a}\\
&M_b + M_c \geq 12 && \text{Fig. 10-3b}\\
&M_b + M_c \geq 5 && \text{Fig. 10-3c}\\
&M_b \geq 2.5 && \text{Fig. 10-3d}\\
&M_b + 2M_c \geq 17 && \text{Fig. 10-3e}\\
&2M_b + M_c \geq 17 && \text{Fig. 10-3f}
\end{aligned} \qquad (10\text{-}8)$$

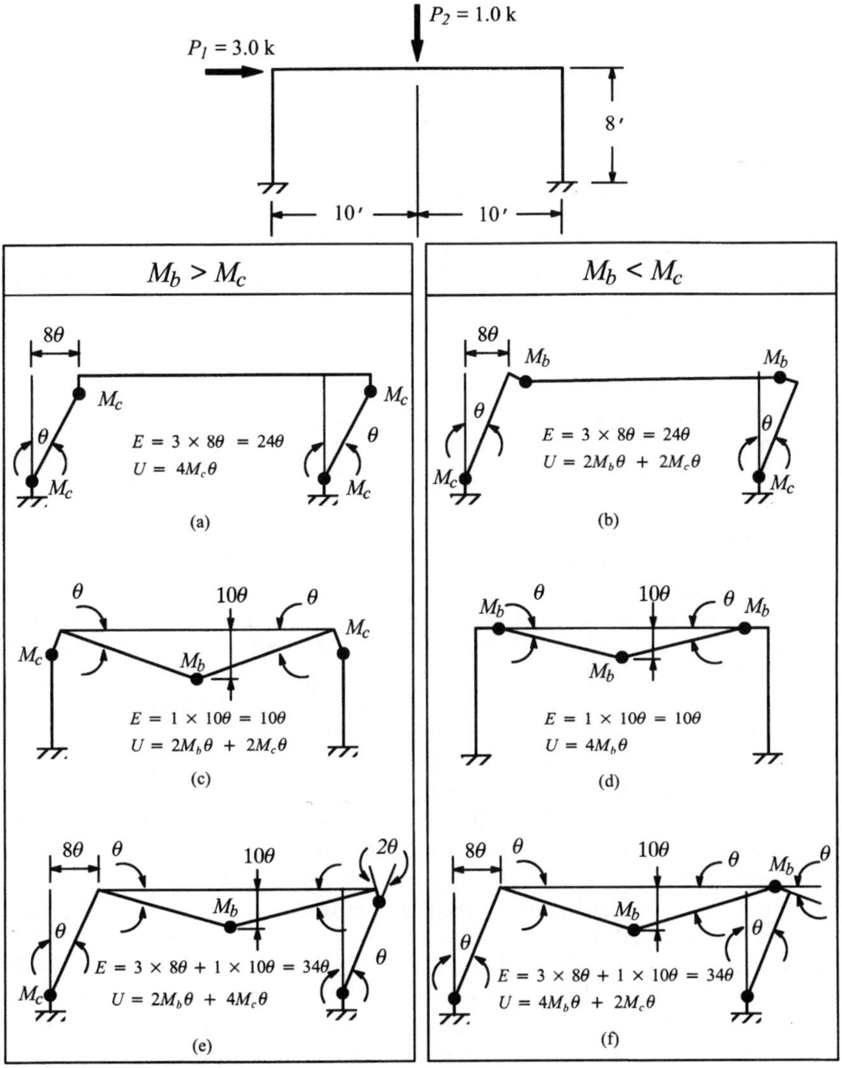

FIG. 10-3. Collapse Mechanisms: (a), (b) Sway Mechanisms; (c), (d) Beam Mechanisms; and (e), (f) Combined Mechanisms.

It is obvious that the third constraint in Eq. (10-8) is satisfied if the second constraint is satisfied. Therefore we drop the third constraint in Eq. (10-8). The constraint set given in Eq. (10-8) can be written in matrix form as

$$[a]\begin{Bmatrix} M_b \\ M_c \end{Bmatrix} \geq \{b\} \tag{10-9}$$

in which

$$[a] = \begin{bmatrix} 0 & 1 \\ 1 & 1 \\ 1 & 0 \\ 1 & 2 \\ 2 & 1 \end{bmatrix} \qquad \{b\} = \begin{Bmatrix} 6 \\ 12 \\ 2.5 \\ 17 \\ 17 \end{Bmatrix} \tag{10-10}$$

These work equations may be represented as straight lines on a diagram in which M_b and M_c are rectangular coordinate variables, as in Fig. 10-4. For each of the three types of mechanism there are two lines, representing the work equations for the cases $M_b \geq M_c$ and $M_b \leq M_c$. Thus for the sway mechanism, lines (a) and (b) represent the two cases. The other work equations are shown similarly as the lines (c) through (f), the lettering corresponding to that of Fig. 10-3. Each inequality given in Eq. (10-8) represents a region on one side of a straight line in a plane with coordinates M_b and M_c. For instance, the inequality $M_c \geq 6$ ft kips represents the region to the right of line (a). The region that satisfies all the inequalities is called the "region of admissible solutions" because only values of M_b and M_c that are within this region will satisfy all of the inequalities. The objective is then to find the point with coordinates M_b and M_c in the region of admissible solutions that will minimize the objective function, Eq. (10-7). It can be shown that the minimum value is at one of the intersections of the lines defining the region of admissible solutions. These interactions are called "extreme points" and are marked by A, B, and C in Fig. 10-4.

One way to establish which extreme point represents the solution is to substitute into the objective function Eq. (10-7) the values of M_b and M_c at each extreme point, and select the point for which the objective function W is minimum. In the present example we have

$$\begin{aligned} &\text{At point A,} && M_b = 6, && M_c = 6 \rightarrow W = 216\lambda \\ &\text{At point B,} && M_b = 5, && M_c = 7 \rightarrow W = 212\lambda \\ &\text{At point C,} && M_b = 2.5, && M_c = 12 \rightarrow W = 242\lambda \end{aligned} \tag{10-11}$$

Consequently, the objective function W is minimum at point B, and

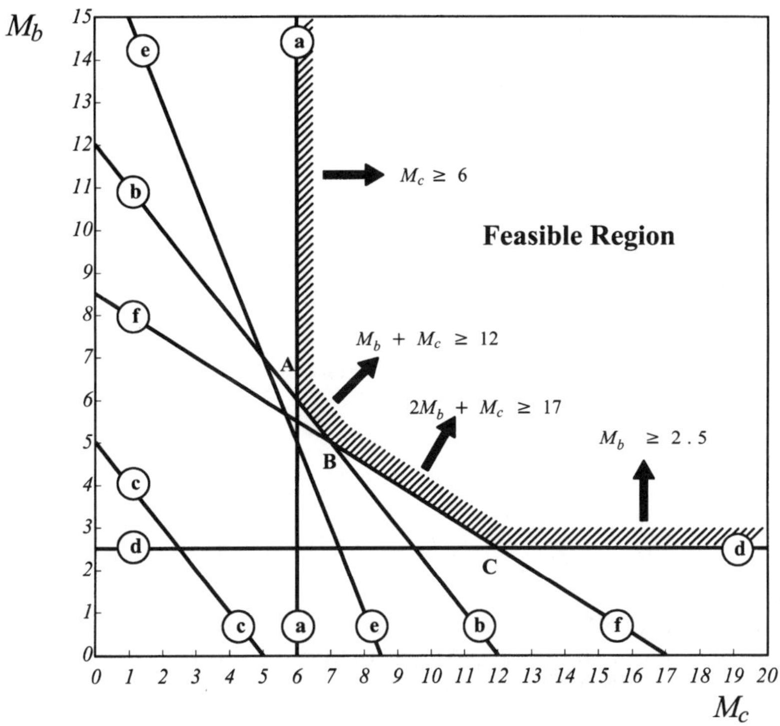

FIG. 10-4. Region of Admissible Solution.

the minimum weight design calls for a beam section and a column section with plastic moment capacities of 5 ft kips and 7 ft kips, respectively.

Design Solution Using a Linear Programming Approach

For problems in which more than three variable plastic moment capacities are involved, the graphical solution shown in Fig. 10-4 is impossible. In such cases the Simplex Method (Ficken 1961) is used. The simplex method is a systematic numerical procedure for the solution of linear programming problems.

The optimization problem as it is given by Eqs. (10-7) and (10-8) is called a *primal problem,* which can be represented in a general form as

Minimize

$$W = \sum_{j=1}^{g} M_{pj} L_j$$

Subject to (10-12)

$$\sum_{j=1}^{g} a_{ij} M_{pj} \geq b_i \qquad i = 1, \ldots, m$$

For each primal problem, a *dual problem,* with variable ω_i, can be stated as (Kirsch 1981):

Maximize

$$Z = \sum_{i=1}^{m} b_i \omega_i$$

Subject to (10-13)

$$\sum_{i=1}^{m} a_{ij} \omega_i \leq L_j \qquad j = 1, \ldots, g$$

in which the coefficients a_{ij}, b_i, and L_j are identical to those of the primal problem. Therefore, for the primal optimization problem defined by Eqs. (10-7) and (10-8), we can state the dual problem as

Maximize

$$Z = 6\omega_1 + 12\omega_2 + 2.5\omega_3 + 17\omega_4 + 17\omega_5$$

Subject to (10-14)

$$\omega_2 + \omega_3 + \omega_4 + 2\omega_5 \leq 20$$

$$\omega_1 + \omega_2 + 2\omega_4 + \omega_5 \leq 16$$

It can be shown that the solutions of the primal and dual problems are identical, namely, min W = max Z. If one of the problems is solved, we also have the solution of the other problem. Since the computational effort in solving linear programming problems is a function of constraints, it is desirable to reduce this number (i.e., for our case, the primal problem has five constraints and the dual problem has only two constraints). In addition, handling constraints with "$\geq$" in the simplex method is more complex than constraints with "$\leq$" Therefore for the example problem it is advantageous to solve the dual problem.

The dual problem given in Eq. (10-14) can be easily solved using many commercially available linear programming software packages such as CPLEX (Optimization, Inc.), LINDO (Lindo Systems, Inc.), or MPL (Maximal Software). Using any of the mentioned linear programming software packages, the solution to the optimization problem, given in Eq. (10-14), can be found to be 212λ for the objective function with $M_b = 5$ and $M_c = 7$.

SELECTION OF DISCRETE SECTIONS

The use of the linearized relationship between weight and plastic moment [Eq. (10-3)] has to be followed by the selection of a suitable section from the Manual of Steel Construction (AISC-ASD 1989 or AISC-LRFD 1986) among the list of available sections. A safe design results if a section is chosen giving a plastic moment not less than that which emerges from the linear optimization design routines. Consequently, the overall design will then be stronger than required and ideally it may be possible to reduce some of the sections to a plastic moment less than that emerging from the linear analysis. In practice, because the linear optimization process ignores various factors all tending to overestimate the carrying capacity of the structure as designed, it is found advantageous to accept the degree of overdesign inherent in selecting the discrete sections in the manner described. The designer or the computer program that selects the discrete sections must choose the most economical section for the given axial thrust and reduced plastic moment; that is, it must choose the appropriate "economy" section. An economical section is defined as being the one that, for a certain range of plastic moment capacity, has a weight equal to or less than that of any other available sections. The AISC steel manuals (AISC-ASD 1989 and AISC-LRFD 1986) provide a series of tables that can be used to achieve this goal.

HOW TO IDENTIFY ALL OF THE COLLAPSE MECHANISMS

The number of collapse mechanisms rises extremely quickly with the complexity and size of the structure. However, recently Tam and Jennings (1992) developed a technique that finds all the collapse mechanisms systematically even in the presence of tapered members, pinned connections, plastic moment bounds, or multiple loading cases.

OPTIMAL PLASTIC DESIGN FOR TALL BUILDINGS

Regarding *unbraced* multistory frames, plastic procedures are continuing to be developed. Load and Resistance Factor Design (AISC-LRFD 1986) has no restriction on the number of stories, and the $P - \Delta$ effect *must* be considered. Plastic Design in Chapter N of the 1989 AISC-ASD Specification (1989) has no restriction on the number of stories; however, for one- and two-story frames, the frame stability effect $P - \Delta$ may be ignored (ASD-N3). Chapter N does require that

"The strength of an unbraced multi-story frame shall be determined by a rational analysis which includes the effect of frame instability. . . ." Therefore the classical optimal plastic design approach discussed in this chapter can be applied only to one- or two-story frames. For slender and taller buildings, we need to include $P - \Delta$ in the design process. Otherwise, the strength of slender structures such as tall buildings can be grossly underdesigned if the stability ($P - \Delta$) effect is not considered (Lu 1984).

The problem of member stability is taken into account in the member selection phase of the design process. The effect of overall instability, however, tends to change the failure mode of the structure from plastic collapse to inelastic instability. The maximum strength of the structure is represented in the latter case by the instability limit load, which can be determined by performing a second-order analysis.

Optimal limit design of tall buildings is a current topic of research and is beyond the scope of this chapter. Interested readers may consult Hjelmstad and Pezeshk (1991), Horne and Morris (1974), Pezeshk (1992), and Pezeshk and Hjelmstad (1991) for more information.

SUMMARY

An optimal design procedure using plastic design concepts has been presented. For low-rise building frames, optimal plastic designs can be made using the plastic limit load as the basis of design. However, the design of multistory building frames is complicated by the incidence of frame instability, and rigid-plastic methods of design must be modified to allow for this.

Inelastic behavior is recognized in different ways in current design practice. More extensive application of optimal plastic design concepts for tall buildings is expected in the future through continuing research and updating of design specifications.

APPENDIX 10-A. NOTATION

a_i and b_i = constants
E = the external energy
J = number of plastic hinges to develop the collapse mechanism i
M_p = fully plastic moment
M_{pj} = plastic moment for plastic hinge j
U = energy absorption capacity
Z = objective function
δ = virtual displacement
λ = constant representing the weight of a linear foot of a member with a plastic moment capacity of 1 ft kip
θ_{ij} = rotation for plastic hinge j

REFERENCES

AISC-ASD (1989). *Manual of steel construction: Allowable stress design*. Ninth edition, American Institute of Steel Construction, Chicago, IL.

AISC-LRFD (1986). *Manual of steel construction: Load and resistance factor design*. First edition, American Institute of Steel Construction, Chicago, IL.

Ficken, F. A. (1961). *The simplex method of linear programming*. Holt, Rinehart, and Winston, New York.

Hjelmstad, K. D. and Pezeshk, S. (1991). "Optimal design of frames to resist buckling under multiple loading cases." *ASCE J. of Struct. Engrg.* 117(3) March, 914–935.

Horne, M. R. and Morris, L. J. (1974). *Optimum structural design*. R. H. Gallagher and O. C. Zienkiewicz, eds., John Wiley, New York, chapter 14.

Kirsch, U. (1981). *Optimum structural design*. McGraw-Hill Company, New York.

Lu, L. - W. (1984). "The present status of application of the plastic concepts in structural design." *Proceedings of Structural Congress III*, Session ST-25, San Francisco, CA.

Neal, B. G. (1985). *The plastic methods of structural analysis*. Chapman and Hall, New York.

Pezeshk, S. (1992). "Optimal design of structures with kinematic nonlinear behavior." *ASCE J. of Mech. Div.*, 118(4) April, 702–720.

Pezeshk, S. and Hjelmstad, K. D. (1991). "Optimal design of planar frames based on a stability criterion." *ASCE J. of Struct. Engrg.* 117(3) March, 896–913.

Rubinstein, M. F. and Karagozian, J. (1966). "Building design using linear programming." *ASCE J. of Struct. Engrg.* 92(ST6), Dec. 223–245.

Tam, T. K. H. and Jennings, A. (1992). "Computing alternative optimal collapse mechanisms." *Int. J. for Numerical Methods in Engrg.* 33, 1197–1216.

Chapter 11

HOW TO INCORPORATE RELIABILITY IN STRUCTURAL OPTIMIZATION

Structural optimization is a process made up of design cycles aimed at finding the best possible solution. At each design cycle, ultimate and/or serviceability limit states must be checked according to the code requirements. In the United States, these requirements are generally based on the Load and Resistance Factor Design (LRFD) methodology. This methodology uses load and resistance factors that have been developed from the modern theory of structural reliability. Therefore structural optimization and structural reliability have to be coupled in order to achieve practical optimal solutions. This chapter describes how to incorporate reliability into the structural optimization process. A framework for controlling structural reliability during the optimization process is provided. Numerical examples are presented for illustration purposes.

INTRODUCTION

Traditionally, structural optimization has been based on deterministic design philosophy. Using this philosophy, the general formulation of a structural optimization problem is as follows [e.g., Grierson and Schmit (1982)]. Find the design variable vector

$$\mathbf{X} = (X_1, X_2, \ldots, X_n)^t \tag{11-1}$$

to minimize the objective function

$$F = F(\mathbf{X}) \tag{11-2}$$

Dan M. Frangopol, Fellow, ASCE, Professor of Civil Engineering, Department of Civil, Environmental and Architectural Engineering, University of Colorado at Boulder, P.O. Box 428, Boulder, CO 80309-0428; email: ⟨Dan.Frangopol@colorado.edu⟩.

subject to the side constraints

$$X_i^L \le X_i \le X_i^U \qquad i = 1, 2, \ldots, n \tag{11-3}$$

and to the performance constraints

$$\delta_j^L \le \delta_j \le \delta_j^U \qquad j = 1, 2, \ldots, p \tag{11-4}$$

$$\sigma_k^L \le \sigma_k \le \sigma_k^U \qquad k = 1, 2, \ldots, q \tag{11-5}$$

$$\lambda_l^L \le \lambda_l \le \lambda_l^U \qquad l = 1, 2, \ldots, r \tag{11-6}$$

where Eq. (11-2) defines the objective function F (also denoted as f, C, or W) which is a scalar function, such as total expected cost C, or total structural weight W, of the design variable vector $\mathbf{X}$; Eq. (11-3) defines the n side constraints on the element sizes X_i to satisfy fabrication and technological requirements; Eq. (11-4) defines the p service performance constraints (e.g., acceptable elastic deflections) on deformations δ_j; Eq. (11-5) defines the q service performance constraints (e.g., acceptable safety level against the occurrence of the first plastic hinge) on stresses σ_k; Eq. (11-6) defines the r ultimate performance constraints (e.g., acceptable safety level against the occurrence of plastic collapse) on plastic collapse load factors λ_l; and L and U are superscripts denoting specified lower and upper bounds, respectively.

In general, the service and the ultimate performance constraints are implicit functions of the design variables X_i that must be determined, and the objective function F is an explicit function of the design variables. All the (deterministic) constraints defined in Eqs. (11-4) through (11-6) include fixed and predetermined safety factors that limit the deformations δ_j, stresses σ_k, and collapse load factors λ_l to the pair of allowable bounds (δ_j^L, δ_j^U), (σ_k^L, σ_k^U), and $(\lambda_l^L, \lambda_l^U)$, respectively.

As shown by Feng and Moses (1986) and Frangopol and Klisinski (1989), deterministic optimized structures will usually have higher failure probabilities and less load-path abilities than code-design (unoptimized) structures, the reason being that the safety factors specified in design codes usually refer to the design of elements, such as beams and columns, and do not allow for a near simultaneous occurrence of many failure modes (Frangopol and Moses 1994).

The structural optimization process must be accomplished under realistic conditions. These conditions have to include uncertainties associated with inherent variabilities in loads and strengths, and model imperfections (Ang and Cornell 1974). In order to obtain an optimal solution that recognizes these uncertainties, the structural optimization process should be based on modern structural reliability theory. Such a process is called Reliability-Based Structural Optimization (RBSO).

RBSO has been the focus of research for more than 30 years. Early attempts concentrated on the feasibility of incorporating reliability analysis into optimum design of structural systems (Moses 1969). Subsequently, significant contributions to RBSO were made. For a review of these contributions the reader is referred to Frangopol (1985), Enevoldsen (1991), Thoft-Christensen (1991), Enevoldsen and Sørensen (1993), Frangopol and Moses (1994), Frangopol (1995), and Frangopol and Corotis (1996). Today the RBSO concepts and methods are beginning to be used in realistic design applications.

This chapter describes how to incorporate reliability into the structural optimization process. Numerical examples are presented for illustration purposes.

STRUCTURAL RELIABILITY BACKGROUND

General

Considerable background material on structural reliability concepts and methods may be found in books by Thoft-Christensen and Baker (1982), Ang and Tang (1984), Madsen et al. (1986), Thoft-Christensen and Murotsu (1986), and Melchers (1987), among others. Also, several international [e.g., International Conference on Structural Safety and Reliability (ICOSSAR), International Conference on Applications of Statistics and Probability in Civil Engineering (CERRA-ICASP)] and national (ASCE Specialty Conference on Probabilistic Mechanics and Structural Reliability) conferences are held every four years in different locations. In addition, two international journals (*Structural Safety* and *Probabilistic Engineering Mechanics*) deal extensively and almost exclusively with structural reliability concepts, methods, and applications. Owing to space limitations, it is not possible to include in this section a detailed review of structural reliability. However, several basic reliability concepts and methods that are applied in RBSO are indicated.

Single Failure Mode Reliability

Consider first the fundamental single failure mode reliability problem in which the random resistance R and the random load effect (or load) S are known (Ang and Cornell 1974) by their probability density distributions, $f_R(r)$ and $f_S(s)$, respectively. The reliability may be measured in terms of the probability of the "safe" event

$$R > S \tag{11-7}$$

Therefore the measure of reliability is

$$\text{Probability of survival} = P(R > S) \tag{11-8}$$

and, alternatively, the measure of unreliability (risk) is the probability of occurrence of the "failure" event $R \leq S$

$$\text{Probability of failure} = P_f = P(R \leq S) = 1 - P(R > S) \quad (11\text{-}9)$$

For statistically independent R and S, precise calculation of the probability of failure is obtained by solving the convolution integrals

$$P_f = \int_0^\infty F_R(s) f_S(s)\, ds = 1 - \int_0^\infty F_S(r) f_R(r)\, dr \quad (11\text{-}10)$$

in which $F_R(.)$ and $F_S(.)$ are the cumulative distribution functions of R and S, respectively, and $f_R(.)$ and $f_S(.)$ are the probability density functions of R and S, respectively. Clearly, P_f depends on the type of probability density functions (e.g., normal, log-normal, Poisson, beta, gamma) of R and S, and on the descriptors (e.g., means, variances, coefficients of variation, skewness coefficients) of these random variables. In practical applications, information on $f_R(r)$ and $f_S(s)$ is often limited to the first moment (i.e., mean) and the second central moment (i.e., variance) of these density functions.

Some closed-form solutions of Eq. (11-10) are available when R and S are both normal (Gaussian) or log-normal. For normal distributed variables, Eq. (11-10) yields

$$P_f = 1 - \Phi\left(\frac{\bar{R} - \bar{S}}{\sqrt{\sigma^2(R) + \sigma^2(S)}}\right) \quad (11\text{-}11)$$

where $\bar{R}$ and $\bar{S}$ are the mean values of R and S, respectively, $\sigma^2(R)$ and $\sigma^2(S)$ are the variances of R and S, respectively, and $\Phi(.)$ = the standard normal distribution function.

When both R and S are log-normal (Ang and Cornell 1974) and the coefficients of variation of R and S, $V(R)$ and $V(S)$, are not large (say, ≤ 0.30), the approximate failure probability is

$$P_f \simeq 1 - \Phi\left(\frac{ln(\bar{R}/\bar{S})}{\sqrt{V^2(R) + V^2(S)}}\right) \quad (11\text{-}12)$$

In terms of the central safety factor $\theta = \bar{R}/\bar{S}$, Eqs. (11-11) and (11-12) are, respectively, as follows.

$$P_f = 1 - \Phi\left(\frac{\theta - 1}{\sqrt{\theta^2 V^2(R) + V^2(S)}}\right) \quad (11\text{-}13)$$

$$P_f \simeq 1 - \Phi\left(\frac{ln\ \theta}{\sqrt{V^2(R) + V^2(S)}}\right) \tag{11-14}$$

In many cases, the probability density functions of R and S are difficult to define or simply unknown. In these cases simplified reliability-based analysis and design formats have to be considered. These simplified formats were proposed by Cornell (1969), Rosenblueth and Esteva (1972), Ang and Cornell (1974), and Hasofer and Lind (1978), among others.

For example, in the reliability index format (also called "second moment format"), a design scheme is proposed in which a minimum safety margin is maintained (Cornell 1969; Ang and Cornell 1974). Therefore, instead of requiring that P_f be below a specified probability level P_f^0, it is required that the mean safety margin $\bar{Z}$ be at least a code-specified number of standard deviations $\sigma(Z)$ above zero

$$\bar{Z} \geq \beta^0 \sigma(Z) \tag{11-15}$$

where

$$\bar{Z} = \bar{R} - \bar{S} \tag{11-16}$$

$$\sigma(Z) = \sqrt{\sigma^2(R) + \sigma^2(S)} \tag{11-17}$$

are, respectively, the mean and standard deviation of the safety margin

$$Z = R - S \tag{11-18}$$

If R and S are both assumed normal or log-normal distributed, the code-specified value of the reliability index β^0 may be related to the accepted risk level P_f^0. For example, for normal distributed R and S the relation between β^0 and P_f^0 is given in Table 11-1 (excerpted from any Standard Normal Probability Table).

On the basis of the preceding format, the design checking equation is

$$\beta = \frac{\bar{Z}}{\sigma(Z)} = \frac{\bar{R} - \bar{S}}{\sqrt{\sigma^2(R) + \sigma^2(S)}} \geq \beta^0 \tag{11-19}$$

where β is the actual reliability index.

For a structural design problem, the first step is to define the m limit state requirements

$$g_j(Y_1, Y_2, \ldots, Y_i, \ldots, Y_t) = g_j(\mathbf{Y}) = 0 \qquad j = 1, 2, \ldots, m \tag{11-20}$$

TABLE 11-1 Reliability Index β^0 and Failure Probability P_f^0

β^0 (1)	$P_f^0 = \Phi(-\beta^0)$ (2)
0.00	0.500000
0.50	0.308537
1.00	0.158655
1.50	0.066807
2.00	0.022750
2.50	0.006210
3.00	0.001350
3.50	0.000233
4.00	0.316712E-04
4.50	0.339767E-05
5.00	0.286652E-06
5.50	0.189896E-07
6.00	0.986588E-09
6.50	0.402E-10
7.00	0.128E-11

where Y_i = the ith basic random variable (load, resistance), $g_j(\mathbf{Y})$ = the performance function which is defined such as $g_j(\mathbf{Y}) > 0$ means "survival state" and $g_j(\mathbf{Y}) \leq 0$ means "failure state." Hence, the probability of occurrence of limit state j is

$$P_{f_j} = \int_{g_j(\mathbf{Y}) \leq 0} f_{\mathbf{Y}}(\mathbf{y})\, d\mathbf{y} \tag{11-21}$$

where $f_{\mathbf{Y}}(\mathbf{y})$ is the joint probability density function of the random variables $Y_1, Y_2, \ldots, Y_t$.

For practical purposes, in general, either first-order (FORM) and second-order (SORM) reliability methods or Monte Carlo simulations (MCS) are used to evaluate the limit state probability P_{f_j}. These methods are implemented in available reliability analysis computer programs [e.g., PROBAN (1986), CALREL (Liu et al. 1989), and RELTRAN (Lee et al. 1993)]. For an excellent review of both traditional and advanced computation methods for reliability analysis, the reader is referred to Bjerager (1989). All these methods are used to find the reliability index

$$\beta_j = \Phi^{-1}(1 - P_{f_j}) \tag{11-22}$$

or the probability of occurrence of a single failure mode defined by the limit state j

$$P_{f_j} = 1 - \Phi(\beta_j) \tag{11-23}$$

of a structural element or system.

System Reliability

The reliability of a structural system is, in general, significantly different from the reliability of its individual components (e.g., beams, columns, connections). This difference depends on many factors including system geometry (e.g., series, parallel), material behavior (e.g., brittle, ductile, semibrittle), statistical correlation (e.g., load correlation, strength correlation), and variability (e.g., load variability, strength variability).

The computation of structural system reliability is important and necessary in evaluation, design, and optimization of structural systems. For example, an evaluation goal may be to determine whether the reliability of a structural system is larger or less than the reliability level of the most critical component, or a design goal may be to optimally proportion a structure for a specified system reliability level. As indicated by Frangopol and Moses (1994), system reliability can, in many cases, indicate whether there are (a) critical locations from which failure may initiate, and (b) failure modes (e.g., flexure, shear, buckling) with large probabilities of occurrence. Also, system reliability can assist structural designers in the best utilization of resources to promote better design solutions and inspection strategies.

The simplest models for system reliability are series ("weakest-link") and parallel ("fail-safe") failure modes. Practical expressions for system reliability include lower and upper bounds for both "weakest-link" and "fail-safe" systems. For example, for failure modes of a "weakest-link" system (e.g., failure of members of a statically determinate truss), the corresponding first-order bounds on the system failure probability are (Cornell 1967)

$$\max_{j=1}^{m} P_{f_j} \leq P_{f_{\text{system}}} \leq 1 - \prod_{j=1}^{m} (1 - P_{f_j}) \tag{11-24}$$

If correlation between pairs of potential failure modes is considered, alternate and more exact system failure probability bounds of a "weakest-link" system are (Ditlevsen 1979)

$$P_{f_{\text{system}}} \geq P_{f_1} + \sum_{j=2}^{m} \max \left(0;\ P_{f_j} - \sum_{i=1}^{j-1} P_{f_{ij}}\right) \tag{11-25}$$

$$P_{f_{\text{system}}} \leq \sum_{j=1}^{m} P_{f_j} - \sum_{j=2}^{m} \max_{i<j} P_{f_{ij}} \qquad (11\text{-}26)$$

where the exact system failure probability is

$$P_{f_{\text{system}}} = P\left(\bigcup_{j=1}^{m} [g_j(\mathbf{Y}) \leq 0]\right)$$

$$= P \text{ (any failure mode occurrence)} \qquad (11\text{-}27)$$

the probability of occurrence of failure mode j is

$$P_{f_j} = P(g_j(\mathbf{Y}) \leq 0) \qquad (11\text{-}28)$$

the probability of occurrence of the dominant (i.e., most probable) failure mode is

$$P_{f_1} = \max_{j=1}^{m} P_{f_j} \qquad (11\text{-}29)$$

and the probability of joint occurrence of failure modes i and j is

$$P_{f_{ij}} = P[\{g_i(\mathbf{Y}) \leq 0\} \cap \{g_j(\mathbf{Y}) \leq 0\}] \qquad (11\text{-}30)$$

Expressions for approximating the system failure probability of parallel (''fail-safe'') and more general systems (e.g., a series system of parallel systems) are also available (Ang and Tang 1984; Enevoldsen 1991). Also, practical methods for system reliability evaluation including advanced Monte Carlo simulations, incremental analysis, fault tree, and event tree models are available [see e.g., Moses (1982), Ang and Tang (1984), Ditlevsen and Bjerager (1986), Thoft-Christensen and Murotsu (1986), Karamchandani (1987), and Frangopol (1989)].

RELIABILITY-BASED STRUCTURAL OPTIMIZATION

Reliability-based structural optimization (RBSO) may be formulated at various levels depending on the nature of the design variables. These levels can be divided into the following groups (Frangopol 1985, Frangopol and Moses 1994, Thoft-Christensen 1991).

1. Sizing design variables (e.g., the cross-section dimensions of structural members).
2. Shape (configuration, geometric layout) design variables (e.g., the coordinates of joints in a framed structure, the length of spans in a continuous bridge).

3. Material design variables (e.g., the type of mechanical and/or physical properties of the materials to be used).
4. Topological (general arrangement) design variables (e.g., the number of spans in a bridge).
5. Structural system types (e.g., truss structure, framed structure, plated structure).

Most RBSO studies deal exclusively with the first and lowest category in this hierarchy of design variables (i.e., the selection of cross-sectional sizes) because of the relative simplicity of this problem. However, it should be noted that extending optimization capabilities upward into the design variable hierarchy (e.g., consideration of both shape and cross-sectional design variables) may appreciably improve the optimum solution (Frangopol 1985).

The RBSO problem may be formulated in various ways. These include the following formulations (Cohn and Parimi 1972; Frangopol 1985; Cheng and Chang 1985; Chang et al. 1994; Frangopol and Moses 1994).

- Find the vector of design variables $\mathbf{X} = (X_1, X_2, \ldots, X_n)^t$ that will minimize the total expected cost of the structure

$$C = C(\mathbf{X}) = C_0 + \sum_{j=1}^{m} C_{fj} P_{fj} \tag{11-31}$$

where C_0 = the initial cost, which is the function of the vector of design variables $\mathbf{X}$, C_{fj} = the cost of failure associated with the occurrence of limit state j, and P_{fj} = probability of occurrence of limit state j, which also depends on $\mathbf{X}$.

- Find $\mathbf{X}$ to minimize

$$F = F(\mathbf{X}) \tag{11-32}$$

subject to

$$P_{fj} \leq P_{fj}^0 \qquad j = 1, 2, \ldots, m \tag{11-33}$$

in which F = the objective to be minimized (e.g., total weight), P_{fj} = the probability of occurrence of limit state j, P_{fj}^0 = the allowable value of P_{fj}. Side constraints [see Eq. (11-3)] may also be imposed for each limit state under consideration. Alternatively, using a reliability index format the constraints (11-33) are replaced by

$$\beta_j \geq \beta_j^0 \qquad j = 1, 2, \ldots, m \tag{11-34}$$

where β_j = the reliability index associated with the occurrence of limit state j, and β_j^0 = the allowable value of β_j.

- Find $\mathbf{X}$ to minimize

$$F = F(\mathbf{X}) \tag{11-35}$$

subject to

$$P_{f_j} \leq P_{f_j}^0 \qquad j = 1, 2, \ldots, m \tag{11-36}$$

$$P_{f_{\text{system}}} \leq P_{f_{\text{system}}}^0 \tag{11-37}$$

In this formulation, the system reliability constraint (11-37) is added to the m individual limit state constraints. Alternatively, using a reliability index format, the constraints (11-36) are replaced by (11-34) and the system constraint (11-37) is replaced by

$$\beta_{\text{system}} \geq \beta_{\text{system}}^0 \tag{11-38}$$

where β_{system} = the system reliability index, and β_{system}^0 = the allowable value of β_{system}. Of course, other RBSO formulations are possible including minimizing the probability of system failure (or, equivalently, maximizing the system reliability index) for a total expected cost. Also, multicriteria RBSO formulations were proposed (Fu and Frangopol 1990; Frangopol and Moses 1994; Frangopol 1995) in which the purpose is to minimize a vector objective

$$\mathbf{F}(\mathbf{X}) = [F_1(\mathbf{X}), F_2(\mathbf{X}), \ldots, F_p(\mathbf{X})]^t \tag{11-39}$$

whose components are certain characteristic properties of the structural system to be designed such as failure cost, material volume, system probabilities of collapse, first yielding, or plastic hinge occurrences. For a review of RBSO problem types and solution methods the reader is referred to Frangopol (1985), Enevoldsen (1991), Enevoldsen and Sørensen (1993), Frangopol and Moses (1994), and Frangopol (1995). It should be noted that computer software for reliability analysis of structural elements and systems [such as PROBAN (1986), CALREL (Liu et al. 1989), RELTRAN (Lee et al. 1993)] may be coupled with optimizers [such as ADS (Vanderplaats 1986), IDESIGN (Arora and Tseng 1988)] to produce efficient software for RBSO. The RELTRAN-ADS linkage was created and successfully used at the University of Colorado by Hendawi (1994), among others. Results of this linkage are presented in the next section.

NUMERICAL EXAMPLES

RBSO of a Steel Portal Frame

This example is the rectangular fix-ended single-story rigid steel portal frame with deterministic geometry, random strength (M_B is the plastic moment of the critical sections of the beam, and M_C is the plastic moment of the critical sections of the columns), and random loading shown in Fig. 11-1. This frame is a typical example used in RBSO (Frangopol 1985). It is also similar in geometry and loading to the frame optimized deterministically by Pezeshk (1996).

The statistical information is as follows: (a) mean loads $\bar{H} = 20\ kN$, $\bar{P} = 80\ kN$; (b) coefficients of variation of loads $V(H) = 0.15$, $V(P) = 0.20$; (c) mean plastic moment of critical sections of the beam $\bar{M}_3 = \bar{M}_4 = \bar{M}_5 = \bar{M}_B$; (d) mean plastic moment of critical sections of columns $\bar{M}_1 = \bar{M}_2 = \bar{M}_6 = \bar{M}_7 = \bar{M}_C$; (e) coefficient of variation of plastic moments $V(M_1) = V(M_2) = V(M_3) = V(M_4) = V(M_5) = V(M_6) = V(M_7) = 0.10$; (f) load correlation coefficient $\rho(P, H) = 0$; (g) plastic moment correlation coefficients $\rho(M_i, M_j) = \rho(M_B, M_C) = 1$; and (h) normal distribution is assumed for all nine (two loads and seven plastic moments) random variables considered.

The total cost of the frame C is approximated by the linear function $W = l\bar{M}_B + 2h\bar{M}_c$, where l = beam length = $10m$ and h = column height = $4m$. The design variable vector $\mathbf{X}$ consists of two mean value variables $\bar{M}_B$ and $\bar{M}_C$. The following three RBSO formulations are considered.

Formulation 1. Find

$$\mathbf{X} = (\bar{M}_B, \bar{M}_C)^t \tag{11-40}$$

to minimize

$$W = 10\bar{M}_B + 8\bar{M}_C \tag{11-41}$$

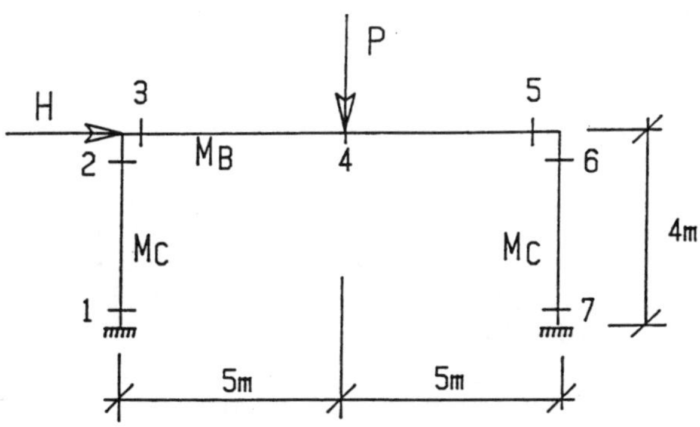

FIG. 11-1. Steel Portal Frame (1 kN = 0.22 kip; 1 m = 3.28 ft).

subject to

$$P_{f_1} = P_{f\text{collapse}} \leq P_{f_1}^0 \tag{11-42}$$

$$P_{f_2} = P_{f\text{hinge}} \leq P_{f_2}^0 \tag{11-43}$$

$$\bar{M}_B > 0, \qquad \bar{M}_C > 0 \tag{11-44}$$

Formulation 2. Find

$$\mathbf{X} = (\bar{M}_B, \bar{M}_C)^t \tag{11-45}$$

to minimize

$$W = 10\bar{M}_B + 8\bar{M}_C \tag{11-46}$$

subject to

$$P_{f_1} = P_{f\text{collapse}} \leq P_{f_1}^0 \tag{11-47}$$

$$\bar{M}_B > 0, \qquad \bar{M}_C > 0 \tag{11-48}$$

Formulation 3. Find

$$\mathbf{X} = (\bar{M}_B, \bar{M}_C)^t \tag{11-49}$$

to minimize

$$W = 10\bar{M}_B + 8\bar{M}_C \tag{11-50}$$

subject to

$$P_{f_2} = P_{f\text{hinge}} \leq P_{f_2}^0 \tag{11-51}$$

$$\bar{M}_B > 0, \qquad \bar{M}_C > 0 \tag{11-52}$$

where W = objective function, $P_{f_1} = P_{f\text{collapse}}$ = probability of plastic collapse of frame, $P_{f_2} = P_{f\text{hinge}}$ = probability of frame unserviceability by first plastic hinge occurrence, and $P_{f_1}^0$ and $P_{f_2}^0$ are allowable values of P_{f_1} and P_{f_2}, respectively. Probabilities P_{f_1} and P_{f_2} are both computed as system probabilities as follows.

$$P_{f_1} = P \text{ (any collapse mechanism will occur)} \tag{11-53}$$

$$P_{f_2} = P \text{ (first plastic hinge will occur at any critical section)} \tag{11-54}$$

The frame in Fig. 11-1 has 10 possible collapse mechanisms (see Fig. 11-2) and seven critical sections (see Fig. 11-1). Therefore

$$P_{f_1} = P\left(\bigcup_{u=1}^{10} (g_u \leq 0)\right) \tag{11-55}$$

$$P_{f_2} = P\left(\bigcup_{v=1}^{7} (g_v \leq 0)\right) \tag{11-56}$$

where $g_u \leq 0$ defines the occurrence of plastic collapse mechanism u, and $g_v \leq 0$ defines the occurrence of the first plastic hinge at the critical section v. In this example, the probabilities P_{f_1} and P_{f_2} are conservatively approximated by the upper bound system failure probability defined by the right-hand side of Eq. (11-24). Therefore

$$P_{f_1} = P_{f\text{collapse}} = 1 - \prod_{u=1}^{10} (1 - P_{f_u}) \tag{11-57}$$

$$P_{f_2} = P_{f\text{hinge}} = 1 - \prod_{v=1}^{7} (1 - P_{f_v}) \tag{11-58}$$

where P_{f_u} = the probability of occurrence of plastic collapse mechanism u, and P_{f_v} = the probability of occurrence of the first plastic hinge at critical section v.

Figure 11-3 contains a comparison of the results associated with Formulations 1, 2, and 3. The allowable probabilities of collapse and unserviceability by first plastic hinge occurrence are, respectively, specified as $P_{f_1}^0 = 10^{-5}$ and $P_{f_2}^0 = 10^{-2}$. The two-dimensional mean design space shown in Fig. 11-3 has two reliability constraint curves [i.e., Eqs. (11-42) and (11-43) considered as equalities] such that all points on the curves $P_{f_1}^0 = 10^{-5}$ and $P_{f_2}^0 = 10^{-2}$ have the probabilities of plastic collapse and unserviceability equal to the prescribed values 10^{-5} and 10^{-2}, respectively. Also shown is the linear objective function W that should be minimized.

The boundary of the feasible design space (containing all points with maximum acceptable probabilities of failure) depends on the optimization formulation. The boundary $a_2bc_1de_2$ corresponds to the multiconstraint Formulation 1, and boundaries $a_1bc_1de_1$ and $a_2bc_2de_2$ correspond to the single-constraint optimization Formulations 2 and 3, respectively. Consequently, if Formulations 1 or 2 are chosen, the optimum solution (the point at which W is tangent to the P_f constraint) is the design point c_1 (collapse constraint is critical at the optimum). On the other hand, if Formulation 3 is chosen, the optimum solution is the design point c_2. It is interesting to note that both reliability constraints are satisfied as equalities at two design points

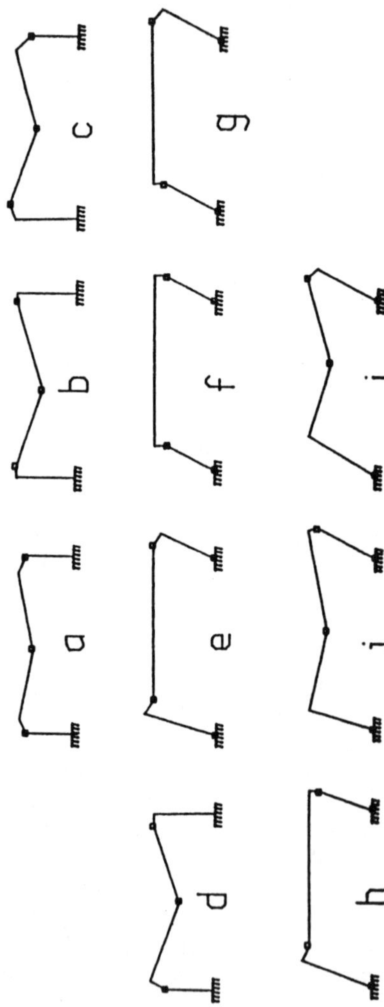

FIG. 11-2. Possible Collapse Modes of Portal Frame in Fig. 11-1.

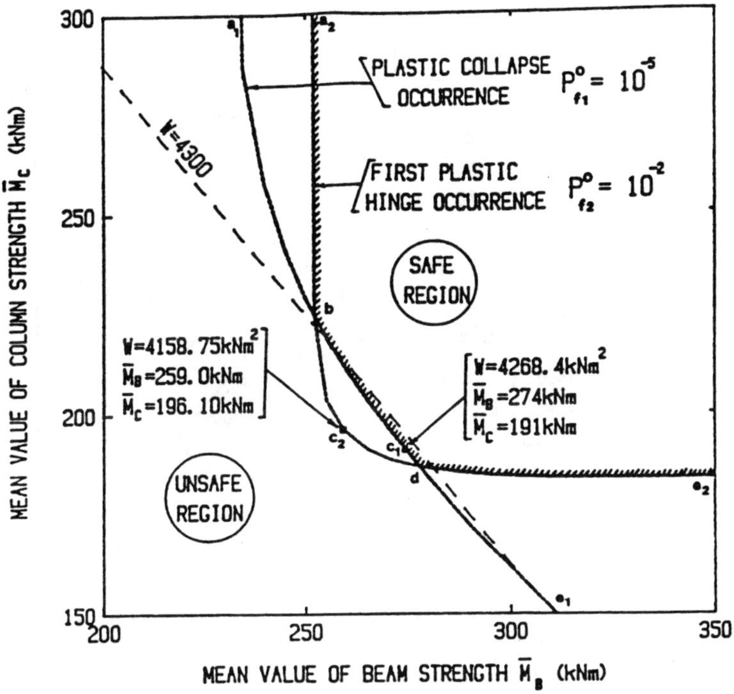

FIG. 11-3. Design Space with Two Reliability Constraints (1 kN = 0.22 kip; 1 m = 3.28 ft).

(*b* and *d*); the points lying on the curve bc_2d have unacceptable reliabilities with regard to plastic collapse ($P_{f_1} > 10^{-5}$); and the points lying on the curves a_1b and de_1 have unacceptable reliabilities with regard to unserviceability ($P_{f_2} > 10^{-2}$).

Figures 11-4 and 11-5 show optimum solutions for various sets of specified risk levels against plastic collapse and unserviceability, respectively. Figure 11-4 presents solutions for different values of $P^0_{f_1}$ when $P^0_{f_2}$ is kept constant, whereas Fig. 11-5 presents solutions for different values of $P^0_{f_2}$ when $P^0_{f_1}$ is kept constant. The optimum solutions presented in Figs. 11-3 to 11-5 were all obtained by means of the usable feasible direction method (Kirsch 1981). These figures indicate the possible fallacy in a single-constraint optimization approach. Selection of discrete sections corresponding to the continuous RBSO solutions shown in Figs. 11-3 to 11-5 should be performed as indicated for deterministic optimization by Pezeshk (1996).

RBSO of a Composite Hybrid-Plate Girder

In this second example the deterministic optimization approach of composite hybrid-plate girders presented by Dhillon and Kuo (1996)

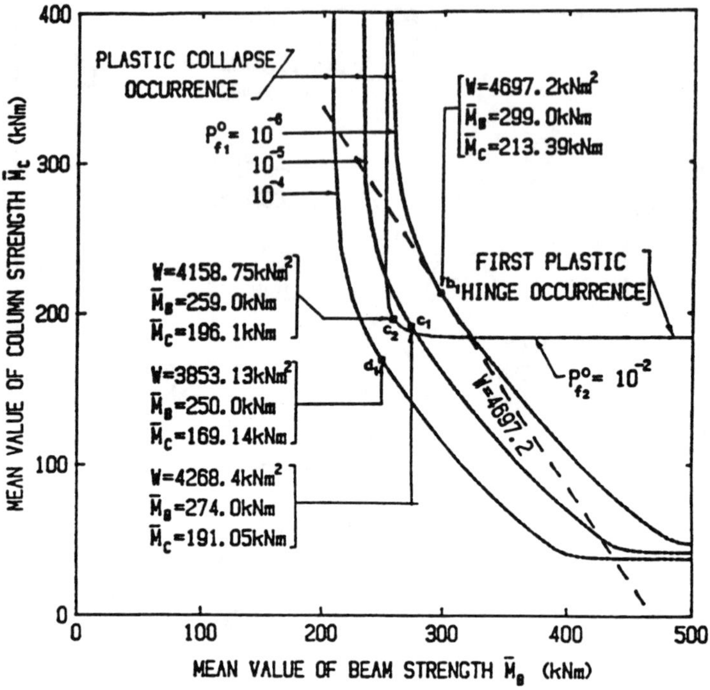

FIG. 11-4. *Effect of Allowable Probability of Plastic Collapse P^0_{f1} on Optimum Solution (1 kN = 0.22 kip; 1 m = 3.28 ft).*

[see also Dhillon and Kuo (1991)], is replaced by a reliability-based optimization approach (Hendawi 1994). The notations herein are almost identical to those used by Dhillon and Kuo (1991, 1996). The only difference is that both independent and dependent variables are denoted in this RBSO example as X_i instead of the notation x_i used in the corresponding deterministic example.

Nine independent variables (X_1 to X_6 and X_{19} to X_{21}) are indicated in Fig. 1 of Chapter 3 (Dhillon and Kuo 1996). As shown by Dhillon and Kuo (1991, 1996), who based their approach on all the deterministic constraints specified in the AASHTO bridge code (Standard Specifications for Highway Bridges 1983 and 1992), the dependent variables may be expressed in the function of (a) independent variables only or (b) a combination of independent variables and design parameters Y_i. Table 11-2 based on the studies carried out by Dhillon and Kuo (1991, 1996) summarizes the variables X_i and parameters Y_i considered in this RBSO example.

The main difference between the deterministic and the reliability-based optimization approaches is that uncertainties are considered

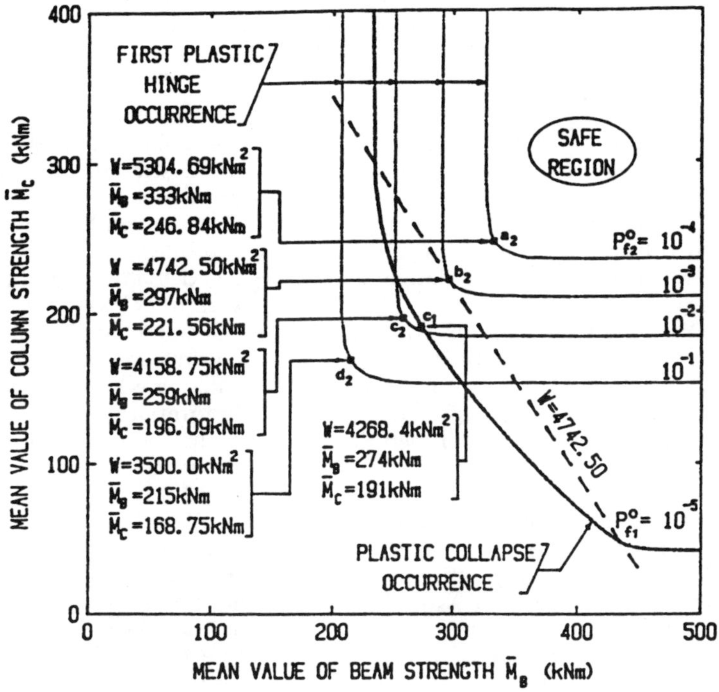

FIG. 11-5. Effect of Allowable Probability of First Plastic Hinge Occurrence, P^0_{f2} on Optimum Solution (1 kN = 0.22 kip; 1 m = 3.28 ft).

explicitly in the latter. In fact, as shown in Table 11-3, eight of the design parameters (i.e., Y_2, Y_5 to Y_{11}) are treated as random variables. Objective functions representing the weight of the unstiffened plate girder W_1 and the weight of the stiffened plate girder W_2 are the same in both the deterministic and the reliability-based optimization approaches.

Dhillon and Kuo (1991, 1996) specify all deterministic equality and inequality constraints. The RBSO approach used in this example replaces the deterministic behavior constraints specified in Dhillon and Kuo (1991, 1996) by reliability-based constraints of the following form:

$$\beta_j \geq \beta_j^0 \qquad j = 1, 2, \ldots, m \tag{11-59}$$

where β_j = the reliability index of the hybird-plate girder with re-

TABLE 11-2 Definition of Design Variables and Parameters*

Design Variables (1)	Design Parameters (2)
X_1 = Width of top flange (in.)	Y_1 = Deck slab thickness (in.)
X_2 = Thickness of top flange (in.)	Y_2 = Yield strength of top flange steel (ksi)
X_3 = Thickness of web (in.)	Y_3 = Girder span length (ft)
X_4 = Depth of web (in.)	Y_4 = Modular ratio
X_5 = Width of bottom flange (in.)	Y_5 = Dead load excluding girder weight (kips/ft)
X_6 = Thickness of bottom flange (in.)	Y_6 = Max. live load moment including impact (kips/ft)
X_7 = Distance from top of concrete slab to centroid of compression flange (in.)	Y_7 = Max. live load shear including impact (kips)
X_8 = Distance from top of concrete slab to centroid of web (in.)	Y_8 = Yield strength of bottom flange steel (ksi)
X_9 = Distance from top of concrete slab to centroid of tension flange (in.)	Y_9 = Yield strength of web steel (ksi)
X_{10} = Distance from top of concrete slab to neutral axis of composite section (in.)	Y_{10} = Compressive strength of concrete (ksi)
X_{11} = Moment of inertia of composite section (in.^{4})	Y_{11} = Superimposed dead load moment (kips-ft)
X_{12} = Distance from top of concrete slab to neutral axis of plate girder section alone (in.)	Y_{12} = Effective width of deck slab (in.)
X_{13} = Moment of inertia of plate girder alone (in.^{4})	Y_{13} = Unit weight of steel (kips/ft^3)
X_{14} = Dead-load moment (kips-ft)	
X_{15} = Ratio of the neutral axis distance from the outer edge of the tension flange to the depth of the steel section	
X_{16} = Strength reduction factor	
X_{17} = Top flange compression stress caused by noncomposite dead load (ksi)	
X_{18} = Web buckling coefficient for stiffened plate girder	
X_{19} = Width of stiffener (in.)	
X_{20} = Thickness of stiffener (in.)	
X_{21} = Stiffener spacing (in.)	
X_{22} = Variable $J = [2.5(X_4/X_{21})^2 - 2]$ for transverse stiffener stiffness	

Note: 1 in. = 25.4 mm; 1 ft = 304.8 mm; 1 kip = 4.45 kN; 1 ksi = 6.89 MPa.
*As specified by Dhillon and Kuo (1991, 1996).

TABLE 11-3 Deterministic and Random Design Parameters

Design Parameter (1)	Deterministic Value (2)	Random Variable: Mean (3)	Random Variable: Coefficient of Variation, % (4)
Y_1	7.0		
Y_2		37.8	10.0
Y_3	80.0		
Y_4	8.0		
Y_5		0.865	4.0
Y_6		1119.0	21.4
Y_7		59.7	21.4
Y_8		105.0	11.0
Y_9		37.8	10.0
Y_{10}		3.4	18.0
Y_{11}		137.3	9.0
Y_{12}	84.0		
Y_{13}	0.49		

Note: Definition and units of design parameters Y_1 to Y_{13} are listed in Table 2.

spect to the occurrence of limit state j (i.e., bending, shear, concrete crushing), and β_j^0 = the allowable value of β_j [see Eq. (11-34)]. All the random variables are considered statistically independent and normal distributed. Aspect ratio, flange local buckling, and geometric constraints are identical in both the deterministic and RBSO approaches.

For the data presented in Table 11-3, the RBSO was solved for both the unstiffened and stiffened cases by linking the optimizer ADS (Vanderplaats 1986) with the reliability analysis software RELTRAN (Lee et al. 1993). The reliability-based optimization results presented in Table 11-4 were obtained by using a modified method of feasible directions and polynomial interpolation as ADS options, and first order reliability analysis as the RELTRAN option. Allowable reliability indices β_j^0 were assumed the same (β_j^0 = 3.5) for all limit states considered in computations. For comparison purposes, Table 11-4 also presents the deterministic optimization results reported by Dhillon and Kuo (1991, 1996).

Finally, Table 11-5 presents optimum weight results for various values (i.e., 2.0, 3.0, and 4.0) of the allowable reliability index β_j^0. As expected, the optimum girder weight depends on the reliability level. Also, for the same reliability level, use of transverse stiffeners reduces the optimum weight as compared to that associated with the unstiffened case.

TABLE 11-4 Deterministic and Reliability-Based Optimization Results

Independent Design Variables and Objectives (1)	Unstiffened Case: Deterministic Approach (Dhillon and Kuo 1991) (2)	Unstiffened Case: RBSO Approach; Allowable Reliability Level: $\beta_j^0 = 3.5$ (3)	Stiffened Case: Deterministic Approach (Dhillon and Kuo 1996) (4)	Stiffened Case: RBSO Approach; Allowable Reliability Level: $\beta_j^0 = 3.5$ (5)
X_1	10.94	9.5571	11.03	9.7260
X_2	0.4962	0.76835	0.5079	0.42556
X_3	0.4315	0.39984	0.2470	0.31189
X_4	44.05	43.112	47.52	58.2630
X_5	15.42	12.262	16.63	14.2560
X_6	0.4230	0.5695	0.4062	0.32563
X_{19}	—	—	2.836	3.96750
X_{20}	—	—	0.2070	0.10
X_{21}	—	—	33.11	50.092
W_1	8.4177	8.5924	—	—
W_2	—	—	6.940	7.4639

Note: X_1 to X_6 and X_{19} to X_{21} (see Fig. 1 in Dhillon and Kuo 1996) are in inches, and W_1 and W_2 are in kips; 1 in. = 2.54 mm; 1 kip = 4.45 kN.

TABLE 11-5 Effect of Allowable Reliability on Optimum Weight

Optimum Weight for Unstiffened Case W_1				Optimum Weight for Stiffened Case W_2			
Deterministic Approach (Dhillon and Kuo 1991)	RBSO Approach; Allowable Reliability Level β_j^0			Deterministic Approach (Dhillon and Kuo 1991, 1996)	RBSO Approach; Allowable Reliability Level β_j^0		
	2.0	3.0	4.0		2.0	3.0	4.0
(1)	(2)	(3)	(4)	(5)	(6)	(7)	(8)
8.4177	7.2018	8.0671	9.1497	6.940	6.3004	7.0218	8.0048

Note: W_1 and W_2 are in kips, 1 kip = 4.45 kN.

SUMMARY AND DISCUSSION

Problems of structural optimization must be solved in the face of uncertainty. Accordingly, the reliability of structural elements and systems must be included in the optimization formulation. The reliability-based optimization approach (RBSO) is demonstrated herein. Generally, in a RBSO formulation the objective function is minimized under multiple reliability constraints. These constraints require that the probabilities of various limit states being reached are below allowable limits or, alternatively, that the reliability indices with respect to the occurrence of various limit states are above allowable limits. With recent progress in reliability analysis and design methods [e.g., Bjerager (1989)] and nonlinear programming algorithms [e.g., Arora (1990)], the application of the RBSO approach to realistic structural systems is now possible. However, as Frangopol and Corotis (1996) stated, important efforts are needed in order to advance the state-of-the-practice in RBSO. Such efforts must include the development of an effective avenue for informing the structural engineering community of the potential and usefulness of RBSO in engineering practice.

APPENDIX 11-A. NOTATION

The following symbols are used in the main part of this chapter. Additional symbols are defined within the numerical examples.

C = total expected cost
C_0 = initial cost
C_{f_j} = cost of failure associated with the occurrence of limit state j
F = objective function (also denoted as f, C, or W)
$\mathbf{F}$ = objective vector
$F_R(r)$ = cumulative distribution function of R
$F_S(s)$ = cumulative distribution function of S
$f_R(r)$ = probability density distribution of R
$f_S(s)$ = probability density distribution of S
$f_Y(\mathbf{y})$ = joint probability density distribution of the components of $\mathbf{Y}$
$g_j(\mathbf{Y})$ = performance function of limit state j
P_f = probability of failure
P_f^0 = allowable value of P_f
$P_{f_{ij}}$ = probability of joint occurrence of failure modes i and j
P_{f_j} = probability of occurrence of limit state j
$P_{f_j}^0$ = allowable value of P_{f_j}
$P_{f_{system}}$ = probability of system failure
$P_{f_{system}}^0$ = allowable value of $P_{f_{system}}$

R = random resistance variable
$\bar{R}$ = mean value of R
S = random load effect (or load) variable
$\bar{S}$ = mean value of S
$V(R)$ = coefficient of variation of R
$V(S)$ = coefficient of variation of S
W = weight function
$\mathbf{X}$ = vector of design variables
$\mathbf{Y}$ = basic random variable vector
Z = safety margin
$\bar{Z}$ = mean safety margin
β = reliability index
β^0 = allowable value of β
β_j = reliability index with respect to the occurrence of limit state j
β_j^0 = allowable value of β_j
β_k^0 = acceptable value for β_k
β_{system} = reliability index with respect to system failure
β_{system}^0 = allowable value of β_{system}
δ_j = deformation
$\delta_j^L, \sigma_k^L, \lambda_l^L$ = lower limits on $\delta_j, \sigma_k, \lambda_l$
$\delta_j^U, \sigma_k^U, \lambda_l^U$ = upper limits on $\delta_j, \sigma_k, \lambda_l$
$\Phi(\cdot)$ = standard normal distribution function
θ = central safety factor, $\bar{R}/\bar{S}$
λ_l = plastic collapse load factor
σ_k = stress
$\sigma^2(R)$ = variance (dispersion) of R
$\sigma^2(S)$ = variance (dispersion) of S
$\sigma(Z)$ = standard deviation of safety margin

ACKNOWLEDGMENTS

The writer gratefully acknowledges the support of the National Science Foundation in the field of reliability-based structural optimization under Grants Nos. MSM-8618108, MSM-8800882, MSM-9013017, CMS-9506435, and CMS-9522541. Opinions expressed in this chapter are those of the writer and do not necessarily reflect those of the sponsor.

REFERENCES

Ang, A. H. S. and Cornell, C. A. (1974). "Reliability bases of structural safety and design." *J. Struct. Div.*, ASCE, 100(ST9), 1755–1769.
Ang, A. H. S. and Tang, W. H. (1984). *Probabilistic concepts in engineering planning and design*. Vol. II, Wiley, New York.
Arora, J. S. (1990). "Computational design optimization: A review and future directions." *Struct. Safety*, 7(2–4), 131–148.

Arora, J. S. and Tseng, C. H. (1988). *User's manual for program IDESIGN*. Tech. Rep. No. ODL-87.1, Optimal Design Laboratory, College of Engineering, Univ. of Iowa, Iowa City, IA.
Bjerager, P. (1989). "On computation methods for structural reliability analysis." In *New Directions in Structural System Reliability*, D. M. Frangopol, ed., Univ. of Colorado, Boulder, CO, 52–67.
Chang, C. C., Ger, J. F. and Cheng, F. Y. (1994). "Reliability-based optimum design for UBC and nondeterministic seismic spectra." *J. Struct. Engng.*, ASCE, 120(1), 139–160.
Cheng, F. Y. and Chang, C. C. (1985). "Optimum design of steel buildings with consideration of reliability." *Proceedings of 4th International Conference on Structural Safety and Reliability*, Kobe, Japan, Vol. III, 81–90.
Cohn, M. Z. and Parimi, S. R. (1972). "Multi-criteria probabilistic design of reinforced concrete structures." In *Inelasticity and non-linearity in structural concrete*, M. Z. Cohn, ed., SM Study No. 8, Univ. of Waterloo, Waterloo, Ontario, 471–492.
Cornell, C. A. (1967). "Bounds on the reliability of structural systems." *J. Struct. Div.*, ASCE, 93(ST1), 171–200.
Cornell, C. A. (1969). "A probability-based structural code." *J. Amer. Concrete Inst.*, ACI, 66(12), 974–985.
Dhillon, B. S. and Kuo, C. - H. (1991). "Optimum design of composite hybrid plate girders." *J. Struct. Engng.*, ASCE, 117(7), 2088–2098.
Dhillon, B. S. and Kuo, C. - H. (1996). "How to optimize composite members." *Guide to structural optimization*, ASCE, New York (Chapter 3, this publication).
Ditlevsen, O. (1979). "Narrow reliability bounds for structural systems." *J. Struct. Mech.* 7(4), 453–472.
Ditlevsen, O. and Bjerager, P. (1986). "Methods of structural systems reliability." *Struct. Safety*, 3, 195–229.
Enevoldsen, I. (1991). *Reliability-based structural optimization*. Ph.D. Thesis, Dept. of Bldg. Tech. and Struct. Engng., Aalborg Univ., Aalborg, Denmark.
Enevoldsen, I. and Sørensen, J. D. (1993). *Reliability-based optimization in structural engineering*. Struct. Reliability Theory, Paper No. 118, Dept. of Bldg. Tech. and Struct. Engng., Aalborg Univ., Aalborg, Denmark.
Feng, Y. S. and Moses, F. (1986). "Optimum design, redundancy and reliability of structural systems." *Computers and Structures*, 24(2), 239–251.
Frangopol, D. M. (1985). "Structural optimization using reliability concepts." *J. Struct. Engng.*, ASCE, 111(11), 2288–2301.
Frangopol, D. M., ed. (1989). *New directions in structural system reliability*. University of Colorado, Boulder, CO.
Frangopol, D .M. (1995). "Reliability-based optimum structural design." *Probabilistic structural mechanics handbook*, C. Sundararajan, ed., Chapman and Hall, New York, 352–387.
Frangopol, D. M. and Corotis, R. B. (1996). "Reliability-based structural system optimization: State-of-the-art versus state-of-the-practice." *Proceedings of the Twelfth Conference on Analysis and Computation*, F. Y. Cheng, ed., ASCE, New York, 67–78.
Frangopol, D. M. and Klisinski, M. (1989). "Material behavior and optimum design of structural systems." *J. Struct. Engng.*, ASCE, 115(5), 1054–1075.
Frangopol, D. M. and Moses, F. (1994). "Reliability-based structural optimization." *Advances in design optimization*, H. Adeli, ed., Chapman and Hall, London, 492–570.
Fu, G. and Frangopol, D. M., (1990). "Reliability-based vector opimization of structural systems." *J. Struct. Engng.*, ASCE, 116(8), 2143–2161.
Grierson, D. E. and Schmit, L. A. (1982). "Synthesis under service and ultimate performance constraints." *Computers and Structures*, 15, 405–417.
Hasofer, A. M. and Lind, N. C. (1978). "Exact and invariant second-moment code format." *J. Engng. Mech. Div.*, ASCE, 100(EM1), 829–844.
Hendawi, S. (1994). *Structural system reliability with applications to bridge analysis, design and optimization*. Ph.D. Thesis, Department of Civil Engineering, University of Colorado, Boulder, CO.
Karamchandani, A. (1987). *Structural system reliability analysis methods*. Report 83, John A. Blume Earthquake Engineering Center, Stanford University, CA.

Kirsch, U. (1981). *Optimum structural design*. McGraw-Hill, New York.
Lee, Y - H., Hendawi, S., and Frangopol, D. M. (1993). *RELTRAN-A structural reliability analysis program: Version 2.0.*, Report No. CU/SR-93/6, Structural Engineering and Structural Mechanics Research Series, Dept. of Civil Engng., Univ. of Colorado, Boulder, CO.
Liu, P. - L., Lin, H. - Z., and Der Kiureghian, A. (1989). *CALREL user manual*. Report No. UCB/SEMM-89/18, Dept. of Civil Engng., Univ. of California, Berkeley, CA.
Madsen, H. O., Krenk, S., and Lind, N. C. (1986). *Methods of structural safety*. Prentice-Hall, Englewood Cliffs, NJ.
Melchers, R. E. (1987). *Structural reliability analysis and prediction*. Ellis Horwood, Chichester, England.
Moses, F. (1969). "Approaches to structural reliability and optimization." In *SM study no. 1*, M. Z. Cohn, ed., Solid Mechanics Division, Univ. of Waterloo, Ontario, Canada, 81–120.
Moses, F. (1982). "System reliability developments in structural engineering." *Structural Safety*, Elsevier, Amsterdam, 1(1), 3–13.
Pezeshk, S. (1996). "How to optimize frames using plastic design concept." *Guide to structural optimization*, ASCE, New York (Chapter 10, this publication)
PROBAN: The probabilistic analysis program (1986). Manuals, Reports 86-2036, 86-2037, and 86-2038, A. S. Veritas Research, Norway.
Rosenblueth, E. and Esteva, L. (1972). "Reliability basis for some Mexican codes." In *Probabilistic design of reinforced concrete buildings*, ACI Publication SP-31, 1–41.
Standard Specifications for Highway Bridges (1983) and (1992). 13th and 15th eds., AASHTO, Washington, DC.
Thoft-Christensen, P. (1991). "On reliability based structural optimization." In *Lecture notes in engineering*, A. Der Kiureghian and P. Thoft-Christensen, eds., Springer-Verlag, Berlin, Germany, 61, 387–402
Thoft-Christensen, P. and Baker, M. J., (1982). *Structural reliability theory and its applications*. Springer-Verlag, Berlin, Germany.
Thoft-Christensen, P. and Murotsu, Y. (1986). *Applications of structural systems reliability theory*. Springer-Verlag, Berlin, Germany.
Vanderplaats, G. N. (1986). *ADS—A Fortran program for automated design synthesis: Version 1.10*. Engineering Design Optimization Inc., Santa Barbara, CA.

Chapter 12

HOW TO OPTIMIZE FOR EARTHQUAKE LOADS

This chapter describes how to use optimization procedures for the design of structures subject to earthquake loads. A major difference between optimizing for static loads and earthquake loads is the fact that inertia-related seismic loads are dependent upon the natural period of the structure. The fact that the inertia loads are a function of the natural periods, which in turn are a function of the structural stiffness and mass, requires changes in the manner by which the gradients of the dynamic structural responses are determined. These changes are directly affected by the type of seismic analysis performed. Descriptions of the required changes for equivalent lateral force, modal response spectra analysis, and time history analysis procedures are presented. Examples including results for each analysis procedure are presented in order to clarify how structural optimization can lead to better understanding of seismic-resistant designs. A brief introduction and example illustrating the use of probabilistic methods in structural optimization for earthquake loads is also presented.

INTRODUCTION

Structural optimization of dynamically excited structures has been ongoing for several decades. Many times the word "optimization" is taken literally as finding the global optimal design. Realistically,

---

Kevin Z. Truman, M. ASCE, Associate Professor, Department of Civil Engineering, Washington University, 1 Brooks Drive, Campus Box 1130, St. Louis, MO 63130; email: ⟨Ktrum@seas.wustl.edu⟩ and Franklin Y. Cheng, M. ASCE, Curator's Professor, Department of Civil Engineering, Senior Investigator, Intelligent Systems Center, University of Missouri-Rolla, Rolla, MO 65401.

structural optimization for earthquake loads should be viewed as an automated design process. A major advantage of using structural optimization for seismically designed structures is the capability of simultaneously optimizing a structure for multiple load cases and multiple earthquakes. Traditionally, seismic designers design for the anticipated worst earthquake and check other earthquake loads for constraint or design violations. This requires an iterative procedure that can be avoided by using structural optimization procedures that can simultaneously design for multiple earthquake loadings. The optimization procedure will resize members in a fashion providing the most efficient dynamic properties for all the predetermined earthquake events, whereas a traditional design focuses on only one event resulting in an overdesigned structure with respect to a different earthquake.

Structural optimization for seismic loads has several unique features as compared to those for static loads. Earthquake loads and structural costs can be either deterministic (fixed earthquake properties, material properties, and structural cost coefficients) or probabilistic (random loads, costs associated with risk levels, and costs associated with possible failures). Optimum design of probabilistic systems implies that the seismic excitations take into account both the random nature of the seismic loads as well as the use of reliability-based analysis methods. The aforementioned random nature means that no uncertainties in structural material properties, external loads, and cost functions are considered. Reliability methods do incorporate probability distributions for response and resistance and can include different variance approaches for studying the sensitivity of specific parameters on the optimal designs. Deterministic systems have seismic loads with specific load equations or accelerograms that are used to reflect the possible seismic events. A major difference in optimizing a dynamically excited structure versus a statically loaded structure is that the inertia-generated earthquake loads and responses change each time the structure is modified. Therefore each iteration of the optimization procedure requires the regeneration of the seismic-related loads. A statically loaded structure changes with each iteration of the optimization procedure, but the loading remains constant allowing certain simplifications within the optimization algorithms that cannot be made for seismic optimization.

LITERATURE REVIEW

Current research in optimization of structures subject to dynamic loads covers a wide range of topics from aircraft or automobile design to building frames. Yamakawa (1984), Feng et al. (1977), Bhatti et al. (1981), Balling et al. (1983), Truman and Cheng (1990), Truman and Jan (1988), Truman and Petruska (1991), and Cheng and Juang (1988,

1989) have all contributed to the use of structural optimization which specifically addresses dynamic loads as applied to structural frames.

The research associated with structural optimization for earthquake loads is growing in volume, but is still limited. Cheng and Juang (1988, 1989), Truman and Cheng (1990), and Truman and Jan (1988) studied two- and three-dimensional building structures with respect to various code provisions and cost functions. Cheng and Truman (1983) have studied optimization of structures analyzed by modal response spectrum analyses. Truman and Petruska (1991) have applied optimality criteria techniques for optimization of two-dimensional structures using time history analysis procedures coupled with actual seismic accelerogram base excitations. Structural optimization of systems subjected to random seismic excitations and reliability theory is still in its developmental stage but has been studied by Cheng and Chang (1985, 1988) and Austin et al. (1987a,b).

TYPICAL SEISMIC DESIGN LOADS

Deterministic, seismically designed structures are typically loaded either through equivalent static forces based on design code formulae, a response spectrum, or a time-dependent accelerogram. Equivalent static force procedures can be a single set of forces or multiple sets of modal forces used as a means of approximating a full modal response spectrum analysis. The equivalent static force method is a common approach for designers when actual earthquake data are scarce or their use is unwarranted (noncritical structures). Modal response spectrum and time history analysis procedures are very common when earthquake data for a specific site are known or when the structure is deemed a critical facility. Many times these procedures are used for structures located in a region of frequent seismic activity or a region of potentially high magnitude seismic events.

Equivalent Static Forces

The Uniform Building Code (UBC 1991) equivalent static design forces are based on finding a total base shear and then distributing this base shear to each floor according to the mass distribution and distance above the ground. The UBC equivalent static base shear equation is of the form:

$$V = \frac{1.25ZIS}{R_w T^{2/3}} W \tag{12-1}$$

where Z is a seismic zone factor, I is an importance factor, S is a soil factor, R_w is a structural framing related ductility factor, T is the fundamental period of the structure, and W is the seismic dead load.

Typically, the lateral forces are a percentage of the base shear distributed according to height and mass for a certain level. The UBC lateral force distribution is of the form:

$$F_x = \frac{(V - F_t)w_x h_x}{\sum_{i=1}^{n} w_i h_i} = (V - F_t)\rho_x \tag{12-2}$$

where F_x is the xth level horizontal force, $F_t = 0.07TV$ and is the top-story horizontal lateral force, w_x is the xth level portion of the weight W used in Eq. (12-1), h_x is the height above the base to level x, and ρ_x is the xth story distribution percentage. Note that the base shear and lateral forces are dependent upon the natural period of the structure.

Modal Response Spectra

The modal response spectrum procedure suggested by the NEHRP Provisions (1992) is also based on finding a set of equivalent static forces based on modal base shear forces for a prescribed number of modes. The displacements, stresses, and drifts resulting from individual modes are then combined through the use of a statistically based procedure such as the square-root-of-the-sum-of-the-squares (SRSS) or complete-quadratic-combination (CQC). The jth modal base shear for a response spectrum analysis is of the form:

$$V_j = \frac{\overline{W_j}}{g} S_{Aj} \tag{12-3}$$

where the effective weight $\overline{W_j}$ is defined as:

$$\overline{W_j} = \frac{[\{\phi\}_j^T[\mathbf{M}]\{\mathbf{I}\}]^2}{\{\phi\}_j^T[\mathbf{M}]\{\phi\}_j} g \tag{12-4}$$

where $\{\phi\}_j$ is the jth mode shape, $[\mathbf{M}]$ is the mass matrix, S_{Aj} is the jth mode spectral acceleration, g is the gravitational constant, and $\{\mathbf{I}\}$ is a unity vector. Note that the mode shapes and the spectral acceleration are dependent upon the natural period of the structure. A classical response spectrum procedure where the displacements are found directly from the response spectrum analysis coupled with optimization procedures is presented in Cheng and Truman (1985).

Ground Motion Accelerogram

Time history analyses are based on the solution of the classic differential equation for seismic analysis using a ground motion accelerogram. The differential equation can be solved by numerous techniques as described in Bathe and Wilson (1976). Truman and Petruska (1991) have successfully used Newmark integration procedures within the optimization algorithms. The classical differential equations are of the form:

$$[\mathbf{M}]\{\ddot{\mathbf{u}}\} + [\mathbf{C}]\{\dot{\mathbf{u}}\} + [\mathbf{K}]\{\mathbf{u}\} = [\mathbf{M}]\{\mathbf{I}\}\ddot{x} = \{\mathbf{R}\} \tag{12-5}$$

where $[\mathbf{C}]$ is the damping matrix, $[\mathbf{K}]$ is the stiffness matrix, $\{\mathbf{u}\}$ is the relative displacement vector, $\{\mathbf{R}\}$ is the load vector, x is the ground displacement, and each dot is one derivative with respect to time. Note that the inertia force on the right side of (12-5) is mass and time dependent; therefore the maximum response for each iterative design typically occurs at different times within a single earthquake loading. Using a Newmark integration scheme allows direct formulation of the gradients for the velocity and acceleration in terms of the displacement gradients (Truman and Petruska 1991).

The algorithm for using the Newmark Integration Method (Bathe and Wilson 1976) can be broken into two parts: initial calculations and calculations for each time step. The initial calculations involve forming the mass, damping and stiffness matrices, initial displacement, velocity and acceleration at time zero, selecting a suitable time step Δt, calculating the integration constants a_{0-7} and forming the effective stiffness matrix given as:

$$[\hat{\mathbf{K}}] = [\mathbf{K}] + a_0[\mathbf{M}] + a_1[\mathbf{C}] \tag{12-6}$$

For each time step, the effective loads at time $t + \Delta t$ are calculated as:

$$\{\hat{R}_{t+\Delta t}\} = \{R_{t+\Delta t}\} + [\mathbf{M}](a_0\{\mathbf{U}_t\} + a_2\{\dot{\mathbf{U}}_t\} + a_3\{\ddot{\mathbf{U}}_t\}) + [\mathbf{C}](a_1\{\mathbf{U}_t\} + a_4\{\dot{\mathbf{U}}_t\} + a_5\{\ddot{\mathbf{U}}_t\}) \tag{12-7}$$

thus the displacements at time $t + \Delta t$ can be solved using

$$[\hat{\mathbf{K}}]\{\mathbf{U}_{t+\Delta t}\} = \{\hat{\mathbf{R}}_{t+\Delta t}\} \tag{12-8}$$

and the accelerations and velocities at time $t + \Delta t$ are given as

$$\{\ddot{\mathbf{U}}_{t+\Delta t}\} = a_0(\{\mathbf{U}_{t+\Delta t}\} - \{\mathbf{U}_t\}) - a_2\{\dot{\mathbf{U}}_t\} - a_3\{\ddot{\mathbf{U}}_t\} \tag{12-9}$$

$$\{\dot{\mathbf{U}}_{t+\Delta t}\} = \{\dot{\mathbf{U}}_t\} + a_6\{\ddot{\mathbf{U}}_t\} + a_7\{\ddot{\mathbf{U}}_{t+\Delta t}\} \tag{12-10}$$

Equation (12-5) indicates that the solution to the differential equation is highly dependent on $[\mathbf{M}]$ and $[\mathbf{K}]$. This is equivalent to forcing the period and resonance effects of the structure to change with respect to the inertia loads {**R**} for each iteration of the structural optimization.

Probabilistic Systems

Optimum design of probabilistic-based systems subjected to seismic excitations takes into account the random nature of seismic events as well as advances in reliability analyses of structures. Much of the research work to date has focused on the safety levels of a system with respect to its various failure modes, uncertainties in seismic forces such as equivalent seismic loads in the UBC and Newmark's nondeterministic response spectra, and random parameters in response and resistance (Cheng and Chang 1988). Newmark's response spectra includes random parameters in both vertical and horizontal earthquake excitations (Mohraz et al. 1972). The complete description of how to incorporate reliability into structural optimization for the previously presented analysis methods is given in Chang et al. (1994).

OPTIMIZATION PROCEDURES

General

Any structural optimization subroutine or program that can account for the gradient of the loads with respect to the design variables can be used for structural optimization for earthquake loads. The gradients of these loads with respect to the design variables can be found explicitly or implicitly, as seen in Chapter 5 and Appendix C, but are typically necessary for stable convergence of the optimization algorithms. The need for the gradients of the loads can easily be seen by exploring the use of an optimality criteria approach which is generally described in Appendix C, but specifically modified for earthquake loads.

Optimality Criteria

Truman et al. (1988, 1990, 1991) have had much success using optimality criteria methods. Their major advantage is the significant reduction of iterations required for convergence to an optimum design. The optimality criteria used in the optimization algorithm is derived from the Kuhn-Tucker conditions of optimality and can be stated as

$$\tau_i = -\sum_{j=1}^{l} \lambda_j \left(\frac{\partial g_j}{\partial X_i}\right) \div \left(\frac{\partial f}{\partial X_i}\right) = 1 \qquad i = 1, \ldots, n \qquad (12\text{-}11)$$

where λ_j is the jth constraint Lagrange multiplier, g_j is the jth constraint, X_i is the ith design variable (member size), f is the function being minimized (objective function), l is the number of active constraints, and n is the number of structural members. As the structure reaches its optimal design, τ_i approaches unity. Therefore a recurrence relationship using (12-11) can be derived as shown in Truman and Cheng (1990). Prior to the use of (12-11) in a recurrence function for resizing of the structural elements, an estimate of the optimal Lagrange multipliers must be determined as described in detail in Truman and Cheng (1983), Grierson and Chan (1991), and briefly in Appendix C. The gradient g_j requires the gradient of the different constraint responses such as displacements, stresses, natural frequencies, and drifts.

SENSITIVITY ANALYSES

Equation (12-11) indicates that the gradients for the constraints are required, which is where the major difference occurs between statically and dynamically loaded structural optimization algorithms. Using the simplest case of finding the gradients for the displacements of a statically loaded system, differentiating the equation $[\mathbf{K}]\{\mathbf{u}\} = \{\mathbf{P}\}$, the general formulation becomes:

$$\frac{\partial\{\mathbf{u}\}}{\partial X_i} = [\mathbf{K}]^{-1}\left[\frac{\partial\{\mathbf{P}\}}{\partial X_i} - \frac{\partial[\mathbf{K}]}{\partial X_i}\{\mathbf{u}\}\right] \tag{12-12}$$

where $\{\mathbf{P}\}$ is the vector of applied forces. For a statically loaded system, each term of the gradient of $\{\mathbf{P}\}$ is zero.

Equivalent Static Forces

For the UBC equivalent static case the gradient of load vector $\{\mathbf{P}\}$ from Eqs. (12-1) and (12-2) becomes:

$$\frac{\partial\{\mathbf{P}\}}{\partial X_i} = (1 - 0.07T)\,\frac{1.25ZIS}{R_w}\left[T^{-2/3}\frac{\partial W}{\partial X_i} + W\frac{\partial T^{-2/3}}{\partial X_i}\right]\{\rho\} - 0.07V\frac{\partial T}{\partial X_i} \tag{12-13}$$

where $\{\rho\}$ is the vector of story percentages for the distribution of the base shear according to story heights and mass. The gradient of the period can be found by using the chain rule (Cheng and Truman 1983) and the well known natural frequency gradient:

$$\frac{\partial\omega_j^2}{\partial X_i} = \frac{\{\phi\}_j^T\left[\dfrac{\partial[\mathbf{K}]}{\partial X_i} - \omega_j^2\dfrac{\partial[\mathbf{M}]}{\partial X_i}\right]\{\phi\}_j}{\{\phi\}_j^T[\mathbf{M}]\{\phi\}_j} \tag{12-14}$$

Modal Response Spectra

The gradients for the modal response spectra displacements, stresses, and drifts are also derived by direct differentiation and application of the chain rule (Truman and Cheng 1990). Therefore the gradient for the *j*th modal base shear (12-3) can be written as

$$\frac{\partial V_j}{\partial X_i} = \frac{1}{g}\left[\frac{\partial \overline{W}_j}{\partial X_i} S_{Aj} + \overline{W}_j \frac{\partial S_{Aj}}{\partial X_i}\right] \tag{12-15}$$

where the partial derivative of the effective weight $\overline{W}_j$ from Eq. (12-4) requires the partial derivatives of the mode shapes, the mass, and the spectral acceleration as developed in Truman and Cheng (1990). The gradient of the load vector {**P**} is found by using the appropriate lateral force distribution equations which would be similar to Eq. (12-2).

Ground Motion Accelerogram

The equilibrium equations of motion are approximately solved for by using Newmark's numerical integration method outlined in Eqs. (12-6) through (12-10). Differentiating (12-8) with respect to the design variable X and noting that the derivatives of the integration constants with respect to the design variables X are zero, gives the gradient of the dynamic displacement for each time step as

$$\frac{\partial\{\mathbf{U}_{t+\Delta t}\}}{\partial X_i} = [\hat{\mathbf{K}}]^{-1}\left(\frac{\partial\{\hat{\mathbf{R}}_{t+\Delta t}\}}{\partial X_i} - \left[\frac{\partial[\mathbf{K}]}{\partial X_i} + a_0\frac{\partial[\mathbf{M}]}{\partial X_i} + a_3\frac{\partial[\mathbf{C}]}{\partial X_i}\right]\{\mathbf{U}_{t+\Delta t}\}\right) \tag{12-16}$$

where

$$\frac{\partial\{\hat{\mathbf{R}}_{t+\Delta t}\}}{\partial X_i} = \frac{\partial\{\mathbf{R}_{t+\Delta t}\}}{\partial X_i} + \frac{\partial[\mathbf{M}]}{\partial X_i}(a_0\{\mathbf{U}_t\} + a_2\{\dot{\mathbf{U}}_t\} + a_3\{\ddot{\mathbf{U}}_t\}$$
$$+ [\mathbf{M}]\left(a_0\frac{\partial\{\mathbf{U}_t\}}{\partial X_i} + a_2\frac{\partial\{\dot{\mathbf{U}}_t\}}{\partial X_i} + a_3\frac{\partial\{\ddot{\mathbf{U}}_t\}}{\partial X_i}\right)$$
$$+ [\mathbf{C}]\left(a_1\frac{\partial\{\mathbf{U}_t\}}{\partial X_i} + a_4\frac{\partial\{\dot{\mathbf{U}}_t\}}{\partial X_i} + a_5\frac{\partial\{\ddot{\mathbf{U}}_t\}}{\partial X_i}\right) + \frac{\partial[\mathbf{C}]}{\partial X_i}(a_1\{\mathbf{U}_t\} + a_4\{\dot{\mathbf{U}}_t\} + a_5\{\ddot{\mathbf{U}}_t\}) \tag{12-17}$$

and the derivatives of the acceleration and velocity can be found by directly differentiating Eqs. (12-9) and (12-10). With the displacement gradients for each time step calculated, the story drift gradients can be obtained and so can the stress gradients (Truman and Cheng 1985).

TYPICAL CONSTRAINTS

From a practical point of view, constraints are used to produce a feasible design. The constraints can be used to control drift, displacements, frequencies, stresses, stability, reliability, and maximum and minimum member sizes. Note that each constraint must have quantifiable gradients to be used in Eq. (12-11) as shown for displacements in (12-12). Once these gradients are derived, the structural optimization, using the Kuhn-Tucker relationship, is able to control these responses while finding an optimal structural system. To be able to control these responses simultaneously for multiple earthquake loads is extremely useful for seismic-resistant design.

OBJECTIVE FUNCTIONS

Deterministic seismic-related objective functions are typically expressed in terms of structural weight or structural cost per geometric property such as volume or surface area. Structural weight is commonly used due to its importance as a major contributor to total construction cost. More importantly, design variables can be clearly defined vis-a-vis structural weight. Other objective functions can be and have been formulated to include any combination of the expense of structural members, nonstructural elements, painting, connections, and maintenance (Cheng and Juang 1988).

For probabilistic-based models, cost functions can include parameters that reflect initial construction costs for structural and nonstructural elements and/or expected costs of failure for various safety levels. Expected failure costs can include parameters reflecting damage, liability, business interruption, and replacement costs. Chang et al. (1994) express estimated structural cost as

$$C_T = C_I + L_f P_{f0} = C_I + C_{VL} C_I P_{f0} \tag{12-18}$$

where C_I is the initial construction cost calculated as C_{in}, ratio of initial cost to structural member cost, multiplied by C_u the unit volume cost of a structural member, and the volume of the members L_f is the expected failure cost, C_{VL} is the ratio of future failure cost to initial cost, and P_{f0} is the system failure probability.

NUMERICAL EXAMPLES

Four examples are presented. Three use deterministic data and one uses probabilistic methods. The examples are used to show how structural optimization can provide automated designs and information regarding the behavior of a structure. The path of optimization can often provide insight that can be used to enhance the designer's understanding of the structure and its interaction with the seismic loading. This enhanced understanding can lead to better seismic-resistant structures.

Multicomponent Modal Response Spectrum Optimization

Nonlinked, Wide-Flange System. A five-story L-shaped, rigid diaphragm, moment-resistant frame structure was optimized with respect to two separate load cases in order to study the effects of multicomponent modal response spectra loadings. Material constants are E = 200,100 MPA (29000 ksi) and ρ = 7820 kg/m^3 (490 lb/ft^3). The beams and columns are wide-flange cross-sections as shown in Fig. 12-1 with the primary design variable as the major axis moment of inertia (Cheng and Truman 1990), and each story height is 3.05 m (10 ft). The initial size of the columns and beams is 395,400 cm^4 (9500 in.^{4}) for all levels. Each member is free (nonlinked) to be any size between 416 cm^4 (10 in.^{4}) and 832,000 cm^4 (20,000 in.^{4}). The constraints consist of displacement constraints of 1.14 cm (0.45 in.) per story. The response spectrum in equation form is:

$$\frac{S_a}{a_{\max}} = -26.14T^2 + 13.94T + 0.935 \tag{12-19}$$

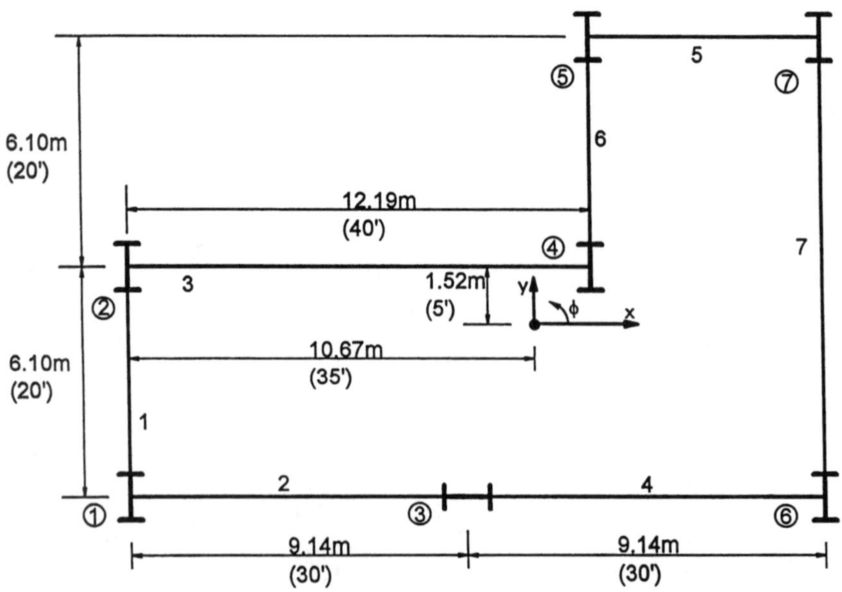

FIG. 12-1. Plan View of Five-Story L-Shaped Steel Structure.

for $T < 0.4$ sec and

$$\frac{S_a}{a_{max}} = 0.1606(T - 0.4)^4 - 1.141(T - 0.4)^3 + 2.996(T - 0.4)^2 - 3.618(T - 0.4) + 2.229 \quad (12\text{-}20)$$

for $T > 0.4$ sec where a_{max} is the maximum ground acceleration.

The first load case uses $0.4g$ as the maximum ground acceleration in the x-direction only and the second load case is $0.4g$ in the x-direction and $0.267g$ in the y-direction. The modal analyses are performed using three modes and the SRSS method of modal superposition. Principal directions of loading are not used in order to simplify the presentation of results.

Initially each load case starts with active x-displacement constraints on levels 2 through 4. Finally, each loading produces active constraints as shown in Table 12-1. Note that the multicomponent cases have additional y-displacement constraints in the active set. Initially, the multicomponent case requires a smaller weight as shown in Fig. 12-2. This is due to the fact that, initially, the rigidity center is located within the third quadrant of the global coordinate system. Therefore the positive y-loading coupled with the rigid diaphragm causes a reduction in the x-displacements. Finally, the multicomponent excited structure produces a final design which is 25% larger than the single component excited structure. It is important to note that the center of rigidity for the nonlinked case is free to relocate on each level for each cycle of optimization. The increase in weight, seen in Fig. 12-2, for the last cycle of the x-excitation case is due to a change in the active set of constraints. The final distributions of stiffnesses are shown in Tables 12-2 and 12-3. The x-only loading produces a strong frame using columns 1, 3, and 6 with beams 2 and 4. For the multicomponent case, the dominant columns are again 1, 3, and 6 with beams 1, 2, 4, and 6. Columns 1, 3, and 6 with beams 2 and 4 are primarily used to control the x-deflections, and columns 1, 2, 4, and 5 with beams 1 and 6 are used to control the y-deflections. Column

TABLE 12-1 Optimal Results for Nonlinked Structure Subject to Multicomponent Seismic Loads and Displacement Constraints (1 kip = 4.45 kN)

a_{max}			Initial Weight (kN)	Final Weight (kN)	Active Constraints (direction$_{level}$)
x	y	z			
$0.40g$	—	—	1220	481.5	x_4, x_5, y_4, y_5
$0.40g$	$0.27g$	—	1165	603.4	$x_4, x_5, y_2–y_5$

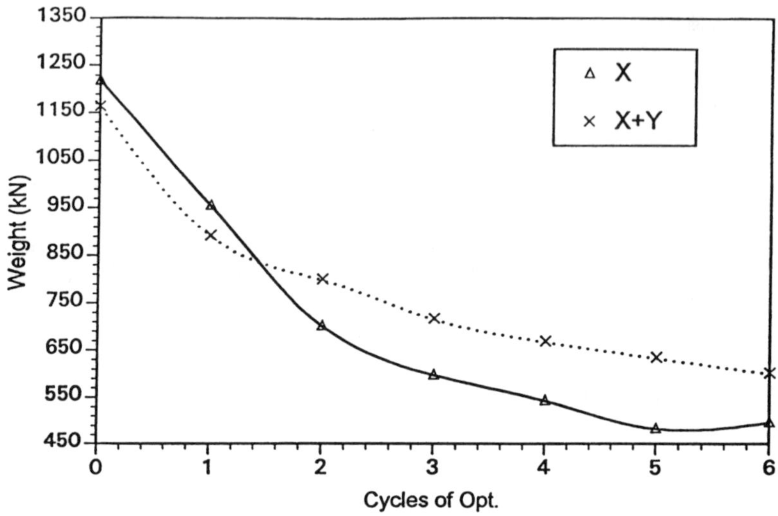

FIG. 12-2. Comparison of Weights for Nonlinked Designs Subjected to Single and Multicomponent Seismic Excitations.

TABLE 12-2 Optimal Stiffness Distribution for Single Component Seismic Excitation with Displacement Constraints ($x = 0.40g$, $y = z = 0$) (1 in.^{4} = 41.6 cm^4)

	Column Moment of Inertia (cm^4)						
Level/Line	1	2	3	4	5	6	7
5	168116	49365	240500	48616	97356	189136	102185
4	327823	51113	285285	91071	182476	452194	138439
3	692234	55400	379020	61061	145223	192590	123995
2	710500	57815	334858	81831	265223	800000	255857
1	109094	40957	832463	43413	86160	99063	65473
	Beam Moments of Inertia (cm^4)						
Level/Bay	1	2	3	4	5	6	7
5	11696	65099	9032	62476	15025	19229	7575
4	4828	181768	16691	195296	37086	90072	8283
3	416	322038	18023	200915	38626	119125	6202
2	416	404369	19896	279583	48866	16649	3996
1	2081	390467	18730	303849	38044	6410	5869

TABLE 12-3 Optimal Stiffness Distribution for Two-Component Seismic Excitation with Displacement Constraints ($x = 0.40g$, $y = 0.27g$, $z = 0$) (1 in.^{4} = 41.6 cm^4)

Column Moment of Inertia (cm^4)							
Level/Line	1	2	3	4	5	6	7
5	384390	119042	280373	136857	148012	282580	79000
4	428926	162871	365867	169323	199333	320373	84037
3	832463	207325	269385	197835	205160	619644	67097
2	832463	312298	535482	228386	217772	832463	86285
1	832463	832463	832463	250571	189260	178813	100603
Beam Moments of Inertia (cm^4)							
Level/Bay	1	2	3	4	5	6	7
5	76295	87950	12362	99687	13070	112757	45577
4	230634	239749	22601	298521	25765	286992	37211
3	359582	278750	22518	308261	25390	400082	36587
2	521496	377314	25848	404743	25265	340644	27638
1	410404	482870	32882	326034	23184	179937	30260

1 is critical since it can be used effectively as part of the frames used to resist both sets of orthogonal deflections and at the same time force the rigidity center to be within the second quadrant.

Linked, Wide-Flange System. This structural system is identical to that used in the preceding example with the exception that linking is used. The initial data are exactly the same except that each level is required to have the same size columns throughout that level and the same beam size throughout each level. The linking is used to stabilize the location of the rigidity center and to provide a slightly more realistic design. Due to the linking and orientation of the columns, the rigidity center is permanently located within the third quadrant of the global coordinate system for each cycle and level.

The final results for the linked systems are given in Figs. 12-3 and 12-4 and Table 12-4. The location of the rigidity center allows the multicomponent excited structure to continually produce results with less weight than the single component excitation as shown in Fig. 12-3. The optimizations of both systems have nearly identical paths which are a result of the column orientation. Most of the major axes are oriented to resist translation in the y-direction. This fact coupled with the linking of columns only produces active constraints in the x-direction. Therefore the only effect of the y-direction loading is to reduce the x-displacements of the mass center due to the rotation about the center of rigidity. The final results and comparison to the nonlinked cases are given in Table 12-4. The single-component linked

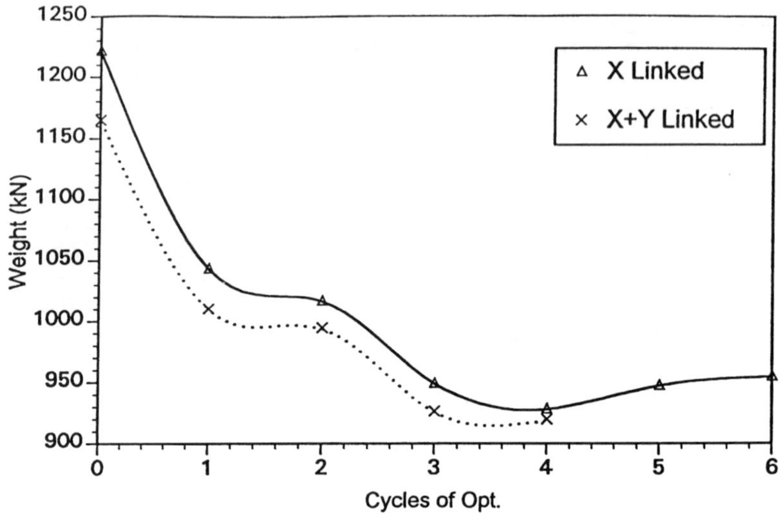

FIG. 12-3. Comparison of Weights for Linked Designs Subjected to Single and Multicomponent Seismic Excitations.

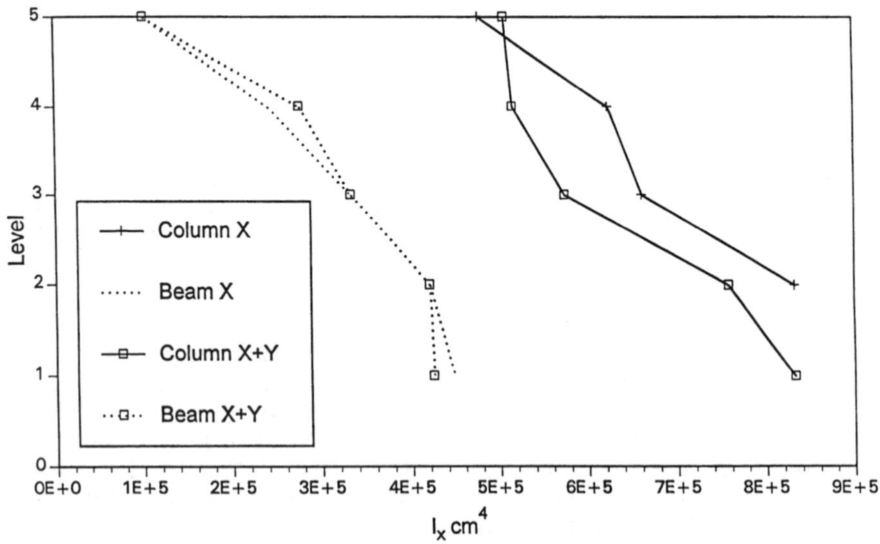

FIG. 12-4. Stiffness Distributions for Optimal Linked Designs for Single and Multicomponent Seismic Excitations.

TABLE 12-4 Comparison of Results for Linked and Nonlinked, Seismically Loaded Five-Story L-Shaped Structure with Displacement Constraints (1 kip = 4.45 kN)

a_{max}			Initial	Final	
x	y	Linkage Constraints	Weight (kN)	Weight (kN)	Active Constraints (direction$_{level}$)
0.40g	—	No	1220	481.5	x_4, x_5, y_4, y_5
0.40g	—	Yes	1220	926.5	x_2–x_5
0.40g	0.27g	No	1165	603.4	x_4, x_5, y_2–y_5
0.40g	0.27g	Yes	1165	918.9	x_2–x_5

case gives a 92% increase in weight over the nonlinked case, whereas the multicomponent linked case provides a 52% increase over the nonlinked case. The additional weight is accumulated since the y-displacements for the linked case are only 7 to 10% of the allowable values, and the nonlinked case allows several y-displacement constraints to become active.

These results suggest that a new design is in order. A lighter structure could be obtained by reorienting several of the columns such that their major axis moment of inertia would help to resist the x-displacements. This reorientation would allow the y-displacement criteria to play a more significant role in the final design, therefore providing a better balance in resistance.

Time History Optimization

The three-story, two-bay frame shown in Fig. 12-5 was analyzed to show the capability of a time history optimization algorithm. The constraints for this design problem are displacement and minimum and maximum member sizes. The absolute value of the horizontal displacement at each floor was limited to 2.54, 4.19, and 5.97 cm (1.00, 1.65, and 2.35 in.) for the first, second, and third floors, respectively. The lower limit on the moment of inertia was 12,070 cm^4 (290 in.^{4}). Initial values for the member sizes were 62,400 cm^4 (1500 in.^{4}). Material constants were E = 200,100 MPa (29000 ksi) and ρ = 7820 kg/m^3 (490 lb/ft^3). A uniformly distributed nonstructural weight totaling 148.3 kN (33.33 kips) was applied to each beam member of the frame. Damping of the structure was neglected. Three different loadings were examined. At time t = 0, the structure is at rest and the base undergoes a transient acceleration. The first base acceleration is defined as

$$\ddot{a}_g(t) = 343 \sin(2\pi t) \text{ cm/sec}^2, \qquad 0 \leq t \leq 1.0 \text{ sec} \qquad (12\text{-}21)$$

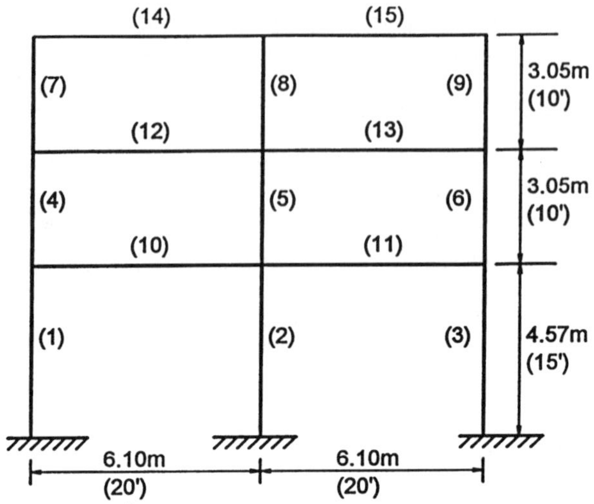

FIG. 12-5. Three-Story, Two-Bay Frame.

and is zero for t greater than 1.0 seconds. The second base acceleration is defined by the accelerogram for the north-south component of the El Centro, California earthquake on May 18, 1940 and is shown in Fig. 12-6. The third base acceleration is for the north-south component of the San Fernando, California earthquake on February 9, 1971 taken at the Caltech Seismological Lab and is shown in Fig. 12-7. The constant of 343 used in the sine function was chosen to match the peak acceleration of the El Centro earthquake record. The San Fernando earthquake was also multiplied by a constant of 1.81 so that both earthquakes would have the same peak acceleration. The time step selected was 0.01 seconds. The examples use different linking schemes as indicated in Table 12-5. For the sin load, the time frame examined was 2.0 seconds and for the El Centro and San Fernando earthquakes, the time frame was 10.0 seconds. Table 12-5 shows the optimum designs and Fig. 12-8 shows the weight versus iteration.

Comparing Case A to Case B and Case C to Case D (defined below Table 12-5) shows that the linking examples produce heavier structures since each member is not able to take on its optimum size. The El Centro earthquake results in a slightly lighter structure than the structures simultaneously optimized for the San Fernando and El Centro earthquakes. It is worth noting that both earthquakes in Case E, simultaneous earthquakes, produce active constraints. This indicates that a design based on only one earthquake would most likely violate the constraints when loaded by the other earthquake. Fig. 12-8 shows for Case C a zigzag pattern between iterations 5 through 9 because at iteration 4, the process violates the displacement con-

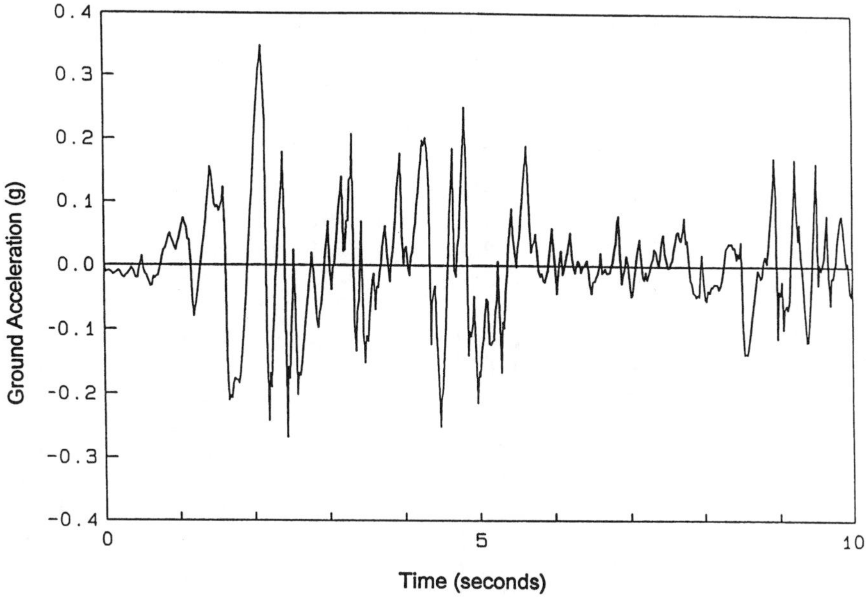

FIG. 12-6. El Centro Earthquake Acceleration Record.

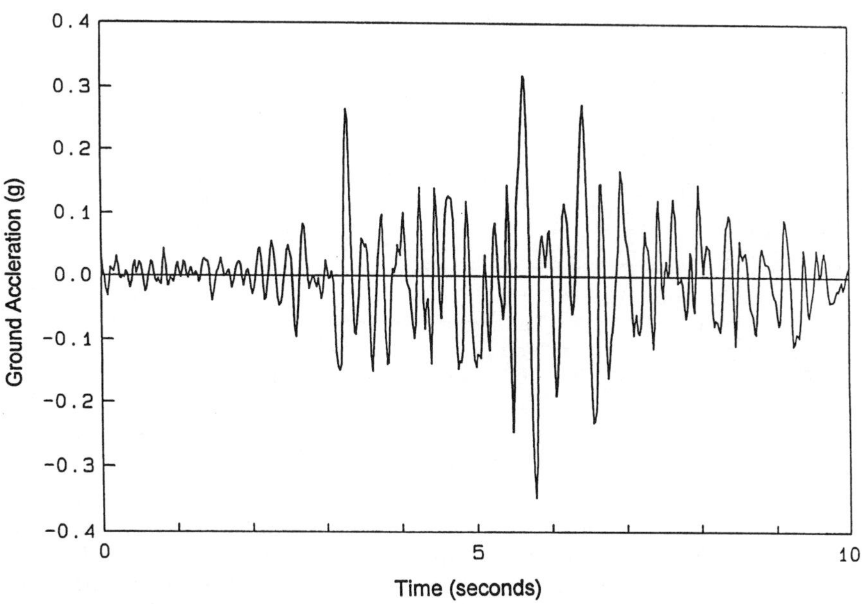

FIG. 12-7. San Fernando Earthquake Acceleration Record.

TABLE 12-5 Optimum Solutions for Ground Acceleration Example Problem

	Moment of Inertia (cm^4)				
Member	Case A	Case B	Case C	Case D	Case E
1	50867	67921	115504	62559	44798
2	94655	67921	115504	237015	404007
3	50867	67921	115504	62485	54473
4	15883	26630	39292	25868	41823
5	44620	26630	39292	76058	71550
6	15883	26630	39292	25960	41952
7	12071	14639	27367	14776	16529
8	22631	14639	27367	43733	52104
9	12071	14639	27367	14918	16204
10	52062	53577	71338	78543	99683
11	52062	53577	71454	78547	109585
12	30834	32566	44820	43080	73631
13	30834	32566	44037	43375	72666
14	12071	12071	16811	12670	14826
15	1207	12071	16358	12799	14859
W^*_{Total} (kN)	43.35	44.26	54.03	52.81	60.15
Iteration	4	5	11	10	12
	Active Constraints				
DOF	X_3	X_3	X_2	X_2	X_3 (El C)
Time (sec)	$t = 0.69$	$t = 0.69$	$t = 8.00$	$t = 3.58$	$t = 4.74$
Value (cm)	5.98	5.97	4.23	4.14	−5.99
DOF			X_3	X_3	X_3 (SF)
Time			$t = 8.02$	$t = 3.59$	$t = 6.81$
Value (cm)			6.09	6.04	5.98
DOF				X_3	
Time				$t = 4.22$	
Value (cm)				−5.91	

Case A. Sine Function (No linking of members).
Case B. Sine Function (Linking of members).
Case C. El Centro Earthquake (Linking of columns).
Case D. El Centro Earthquake (No linking of members).
Case E. El Centro and San Fernando as separate load cases (No linking).

straint, and the optimization process is returning to the constraint in iterations 5 through 9. Case D shows a large decrease in weight at iteration 6 due to a change in the active constraints, thus allowing the optimization process to use a new path to produce a better solution.

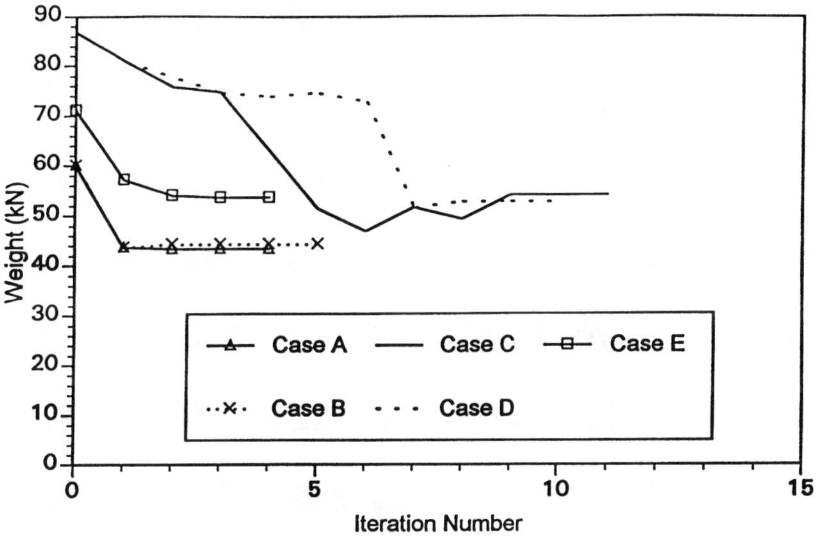

FIG. 12-8. Weight Versus Iteration Number.

Probabilistic Minimum Cost Example

For a structure consisting of a 10-story shear building as shown in Fig. 12-9, the parameters used in the design are: loading area for each is 83.61 m^2 (900 ft^2); mean dead load is 3.82 kPa (80 lb/ft^2) and its allowable variance is 0.12; allowable displacement is 0.005 times the corresponding height relative to the base and allowable variance of displacement is zero; mean elastic modulus is $(2.067)10^8$ kPa (30,000 ksi) and its coefficient of variance is 0.06; and the coefficient of variance for the moment of inertia is 0.05.

The effect of the ratio of initial cost to structural member cost C_{in} is studied by expressing initial cost C_I in terms of the product of C_{in} and member cost. Three values of C_{in} of 2, 5, and 20 are used to study the relationship between structural and nonstructural costs. Unit cost C_u is \$9.1536(10^{-3})/cm^3 (0.15 \$/$in.^3$). Future failure cost as defined in Eq. (12-18) is not considered, and the value of C_{VL}, the ratio of future failure cost to initial cost is assumed to be constant as 1.0. As mentioned earlier, future failure cost can be expressed in terms of expected failure cost and system failure probability. Expected failure cost may include replacement cost, liability due to injury or death, and business interruption.

The optimum solution shown in Fig. 12-10 is based on a modified white noise spectrum (Cheng and Chang 1988). Some difference between total cost and initial cost at low reliability such as 10^{-3} is demonstrated here. This difference becomes very small at high reliability which means that an increase in nonstructural member cost influ-

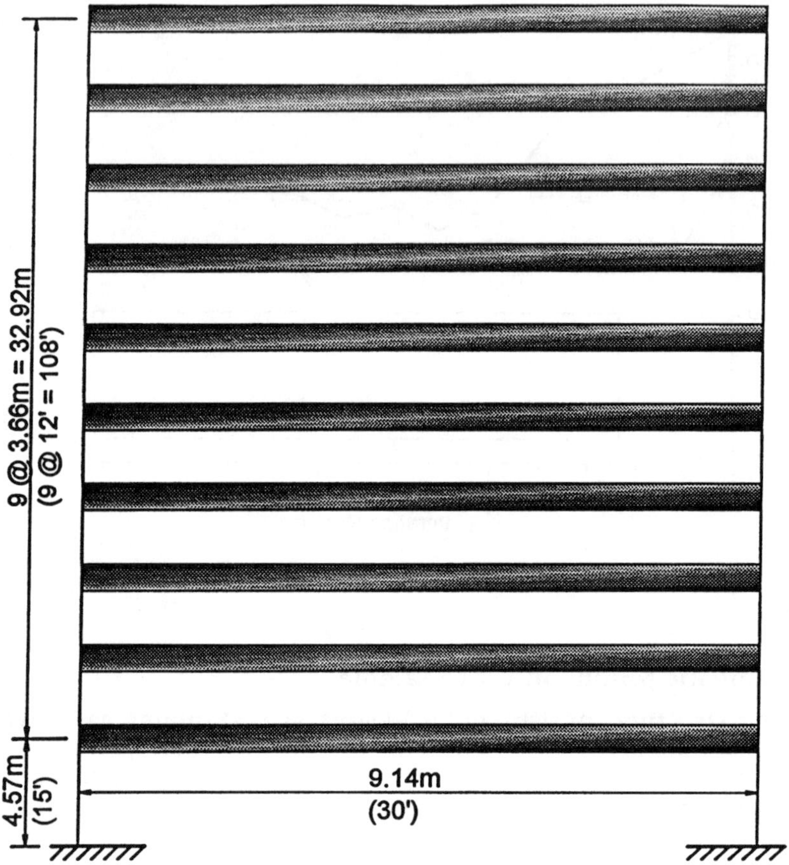

FIG. 12-9. 10-Story Shear Building Structure.

ences the design of structural members. Total cost goes up as C_{in} increases. At higher reliability levels the difference between total cost and initial cost decreases because an increase in reliability leads to a decrease in future failure losses.

The effect of the ratio of expected failure cost to initial cost is studied by expressing expected failure cost in terms of C_{VL} and C_{in}. Three values of C_{VL} of 0.5, 1.0, and 10 are assumed in order to examine the effects of diverse projected failure costs. The value of C_{in} is assumed not to vary and a constant coefficient of 5 is used.

Figure 12-11 shows that total cost goes up as C_{VL} increases at low reliability such as 10^{-5}. As reliability increases, the impact of C_{VL} diminishes because the value of any future failure loss is small at high

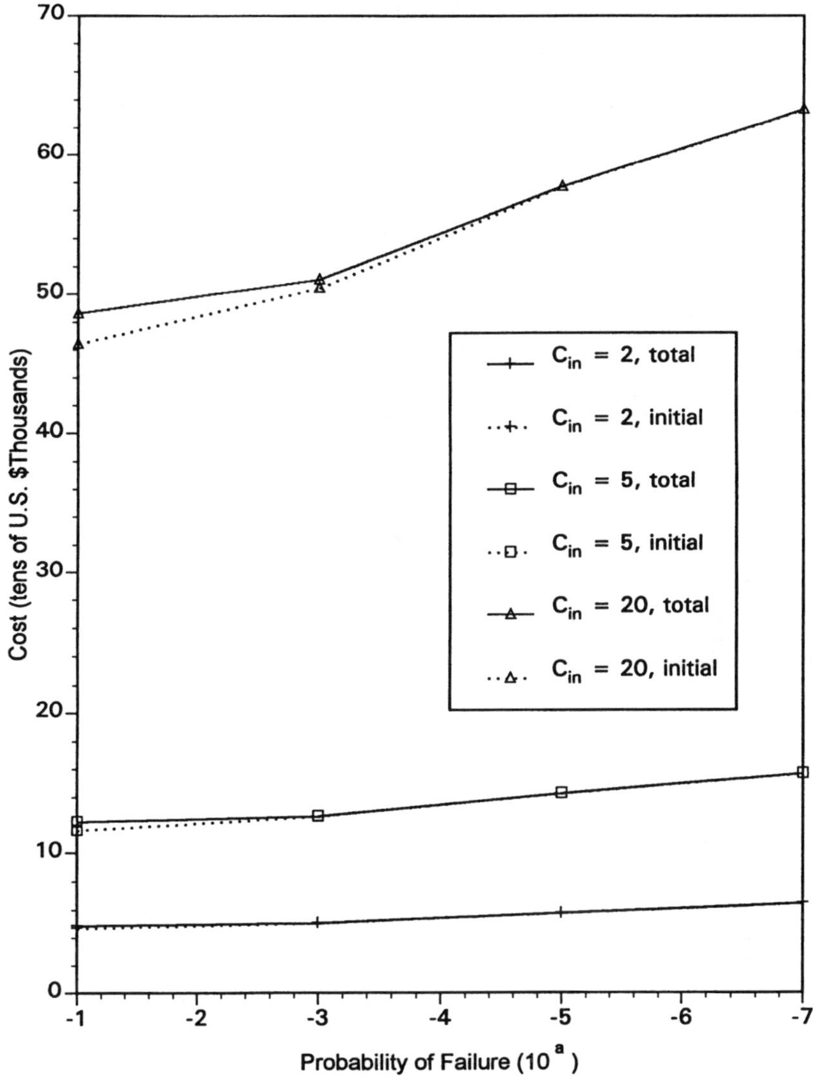

FIG. 12-10. Optimum Cost for Various C_{in} of 10-Story Building.

reliability. Thus the increase of future failure cost at high reliability does not significantly influence the design of structural members. To reiterate, differences between total cost and initial cost decrease at higher reliability levels because an increase in reliability leads to a decrease in future failure losses.

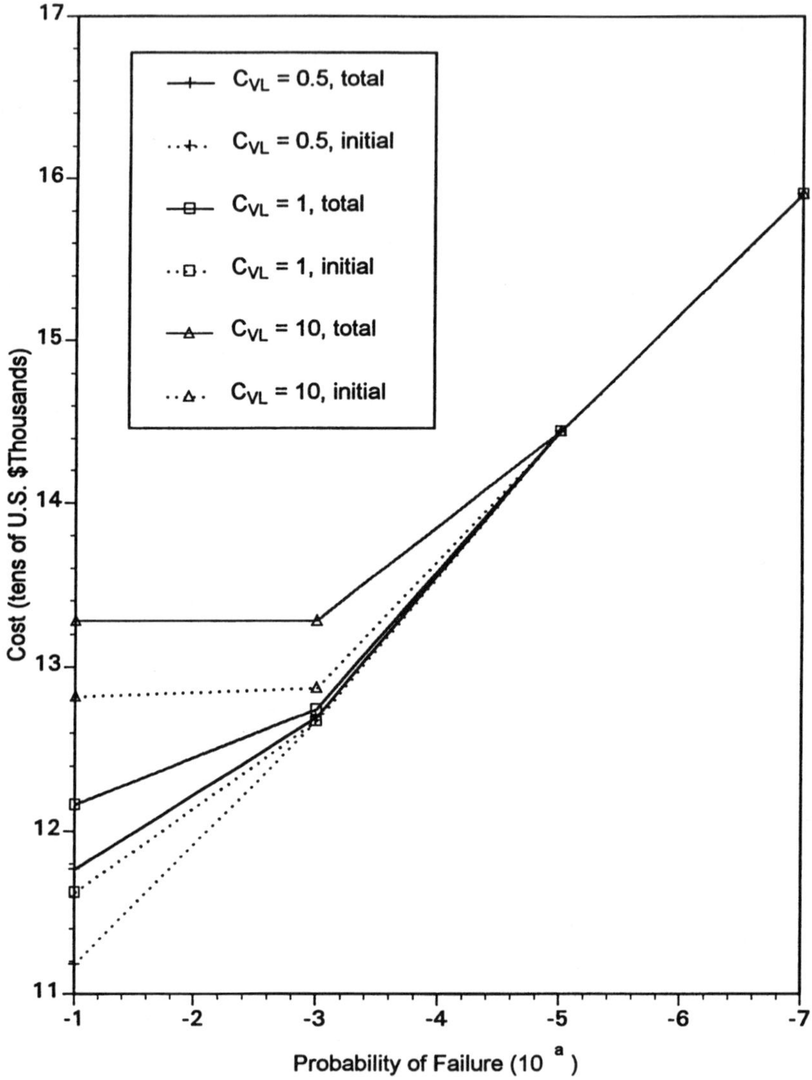

FIG. 12-11. Optimum Cost for Various C_{VL} *of 10-Story Building.*

SUMMARY AND DISCUSSION

Several observations can be made with regard to the use of optimization procedures for earthquake loads. Structural optimization is a viable means of producing seismic-resistant designs. The optimization procedures can produce a series of feasible designs that can be used for the final seismic-resistant design. This series of feasible structures can be extremely useful in enhancing the designer's understanding of the interaction between the structure and the seismic loads. The use of structural optimization procedures for full modal response spectra and time history analyses provides an opportunity to design structures for multiple earthquakes simultaneously. Being able to design a structure for a family of varied seismic accelerograms is of tremendous benefit. Objective functions can be related to cost, weight, damage, energy absorption, or any other quantity that can be related to the actual design of the structure. Nondeterministic procedures are available in order to explore seismic-resistant design from a probabilistic point of view. These techniques are useful in exploring potential structural costs associated with probable events and failures. The use of structural optimization for earthquake loads is not as easily applied as that for static loads, but it is possible. Structural optimization can be an extremely useful and beneficial tool for seismic-resistant system designers.

APPENDIX 12-A. NOTATION

a_0–a_7 = Newmark integration constants
C_I = initial construction cost
C_{in} = ratio of initial cost to structural member cost
C_{VL} = ratio of future failure cost to initial cost
C_u = unit volume cost of a structural element
$[\mathbf{C}]$ = damping matrix
f = objective function
F_t = top story horizontal lateral force
F_x = xth level horizontal force
g = gravitational constant
g_j = jth constraint
h_x = height above the base to level x
I = UBC importance factor
$\{\mathbf{I}\}$ = unity vector
$[\mathbf{K}]$ = stiffness matrix
l = number of active constraints
L_f = expected failure cost
$[\mathbf{M}]$ = mass matrix
n = number of structural members
P_{f0} = system failure probability
$\{\mathbf{P}\}$ = vector of applied forces
R_w = UBC structural framing related ductility factor
$\{\mathbf{R}\}$ = dynamic load vector
S = UBC soil factor
S_{Aj} = jth mode spectral acceleration
T = fundamental period of the structure
$\{\mathbf{u}\}$ = relative displacement vector
V = seismic base shear

V_j = jth modal base shear
W = seismic dead load
w_x = xth level portion of the weight W
$\overline{W}_j$ = effective weight for mode j
x = ground displacement
X_i = ith design variable
Z = UBC seismic zone factor
ρ_x = xth story distribution percentage
λ_j = jth constraint Lagrange multiplier
$\{\phi\}_j$ = jth mode shape

REFERENCES

Austin, M. A., Pister, K. S., and Mahin, S. A. (1978a). "Probabilistic design of moment-resistant structures." *J. of Structural Engng.*, ASCE, 113(8), 1642–1659.

Austin, M. A., Pister, K. S., and Mahin, S. A. (1978b). "Probabilistic design of moment-resistant frames under seismic loading." *J. of Structural Engng.*, ASCE, 118(8), 1660–1677.

Balling, R. J., Pister, K. S., and Ciampi, V. (1983). "Optimal seismic-resistant design of a planner steel frame." *Earthquake Engng. Structural Dynamics*, 11, 541–556.

Bathe, K. and Wilson, E. (1976). *Numerical methods in finite element analysis.* Prentice Hall, Englewood Cliffs, NJ.

Bhatti, M. A. and Pister, K. S. (1981). "A dual criteria approach for optimal design of earthquake-resistant structural systems." *Earthquake Engng. Structural Dynamics*, 9, 557–572.

Chang, C. C., Ger, J. F., and Cheng, F. Y. (1994). "Reliability-based optimum design for UBC and nondeterministic seismic spectra." *J. of Structural Engng.*, ASCE, 120(1), 139–160.

Cheng, F. Y. and Chang, C. C. (1985). "Optimum design of steel building with consideration of reliability." *Proceedings of Fourth International Conference on Structural Safety and Reliability*, Kobe, Japan, Vol. 3, 81–88.

Cheng, F. Y. and Chang, C. C. (1988). *Safety-based optimum design of nondeterministic structures subjected to various types of seismic loads*, NSF Report, U.S. Department of Commerce, VA, NTIS No. PB90-133489/AS (326 pages).

Cheng, F. Y. and Juang, D. S. (1988). "Assessment of various code provisions based on optimum design of steel structures." *J. of Earthquake Engng. and Structural Dynamics*, 16, 46–61.

Cheng, F. Y. and Juang, D. S. (1989). "Recursive optimization for seismic steel frames." *Journal of Structural Engng.*, ASCE, 115(2), 445–466.

Cheng, F. Y. and Truman, K. Z. (1983). "Optimization of 3-D building systems for static and seismic loadings." *Modeling and simulation in engineering*, III, North-Holland Publishing Co., New York, 315–326.

Cheng, F. Y. and Truman, K. Z. (1985). *Optimal design of 3-D reinforced concrete and steel buildings subjected to static and dynamic loads including code provisions.* Final Report Series 85-20 for Nat. Sci. Foundation, U.S. Dept. of Commerce, NTIS Access No. PB87-168564/AS, Washington, DC.

Feng, T. T., Arora, J. S., and Haug, E. J. (1977). "Optimal structural design under dynamic loads." *Int. J. for Numerical Methods in Engng.*, 11(1), 39–52.

Grierson, D. E. and Chan, C. - M. (1991). "Design optimization of tall steel buildings." *OPTI 91—International Conference for Computer Aided Optimum Design of Structures*, Boston, MA, June 25–27, 1991, *Optimization of structural systems and industrial applications*, S. Hernandez and C. A. Brebbia, eds. Elsevier Applied Science, London, 541–551.

Mohraz, B., Hall, W., and Newmark, N. (1972). "A study of vertical and horizontal earthquake spectra." Division of Reactor Standards, U.S. Atomic Energy Commission, Washington, DC.

NEHRP (1992). *Recommended provisions for the development of seismic regulations for new buildings*, Building Seismic Safety Council (for FEMA), Washington, DC.

Truman, K. Z. and Cheng, F. Y. (1990). "Optimum assessment of irregular 3-D buildings." *J. of Structural Engng.*, ASCE, 116(12), 3324–3337.

Truman, K. Z. and Jan, C. T. (1988). "Optimal bracing schemes for structural systems subject to the ATC 3-06, UBC, and BOCA seismic provisions." *Proceedings of the 9th World Conference on Earthquake Engineering*, Tokyo and Kyoto, Japan, August 2–9, 1988, Vol. V, 1149–1155.

Truman, K. Z. and Petruska, D. J. (1991). "Optimum design of dynamically excited structural systems using time history analysis." *OPTI 91—International Conference for Computer Aided Optimum Design of Structures*, Boston, MA, June 25–27, 1991; *Optimization of structural systems and industrial applications*, S. Hernandez and C. A. Brebbia, eds. Elsevier Applied Science, London, 197–207.

UBC (1991). *Uniform building code*. International Conference of Building Officials, Whittier, CA.

Yamakawa, H. (1984). *Optimum structural designs for dynamic response, new directions in optimum structural design*. John Wiley, New York, 249–266.

Chapter 13

HOW TO OPTIMIZE STRUCTURES FOR SHAPE

This chapter deals with how to optimize the shape of a structure so that the design objective is met subject to the performance constraints. It is important to identify meaningful shape design variables that will allow changes in the structural geometry. This chapter gives examples of shape optimization of discrete as well as continuum structures. As stated in earlier chapters, a variety of objective and constraint functions can be chosen that are relevant in the problem formulation. The finite element method is assumed to be suitable for analyzing the structure under consideration. A rudimentary step-by-step shape optimization algorithm is outlined where major computational tasks are explained. Four design examples, one plane stress and the other plane strain along with a transmission tower and solid cantilever beam with a hole are discussed at length to demonstrate the shape optimization methodology presented in this chapter.

INTRODUCTION

The shape optimization process needs to be implemented in the context of finite element analysis to encompass a wide variety of design applications. As stated in earlier chapters, structural design problems are primarily classified into three main categories, namely, sizing, shape, and topology, depending on the choice of the design variables. Design variables that govern shape or geometry of a structure are referred to as shape design variables. For example, coordinates of key nodes in a structure or definition of a certain surface/curve can be cited as typical shape design variables. On the other

P. B. Thanedar, Member ASCE, Ford Motor Company, Dearborn, MI 48126; email: ⟨pthaneda@ford.com⟩.

hand, sizing variables such as the cross-sectional area of a truss member, plate thickness, or cross-sectional dimensions of a beam do not alter the shape of a structure. Thus according to this definition, every possible geometry-related design variable in a three-dimensional continuum structure may be treated as a shape design variable.

In the case of discrete skeletal structures such as a truss or building frame, key nodal coordinates serve as shape design variables. Figure 13-1 shows a 47-member planar transmission tower subject to loads before and after shape optimization (Vanderplaats and Moses 1972). The objective is to minimize structural weight subject to stress and Euler member buckling constraints where x and y coordinates of the key nodes are taken as design variables. Note that the topology of the truss is kept fixed during the shape design process. For continuum structures such as plate/shell, or three-dimensional objects, features of the parameterized geometry such as length or fillet radius may be taken as design variables. Figure 13-2 shows a solid cantilevered beam subject to an end load with a hole where the volume of the beam is to be minimized so that Von Mises stress in the beam is within allowable limits. The shape design variables are beam heights at the base, end, mid-cross-sections, and radius of the hole. Note that

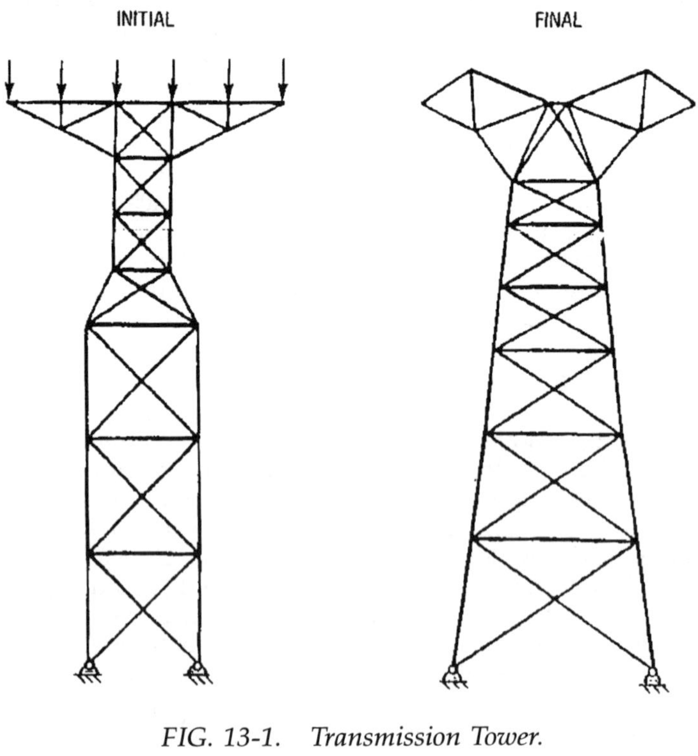

FIG. 13-1. Transmission Tower.

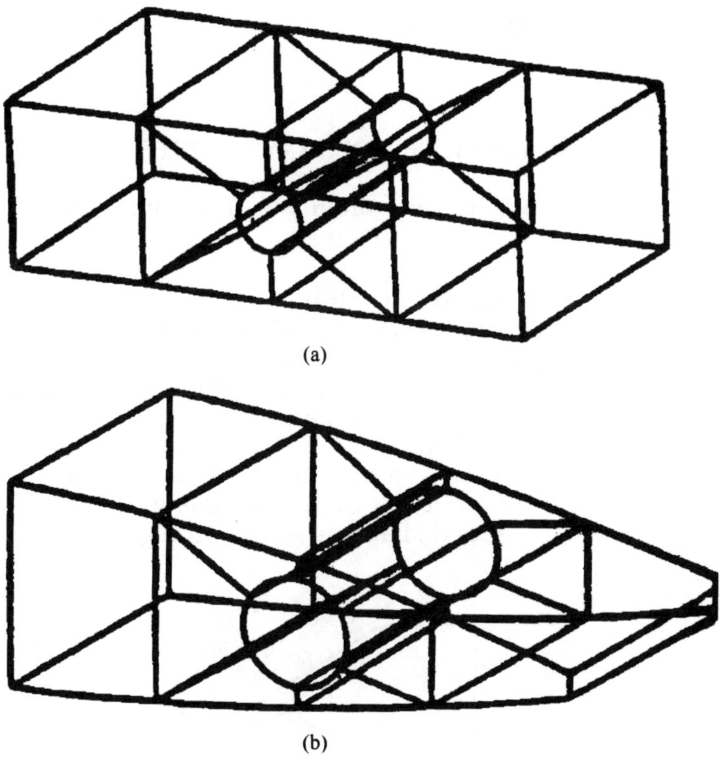

FIG. 13-2. Solid Cantilever Beam: (a) Initial Shape; (b) Final Shape.

both the transmission tower and solid cantilever beam examples are discussed in detail in later sections.

Two alternate approaches exist for shape optimization: parametrized geometry variables are taken as design variables along with automatic meshing capability to automate the design iterations, whereas the alternate approach deals with moving grid locations in the finite element mesh directly instead of changing the underlying geometry. The latter approach (also referred to as the reduced basis approach) involves computation of basis shape or velocity field vectors in addition to the finite element analysis, sensitivity analysis, and design update using an optimization algorithm. In the case of discrete structures such as a frame or truss where nodal coordinates are treated as design variables, there is no need to compute the basis vectors. Discussion in this chapter is limited to the reduced basis approach which is applicable from small to moderate changes in the initial shape. The procedures with automatic remeshing are applicable for moderate to large changes in structural shape and are still in the development stage (Kodiyalam et al. 1992). They are beyond the scope of this chapter.

CLASSIFICATION OF SHAPE DESIGN PROBLEMS

Shape design problems are often classified into the following categories.

1. To decide the shape of an arbitrary open or closed boundary, such as a fillet or dam surface as shown in Fig. 13-3(a).
2. To decide the dimensions of predefined shapes, such as the radius of a circular hole, the major and minor axes of an elliptical curve, dimensions of a slot, or length of a rectangular membrane as shown in Fig. 13-3(b).
3. To decide the location of a surface or curved boundary, such as the coordinates of a circular or elliptic hole which has either an arbitrary or predefined shape as shown in Fig. 13-3(c).

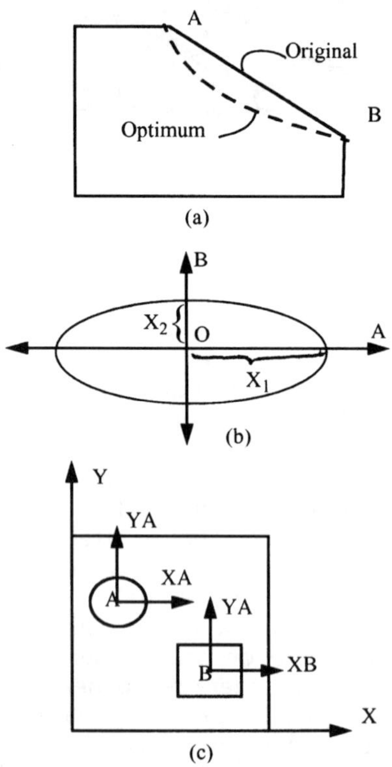

FIG. 13-3. Classification of Shape Design Problems.

REDUCED BASIS APPROACHES

The idea behind the reduced basis approach is to link the movement of free nodes in a finite element model for capturing global shape variations of the designable boundary or surface. The advantage of such a linking is that it is possible to come up with an improved shape by scaling and then combining only a handful of global shape variations of the designable boundary. Thus the reduced basis approach becomes practical and efficient since it deals with only a few independent design variables (or variations) instead of defining hundreds of coordinates of free grid points as design variables.

In shape optimization, geometry of the structure changes from an initial shape to the optimum shape. That is why it is essential that the finite element mesh varies smoothly from one iteration to another. Hierarchical (*h*-type) finite elements are more sensitive to mesh distortion than polynomial (*p*-type) finite elements as demonstrated by Thanedar and King (1993). Highly distorted finite element mesh during the iterative shape design process may yield inaccurate analysis results as well as discontinuous sensitivity derivatives and may lead to instability of the numerical optimization algorithm. Two approaches that belong to the general class of reduced basis method (Pickett et al. 1973), the specified displacements approach (Choi and Chang 1991) and the fictitious loads approach (Belegundu and Rajan 1992), are important in this context. Both approaches are conceptually similar in that they use a linear combination of certain basis vectors to update the structural shape. Note that reduced basis methods, in general, reduce remeshing during shape optimization to a minimum. Figure 13-4 shows three proposed basis vectors for an initial geometry in the form of a circular hole or opening. It is desirable that these basis shape vectors should be linearly independent and orthogonal to each other in the reduced basis method. Thus, in a generic reduced basis method, weighting or scale factors for the basis shape vectors are chosen as design variables instead of the x, y, z coordinates, spline polynomial coefficients, or similar geometric quantities. It is the choice and generation of these basis shape vectors that differ between the specified displacements and fictitious loads approaches.

The quality of finite element mesh maintained during the design process is directly linked to the quality of the basis vectors generated during shape optimization. In the specified displacements approach, basis vectors can be generated by either using a mapped mesh generator or from a static finite element analysis using enforced displacements generated using analytic (as shown in Fig. 13-4) shape functions. In the fictitious loads approach, the deformation pattern of the structure due to some fictitious loads at the key control nodes is used to derive the basis vectors. The eigenmodes of vibration as well as deformation pattern due to thermal loads have also been used as basis vectors.

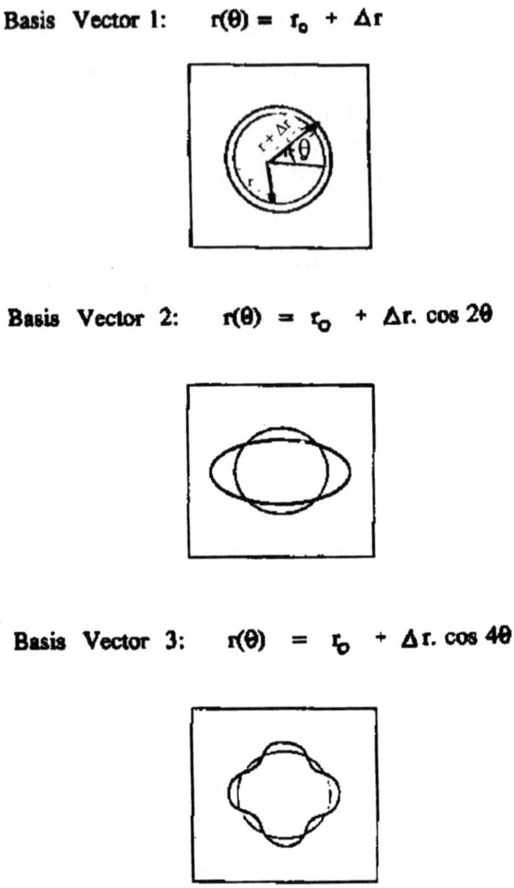

FIG. 13-4. Shape Basis or Velocity Field Vectors.

KEY DEFINITIONS IN SHAPE OPTIMIZATION

In the following, we define the generic sizing or shape optimization problem in a k-dimensional design space with p inequality (as well as equality) constraints and a single real-valued objective function that is minimized.

The problem definition is

Find

$$\mathbf{X} \in \Re^{k} \text{ to}$$

Minimize

$$f(\mathbf{x})$$

Such that

$$g_i(\mathbf{x}, z, \lambda) \leq 0 \qquad i = l, p$$
$$x_i^L \leq x_i \leq x_i^U \qquad i = l, k$$

where $\mathbf{K}(\mathbf{x})\mathbf{z} = \mathbf{F}(\mathbf{x})$ and $\mathbf{K}(\mathbf{x})\mathbf{y} = \mathbf{M}(\mathbf{x})\mathbf{y}$ are the discretized equations of equilibrium via finite element analysis. Note that $\mathbf{x}$ is the vector of design variables, $\mathbf{K}$ is the global stiffness matrix, $\mathbf{M}$ is the global mass matrix, $\mathbf{z}$ is the vector of nodal degrees of freedom, $\mathbf{y}$ is the eigenvector, and λ is the eigenvalue. Similarly, buckling and transient response equations may be included. Basis shape vectors are defined as

$$q^i = \frac{dG}{dx_i}$$

where

$$G(\mathbf{x}) = G_{\text{old}} + \sum_{i=1}^{k} x_i q^i$$

In the preceding equation, $\mathbf{G}(\mathbf{x})$ is the vector of x, y, z coordinates of all the grid points and $df/d\mathbf{x}$ and $dg_i/d\mathbf{x}$ are sensitivity vectors for objective and ith constraint functions, respectively.

The design sensitivity coefficient is interpreted as the slope of structural weight or other response functions such as displacement or stress with respect to a design variable. Chapter 6 gives a flavor of the sensitivity analysis procedure for stress and displacement constraints. A comprehensive background on semi-analytical as well as other methods for sensitivity analysis is given by Haug et al. (1988).

SHAPE OPTIMIZATION ALGORITHM

Let us state a step-by-step algorithm (Belegundu and Rajan 1988) to solve the shape optimization problem defined earlier using the fictitious loads approach. An auxiliary structure is defined to be a modified finite element model where artificial boundary conditions are applied to the nondesignable boundary while generating the basis vectors. Note that the auxiliary structure may have different material properties than the original problem. Thus basis vectors need to be consistent with the designer's intent of changing only that part of the boundary that is designable during shape design. A similar algorithm is programmed in NASTRAN as well as GENESIS programs for shape optimization.

1. Let $\mathbf{G}^{(i)}$ be the ith cycle. Set $\mathbf{x}^{(i)} = 0$ since only change in shape is of interest. Analyze the structure using the finite element method to predict stresses and strains.
2. Apply unit concentrated or pressure loads at control nodes in the auxiliary structure, one at a time, and solve for $\mathbf{q}$ as

$$\mathbf{k}_a \mathbf{q}^j = \mathbf{1}^j$$

Thus establish the [**Q**] matrix as $\Sigma\, \mathbf{q}^j$. Note that $\mathbf{k}_a$ is the global stiffness matrix for the auxiliary structure as exemplified in Fig. 13-7(b) which is obtained by modifying the original global stiffness matrix by appropriate boundary conditions. The right-hand side force vector $\mathbf{1}^j$ is for the jth control node in the appropriate degree of freedom.

3. Compute $df/d\mathbf{x}$ and $dg_i/d\mathbf{x}$ sensitivity coefficients.
4. Construct the linearized subproblem (as described in Appendix B) by making use of finite element and design sensitivity analysis results as

Minimize

$$f(\mathbf{x}^{(i)}) + \left(\frac{df}{d\mathbf{x}}\right) \Delta\mathbf{x}^{(i)}$$

Such that

$$g_i(\mathbf{x}^{(i)}) + \left(\frac{dg_i}{d\mathbf{x}}\right) \Delta\mathbf{x}^{(i)} \leq 0; \qquad (i = 1, p)$$

$$\Delta\mathbf{x}_i^L \leq \Delta\mathbf{x}_i \leq \Delta\mathbf{x}_i^U; \qquad (i = 1, k)$$

Solve for the design variable values from Step 4 and update the shape as

$$G^{(i+1)} = G^{(i)} + \{\mathbf{Q}\}\Delta\mathbf{x}^{(i)}$$

6. If $\Delta\mathbf{x}^{(i)} = \mathbf{0}$ (i.e., no appreciable change in design variables), then convergence has been obtained; otherwise go to Step 1.

NUMERICAL EXAMPLES

Transmission Tower

The 47 member planar tower shown in Fig. 13-1 is optimized for shape by Vanderplaats and Moses (1972) considering three separate loading conditions. The first loading condition represents a load imposed by two power lines coming into the tower at an angle. The second and third load conditions represent cases when one of the two lines has snapped. The weight of the transmission tower needs to be minimized subject to stress and Euler buckling constraints. There are a total of 27 independent member sizing variables and 17 independent coordinate design variables. The geometry variables are linked to maintain symmetry about the *Y*-axis. The weight of the original transmission tower has been reported to be 2450 lbm whereas the optimum transmission tower weighs about 1830 lbm. The SLP method has been used as optimizer. Note that in this example there is no need to implement the reduced basis approach since nodal coordinates of a discrete structure are treated as design variables. In general, discrete structures such as a truss or building frame have nodal coordinates as shape design variables and hence there is no need to generate basis vectors.

Plate with a Square Hole

Let us consider the problem of finding the optimum shape of a hole in a square plate subject to uniformly distributed loads along its edges as shown in Fig. 13-5. The objective is to minimize weight of the plate subject to a limit on the Von Mises stress. The data for the problem are given as:

Plate thickness (t)	= 0.0305 in.
Load (p)	= 10 lb/in.
Young's modulus (E)	= 3.0E + 07 psi
Poisson's ratio (μ)	= 0.3
Mass density (ρ)	= 0.734E-03 lbm/in.^{3}
Von Mises stress limit (σ_a)	= 1000 psi

For the square hole shown in Fig. 13-5(a), the mass of the entire plate is 2.8655E-03 lbm for the original shape. Since the thickness of the plate is very small as compared to in-plane dimensions of the plate, it is appropriate to model this problem as a plane stress approximation. In the plane stress assumption of elasticity, out-of-plane components of the stress tensor are assumed to be zero.

The plate is modeled using quadrilateral plane stress elements. Due to symmetry about both the x and y axes, only a quarter model of the original plate need be used in our study. If we were to choose thickness of the plate as a design variable, then it would become a sizing problem. Since we want to design the shape of the hole, let us keep the thickness fixed at 0.0305 in. and choose radial distances of the boundary nodes of the hole (from 1 to 9) as design variables. There are a total of nine design variables. Note that there is no need to compute basis vectors in this example since the reduced basis approach is not used. Instead, distances of control nodes from the origin are used as design variables.

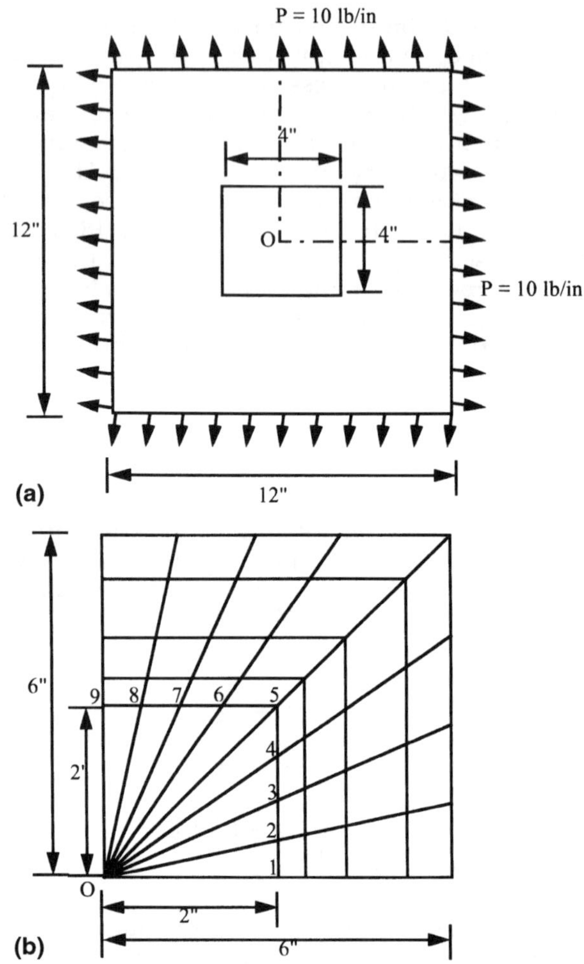

FIG. 13-5. Shape Design of Square Opening: (a) Geometry and Loads for Square Opening; (b) Quarter Symmetry Finite Element Mesh.

The iterative progress of the shape design of the hole is shown in Fig. 13-6 at Iterations 5, 11, and 14. The optimal solution is obtained with a radius of the hole as roughly 3.95 in giving mass of the plate as 2.1382E-03 lbm in 14 iterations. The mass is reduced by 25% during the shape design process. The initial design is feasible but, at the optimum design, the Von Mises stress constraint is active which means Von Mises stress attains the value of 1000 psi at some locations in the plate model. The optimum design variable values (in inches) are given in Table 13-1.

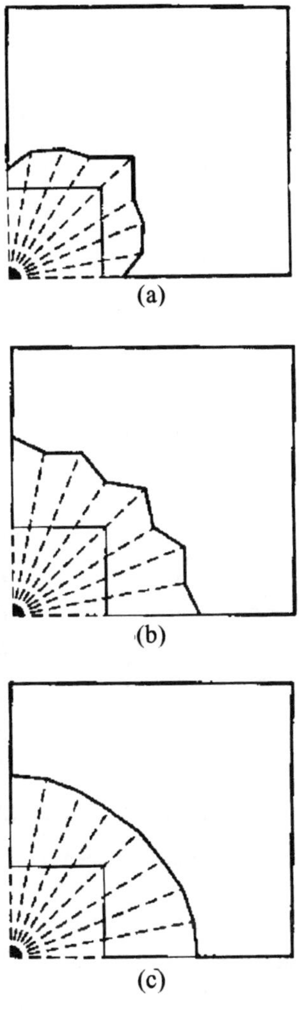

FIG. 13-6. Iterative History of Shape Design of Square Opening: (a) Iteration No. 5; (b) Iteration No. 11; (c) Iteration No. 15.

Culvert

The shape design of a culvert is considered by Belegundu et al. (1991) to demonstrate the usefulness of a limited number of basis shape vectors to achieve an optimum shape. In other words, in this example the reduced basis approach has been implemented. A half-symmetry model of the culvert is shown in Fig. 13-7(a). Pressure loads are applied on top of the structure representing the pavement loads. This example is treated as a plane strain problem. In the plane

TABLE 13-1

Design Variable No.	Value
1	4.000
2	3.976
3	3.929
4	3.902
5	3.913
6	3.902
7	3.929
8	3.976
9	4.000

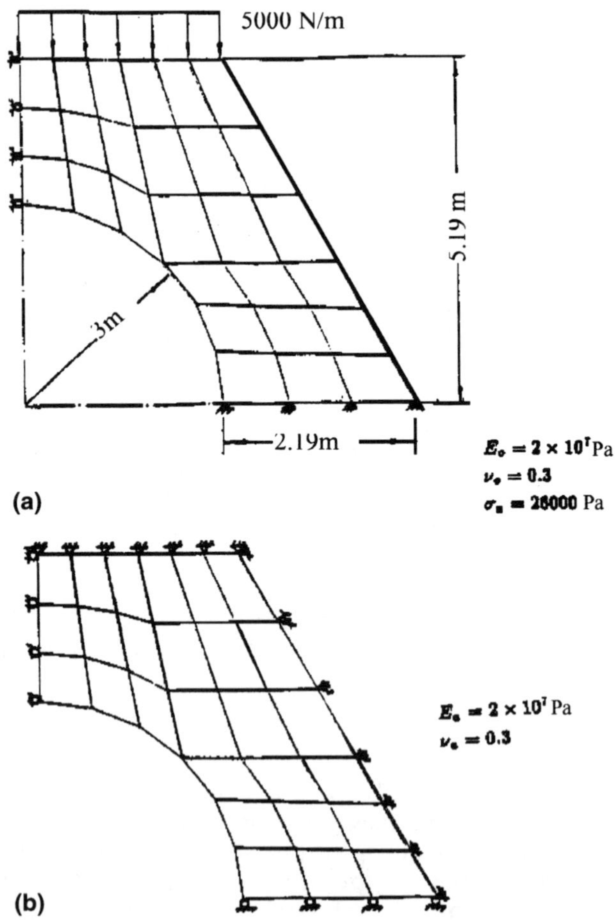

FIG. 13-7. Culvert Design Problem: (a) Primary Structure; (b) Auxiliary Structure.

strain assumption of elasticity, out-of-plane components of the strain tensor are assumed to be zero.

The auxiliary structure that has been used to find the basis shape vectors is shown in Fig. 13-7(b). Note that additional boundary conditions are applied on the nondesignable parts of the original model so that a linear combination of the basis vectors allows only the permissible shape changes in the culvert geometry. Belegundu et al. (1991) have chosen the first six eigenmodes of the auxiliary model of the culvert to be basis shape vectors as shown in Fig. 13-8. Note that it is also valid to generate basis shape vectors as the deformation pattern by applying unit force or pressure load at key nodes of the inner boundary of the culvert. The design formulation is to minimize the volume of the culvert subject to Von Mises stress in the plane strain elements of the finite element model. The inner hole is initially in the form of a circular arc and is free to change, whereas outside edges are fixed. The optimum shape of the culvert is shown in Fig. 13-9.

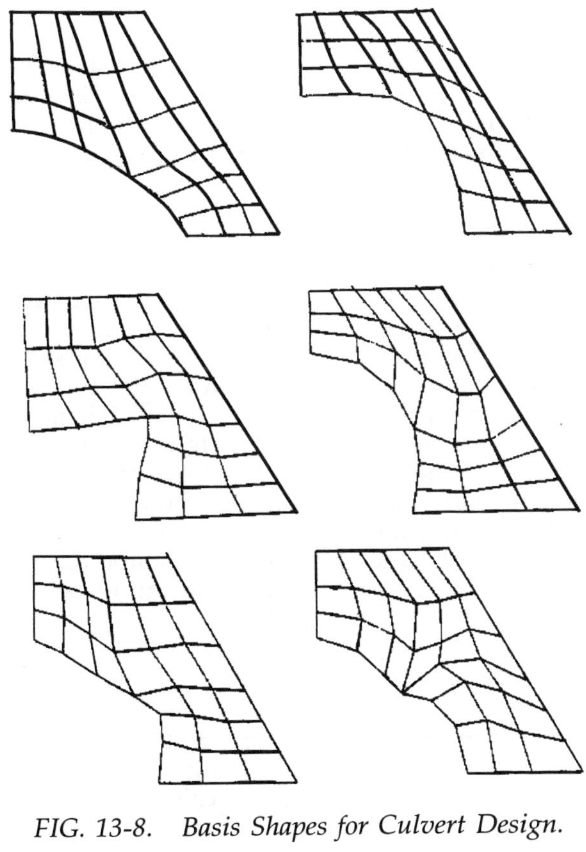

FIG. 13-8. Basis Shapes for Culvert Design.

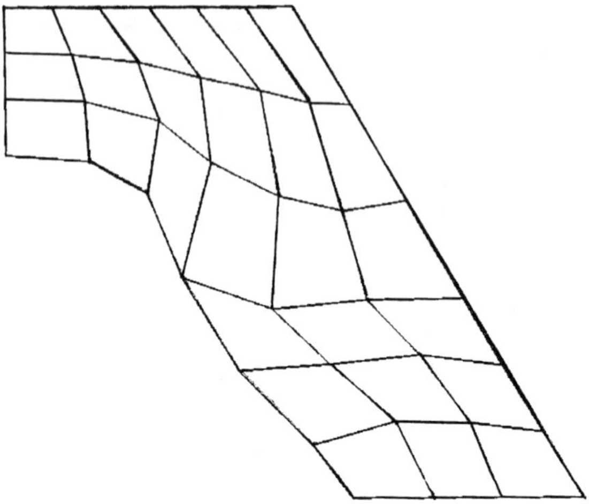

FIG. 13-9. Optimum Shape of Culvert.

The finite element model of half-culvert consists of 27 quadrilateral finite elements and 40 nodal points. The original weight is 0.01198 lbm whereas the final weight is 0.01039 lbm after 12 design iterations, which represents a 13% reduction in weight. The material properties and geometrical dimensions of the culvert are given in Fig. 13-7(a).

Solid Cantilever Beam

A cantilever beam with a circular hole is shown in Fig. 13-2. The shape optimization problem that is considered by Thanedar and King (1993) is to minimize the volume of the beam subject to constraints on the vertical displacements and effective stress at several locations of the finite element model. The design model has a total of four parametric design variables, one each to model height at the fixed and free ends, one for the mid-section height, and the last variable to control the radius of the hole. The initial design is a feasible design with a volume of 18,051 cubic mm whereas the final volume is 13,070 cubic mm. The finite element model consists of eight high-order polynomial-type hexahedral elements. The shape optimization algorithm (SQP method) converged in eight design iterations.

SUMMARY AND DISCUSSION

The problem of finding an optimum structural shape is discussed in this chapter. Along with analysis, design sensitivity analysis, and optimization algorithm computations, calculation of basis vectors may be required in shape optimization. The reduced basis method is presented in the chapter where an improved shape is obtained by linear combination of the basis shape vectors and scale factors for basis vectors are chosen as design variables. A rudimentary shape optimization approach is presented. Nodal coordinates may be chosen as the shape design variables for discrete structures which simplifies their formulation. Four shape design optimization problems are presented in the chapter consisting of a transmission tower, solid cantilever beam with a hole, culvert, and planar plate with a square hole. The example problems are selected so that discrete, continuum, plane stress, and plane strain finite element models are optimized for shape.

APPENDIX 13-A. NOTATION

The following symbols are used in the main part of this chapter. Additional symbols are defined within the numerical examples.

$\mathbf{x}$ = vector of design variables
k = number of design variables
g_i = ith constraint
$\mathbf{z}$ = vector of degrees of freedom in the finite element model
$\mathbf{y}$ = eigenvector
λ = eigenvalue
$\mathbf{G}$ = vector of nodal x, y, z coordinates
$\mathbf{K}$ = global stiffness matrix
$\mathbf{F}$ = vector of nodal loads
$\mathbf{M}$ = global mass matrix
$\mathbf{q}^i$ = ith basis vector
f = objective function
$\mathbf{Q}$ = matrix of basis vectors arranged as columns

REFERENCES

Belegundu, A. D. and Rajan, S. D. (1988). "A shape optimization approach based on natural design variables and shape functions." *Computer Methods in Applied Mechanics and Engng.*, 66, 87–106.

Belegundu, A. D., Zhang, S., Manicka, Y., and Salagame, R. (1991). "The natural approach for shape optimization with mesh distortion control." Technical Report, Pennsylvania State University, University Park, PA.

Choi, K. K. and Chang, K. - H. (1991). "Shape design sensitivity analysis and what-if work-station for elastic solids." *Proceedings of the SDM Conference*, Baltimore, MD, April 8–10, AIAA, Washington, DC, 578–587.

Haug, E. J., Choi, K. K., and Komkov, V. (1988). *Design sensitivity analysis of structural systems.* Academic Press, New York.

Kodiyalam, S., Kumar, V., and Finnigan, P. M. (1992). "Constructive solid geometry approach to 3-dimensional structural shape optimization." *AIAA J.*, 30(5), 1408–1415.

Pickett, R. M., Rubinstein, M. F., and Nelson, R. B. (1973). "Automated structural synthesis using reduced number of design co-ordinates." *AIAA J.*, 11(4), 489–494.
Thanedar, P. B. and King, R. (1993). "Shape optimization using adaptive high order finite elements." *J. of Structural Optimization*, 6(1), 189–194.
Vanderplaats, G. N. and Moses, F. (1972). "Automated design of trusses for optimum geometry." *J. of Structural Engng.*, ASCE, ST3, 671–690.

Appendix A

BASIC CONCEPTS OF OPTIMUM DESIGN

Basic concepts related to optimum design of structural components and systems are described. Local and global minima are defined and existence of solution for the design optimization model is discussed. Optimality conditions for the model are presented. Convexity of the design model is discussed and the question of a global solution for the problem is addressed. Post-optimality analyses are also briefly described.

INTRODUCTION

In this Appendix A, some basic concepts related to optimum design of structural components and systems are presented. (Concepts related to numerical algorithms for optimum design are presented in Appendix B). Local and global minima are defined and existence of solution for the design optimization model is discussed. Optimality conditions for the model are presented and discussed. Convexity of the design model is explained and the question of a global solution for the problem is addressed. Post-optimality analyses are also briefly discussed. Material for the chapter is taken from several books and journal articles, for example, Haug and Arora (1979), Arora (1989, 1990), and several references cited therein.

The concepts presented here are important in practical applications because most of the numerical methods for optimization utilize them. They are also important because they give insights if the optimization process fails to produce a useful solution.

Jasbir S. Arora, M. ASCE, Professor, Civil and Environmental Engineering, The University of Iowa, 1129 EB, Iowa City, IA 52242; email: ⟨Jasbir-Arora@uiowa.edu⟩.

DESIGN OPTIMIZATION MODEL

Formulation of the problem is an important first step in the design optimization process. Problems of static and dynamic response, shape optimization of structural components and systems using finite element and other simulation methods, reliability-based design, optimal control of systems, and many others can be formulated as so-called *nonlinear programming* (NLP) problems. Some of the recent books and articles that can be consulted about formulation of various classes of problems are Haug and Arora (1979), Kirsch (1981), Vanderplaats (1984), Rao (1984), Frangopol (1985), Arora and Thanedar (1986), Rozvany (1988), Arora (1989, 1990), Haftka and Gürdal (1992), and Frangopol and Moses (1994).

The problem formulation process requires identification of design variables (that describe the system), a *cost function* or an *objective function* (that needs to be optimized), and constraints for safe and proper performance of the system (Arora 1989). Depending on the class of problems and system performance requirements, several types of design variables and cost functions can be identified. Constraints usually involve physical and material limitations, buckling and other failure load levels, vibration frequencies, stresses, deflections, and a variety of other response quantities. The design problem can be transcribed into a standard NLP model (i.e., an optimization problem where some of the problem functions are nonlinear) defined as minimization of a function subject to equality constraints and inequality constraints expressed in a "less-than-or-equal-to" form as

Problem P

Find a design variable vector $\mathbf{X} = (X_1, X_2, \ldots, X_n)$ to minimize a cost function $f(\mathbf{X})$ subject to

$$\text{Equality Constraints} \quad g_j(\mathbf{X}) = 0, \qquad j = 1 \text{ to } p \tag{A-1}$$

$$\text{Inequality Constraints} \quad g_j(\mathbf{X}) \leq 0, \qquad j = p + 1 \text{ to } m \tag{A-2}$$

Note that the number of equality constraints must be less than or at the most equal to the number of design variables; otherwise, there may not be any solution for the problem. There is no restriction on the number of inequality constraints.

Collection of all the points that satisfy the constraints of Eqs. (A-1) and (A-2) is called the *feasible set* (also called the *constraint set*). This set is denoted by S. Note that most optimization problems can be transformed into the form of Problem P; that is, maximization of $F(\mathbf{X})$ can be replaced by minimization of $f(\mathbf{X}) = -F(\mathbf{X})$ and $G_i(\mathbf{X}) \geq 0$ can be replaced by $g_i(\mathbf{X}) = -G_i(\mathbf{X}) \leq 0$. In the numerical optimization process, the explicit lower and upper bounds on the design variables are easy to impose. Therefore they are generally treated separately in computer implementation of the algorithms for the sake of numerical efficiency. However, for the sake of simplicity of presentation, the explicit bound constraints are assumed herein to be included in the inequality constraints $g_j(\mathbf{X}) \leq 0$.

An inequality constraint $g_i(\mathbf{X}) \leq 0$ is said to be *active* (critical, tight, or binding) at a design point $\mathbf{X}$ if it is satisfied as an equality at that point, that is, $g_i(\mathbf{X}) = 0$. It is said to be *inactive* if it has negative value at that point, and *violated* if it has positive value. Note that an equality constraint is always either active or violated for any design point.

If all the functions of the problem are linear in terms of the variables, the Problem P is called a *linear programming* (LP) problem. If the cost function is quadratic and the constraints are linear, the problem is called a *quadratic programming* (QP) problem. There are many practical applications that can be formulated as LP or QP problems. In addition, most of the numerical algorithms for nonlinear problems solve either an LP or QP subproblem in their iterative process. Therefore these problem areas are quite important, and it is not surprising that a substantial amount of research has been performed and presented in the literature to develop special methods for their solution.

It is important to note the following points for the foregoing NLP model relative to structural and mechanical system design problems.

1. The model is applicable to all problems with continuous design variables. *Multiobjective* (Osyczka 1984) and *discrete variable* (Arora et al. 1994) problems can also be treated after certain extensions of the model.
2. The functions of the problem are assumed to be *twice differentiable*. Problems having nondifferentiable functions can be treated with additional computational effort (Schittkowski 1985). Also, gradients of active constraints are assumed to be linearly independent at the optimum.
3. The cost and/or constraint functions may be *implicit* as well as *explicit functions* of the design variables; that is, their final form in terms of only the design variables may not be known. The functions, however, can be evaluated using analysis computer programs once a design is specified.
4. *Derivatives* of the functions are needed in numerical methods of optimization. Efficient methods to calculate them taking advantage of the structure of engineering design problems have been developed (Haftka and Adelman 1989).

BASIC DEFINITIONS

Before presenting various concepts related to design optimization, it is prudent to define what is meant by a minimum point for the function $f(\mathbf{X})$, and discuss the existence of the optimum solution for the Problem P.

Local Minimum

The cost function $f(\mathbf{X})$ has a local minimum (also called a relative minimum) at a point $\mathbf{X}^*$ in the feasible set S if the function value is the smallest at the point $\mathbf{X}^*$ compared to all other points $\mathbf{X}$ in a feasible neighborhood of $\mathbf{X}^*$; that is,

$$f(\mathbf{X}^*) \leq f(\mathbf{X}) \tag{A-3}$$

for all $\mathbf{X}$ in a small feasible neighborhood of the point $\mathbf{X}^*$. If strict inequality holds, then $\mathbf{X}^*$ is called the strict or isolated local minimum.

Global Minimum

The cost function $f(\mathbf{X})$ has a global minimum (also called an absolute minimum) at a point $\mathbf{X}^*$ if Inequality (A-3) holds for *all* $\mathbf{X}$ in the feasible set S. If strict inequality holds, then $\mathbf{X}^*$ is called the strict or unique global minimum.

These definitions show that for the local minimum, we test Inequality (A-3) only for a small feasible region around the point $\mathbf{X}^*$, whereas we test it over the entire feasible set S for the global minimum point.

Note that if there are no constraints on the system, then the feasible set for the problem is the entire design space. The preceding definitions hold for such unconstrained problems as well. Also, the cost function $f(\mathbf{X})$ can have many global minimum points (i.e., the function values are the same at all the points). Similarly, there can be several local minima in a small feasible region, and so the local minimum is not isolated in that case.

Generally, it is difficult to know if there is a solution to the Problem P. The question of *existence of a solution* is important, however. It turns out that if $f(\mathbf{X})$ is continuous on a nonempty feasible set S, the constraint functions are continuous, and all inequalities contain their boundary points [i.e., if there are no strict inequalities as $g_j(\mathbf{X}) < 0$], then there is a global minimum point for $f(\mathbf{X})$ in S (Haug and Arora 1979). These conditions guarantee existence of a global minimum point; however, if the conditions are not satisfied, then a global minimum may exist, but its existence cannot be guaranteed.

OPTIMALITY CONDITIONS

The foregoing definitions of local and global minima do not provide any way of finding the minimum points for the Problem P. Moreover, they cannot be used to check optimality of a given point without requiring a large number of calculations because the Inequality (A-3) needs to be checked at an infinite number of points. However, the definitions can be used to derive conditions that characterize a minimum point. These are called the optimality conditions.

There are two types of optimality conditions: necessary and sufficient. A local minimum point must satisfy the necessary conditions; otherwise it cannot be a minimum point. Note, however, that a point satisfying these conditions need not be a minimum point (i.e., points that are not minima may also satisfy the necessary conditions). The points that satisfy the necessary conditions are candidate minima only, or simply stationary points. The sufficient condition, if satisfied, proves the point to be indeed a minimum point. We describe both the necessary and sufficient conditions. Sample problems showing the use of these conditions are presented in Arora (1989).

The optimality conditions are important in these ways: they can be used (1) to develop numerical methods for finding minimum points [this has been done extensively in the literature, e.g., Rozvany (1988)], and (2) to check optimality of a given point [this has been used as a main stopping criterion for iterative numerical algorithms (Arora 1989)]. We first present the optimality conditions for the unconstrained problem and then for the constrained Problem P.

Unconstrained Problem

The problem here is to simply minimize the cost function $f(\mathbf{X})$ without any constraints. The optimality conditions for such problems are derived by analyzing the local behavior of the cost function $f(\mathbf{X})$ at the minimum point $\mathbf{X}^*$ (Arora 1989). These conditions are stated as follows.

Necessary Condition: First Order. If $\mathbf{X}^*$ is a local minimum for the cost function $f(\mathbf{X})$, then its first derivatives at $\mathbf{X}^*$ must vanish; that is,

$$\frac{\partial f}{\partial X_i} = 0; \qquad i = 1, 2, \ldots, n$$

Necessary Condition: Second Order. If $\mathbf{X}^*$ is a local minimum for the function $f(\mathbf{X})$, then its Hessian $\mathbf{H} = [\partial^2 f/\partial X_i\, \partial X_j]$ at $\mathbf{X}^*$ must be at least positive semidefinite; that is, all its eigenvalues must be nonnegative.

Sufficient Condition: Second Order. If the matrix $\mathbf{H}(\mathbf{X}^*)$ is positive definite at the stationary point $\mathbf{X}^*$, then $\mathbf{X}^*$ is an isolated local minimum point. (A matrix is called positive definite if all its eigenvalues are positive.)

Note that the foregoing conditions involve derivatives of the function at the point $\mathbf{X}^*$. As such, the minimum point is not changed if a constant is added to the cost function, or if it is scaled by a positive scalar. The optimum value of the cost function is, of course, changed in the process. Also note that any point satisfying the necessary conditions is called a *stationary point*. If a stationary point is neither a minimum nor a maximum, then it is called an inflection point.

Constrained Problem

The optimality conditions for a constrained problem must involve the constraints in addition to the cost function. Although a constrained problem can have minimum points where no constraints are active (i.e., none of the constraints is binding), this usually does not happen in practical design problems. In case no constraints are active, the optimum point is inside the feasible set S and the optimality conditions for the unconstrained problem apply. In other words, the optimality conditions for the unconstrained problem are a special case of those for the constrained problem. The conditions for the constrained problem can be expressed in several alternate but equivalent ways. We present the conditions that are most commonly used in the literature, known as Karush-Kuhn-Tucker or KKT conditions.

A basic assumption in the derivation of the KKT conditions is that the minimum point is a regular point of the feasible set S. A point $\mathbf{X}$ is called a *regular point* of the feasible set S if gradients of all the active constraints at the point $\mathbf{X}$ are linearly independent and the cost function is differentiable. Since the number of linearly independent vectors cannot be more than n (the number of design variables), the total number of active constraints cannot be more than the number of design variables at the regular point.

Karush-Kuhn-Tucker Necessary Conditions. Let the Lagrangian for the Problem P be defined as

$$L(\mathbf{X}, \mathbf{u}) = f(\mathbf{X}) + \sum_{j=1}^{m} u_j g_j \tag{A-4}$$

where u_j is the Lagrange multiplier for the constraint g_j that needs to be determined. Also, let $\mathbf{X}^*$ be a regular point of the feasible set S that is a local minimum for $f(\mathbf{X})$. Then there exist unique Lagrange multipliers u_j^* such that

$$\frac{\partial L(\mathbf{X}^*)}{\partial \mathbf{X}_i} = 0, \quad \text{or} \quad \frac{\partial f(\mathbf{X}^*)}{\partial \mathbf{X}_i} + \sum_{j=1}^{m} u_j^* \frac{\partial g_j(\mathbf{X}^*)}{\partial \mathbf{X}_i} = 0, \qquad i = 1, 2, \ldots, n \tag{A-5}$$

$$g_j(\mathbf{X}^*) = 0, \quad j = 1 \text{ to } p; \qquad u_j^* g_j(\mathbf{X}^*) = 0, \quad j = (p+1) \text{ to } m \tag{A-6}$$

$$g_j(\mathbf{X}^*) \le 0, \qquad u_j^* \ge 0, \qquad j = (p+1) \text{ to } m \tag{A-7}$$

Note that there are $(n + m)$ unknowns (n design variables and m Lagrange multipliers) and $(n + m)$ equations [n equations in conditions (A-5) and m equations in condition (A-6)]. Therefore the necessary conditions give enough equations for solution of all the unknowns, though they are usually nonlinear. Equations (A-5) show that the Lagrangian L is stationary with respect to $\mathbf{X}$; Eq. (A-6) shows that either the Lagrange multiplier for the jth inequality is zero or the constraint is active at the minimum point (if $u_j = 0$, g_j must be ≤ 0 for feasibility of $\mathbf{X}^*$), and Eq. (A-7) shows that the Lagrange multipliers for the inequality constraints must be nonnegative (the physical meaning of the Lagrange multipliers is explained later). Equation (A-6) is called the *switching condition* or complementary slackness condition because it identifies active and inactive inequality constraints.

Equation (A-5) can also be written differently to gain some physical insight about the necessary condition:

$$-\frac{\partial f(\mathbf{X}^*)}{\partial \mathbf{X}_i} = \sum_{j=1}^{m} u_j^* \frac{\partial g_j(\mathbf{X}^*)}{\partial \mathbf{X}_i}, \qquad i = 1, 2, \ldots, n \tag{A-8}$$

This equation shows that at the minimum point, the negative of the gradient of the cost function (the steepest descent direction) is a linear combination of the gradients of the constraints with Lagrange multipliers as the scalars of the linear combination.

It is important to note that the KKT conditions may or may not be satisfied at points that are not regular. Also, at such points, the Lagrange multipliers may or may not be unique. Stated differently, irregular points can also be local minimum points where the KKT conditions may be violated.

Solutions of the KKT conditions are candidate local minimum designs. Second-order conditions can be used to distinguish the minimum points from others. These conditions also involve Hessians of the functions, as for the unconstrained problems. In the constrained case, we must also consider active constraints at $\mathbf{X}^*$. Thus the second-order conditions involve the Hessian of the Lagrangian function. These conditions are discussed and illustrated in Arora (1989).

GLOBAL OPTIMALITY

Once the optimum design of a system is obtained, the question arises about its global optimality. Usually, the answer to this question is that the solution is only a local minimum. The global solution for the problem can be found by either some exhaustive search of the feasible set, or by showing the problem to be convex. Both procedures can require extensive computations. *If the problem is convex, then any local minimum is also a global minimum and the KKT first-order conditions are necessary as well as sufficient.* The question of convexity of a problem is briefly addressed here.

If the cost function $f(\mathbf{X})$ is convex over the convex feasible set S, then the Problem P is called a convex programming problem. To show convexity of a problem, we need to show convexity of the set S and the function $f(\mathbf{X})$. A set of points S (vectors $\mathbf{X}$) is called a convex set iff for any two points A and B in the set S, the entire line segment AB is also in the set. Graphically this means that a convex set has no holes or re-entrant corners. This definition is difficult to use to check convexity of a set because an infinite pair of points will have to be considered, so a computational procedure is needed. The feasible set S for the Problem P is defined by the constraints $g_i(\mathbf{X}) \le 0$, $i = 1$ to m. It turns out that if all $g_i(\mathbf{X})$ are convex, then the feasible set S is convex. Thus we need to know how to check convexity of a function.

Convex functions are defined over convex sets. *A function of n variables is convex iff its Hessian is at least positive semidefinite everywhere.* If a function $g_i(\mathbf{X})$ is convex, then the set defined by the inequality $g_i(\mathbf{X}) \le e_i$ is convex, where e_i is any constant (note that this is not an "iff" condition). *A linear equality or inequality constraint always defines a convex set.* In contrast, *a nonlinear equality constraint always defines a nonconvex set.* These results can be verified by using the preceding definition of a convex set. The nonlinear inequality constraints $g_i(\mathbf{X}) \le 0$ require a check of their Hessian.

Note that convexity checks for a problem are quite expensive. The Hessian of each function needs to be evaluated and its form needs to be checked over the entire feasible domain. The following points should be noted for convex programming problems.

1. A convex programming problem can have several global minimum points where the cost function has the same numerical value.
2. The convexity check for a constraint function can sometimes fail if the form of the constraint is altered; however, the feasible set defined by the constraint may still be convex. As an example, consider the constraint $g_1 = (a/x_1x_2) - b \le 0$, where $a > 0$ and $b \ge 0$ are specified constants, and $x_1 > 0$ and $x_2 > 0$. The function g_1 satisfies the foregoing convexity check, so the feasible set defined by $g_1 \le 0$ is convex. If the constraint is rewritten as $g_2 = a - bx_1x_2 \le 0$ (which does not alter the feasible set), then g_2 fails the convexity check; however, this may not imply that the set defined by $g_2 \le 0$ is not convex. This discussion shows that the conditions given in the foregoing for checking convexity of the feasible set are only sufficient but not necessary; that is, the condition "if $g_i(\mathbf{X})$ is convex, then the set defined by the inequality $g_i(\mathbf{X}) \le e_i$ is convex" is only a sufficient condition but not necessary.
3. The problems for which convexity checks fail, or problems that are nonconvex, can also have global minimum points. Here it is difficult to know that a global solution has been obtained.

POST-OPTIMALITY ANALYSIS

Many times it is useful to know how the optimum solution changes as some of the parameters in the original problem are changed. Post-optimality analysis implies a study of such variations of the optimum solution. Variations of the optimum cost function, design variables, and Lagrange multipliers due to the variations of many parameters can be studied. Calculation of the new values of the design variables requires further computations that can be time consuming (Vanderplaats and Yoshida 1985). However, sensitivity of the cost function to the variations in the constraint limits can be obtained without any further calculations.

Changes in Constraint Limit

Assume that Problem P has been solved with the current limit values for the constraints as zero. Let e_j be a small variation in the right-hand side of the ith constraint. It is clear that the optimum point for the perturbed problem is a function of the vector $\mathbf{e}$; that is, $\mathbf{X}^* = \mathbf{X}^*(\mathbf{e})$, and $f = f(\mathbf{e})$. However, these are implicit functions of $\mathbf{e}$, and the following result gives a way of obtaining the implicit derivatives $\partial f/\partial e_j$ (Arora 1989; Haug and Arora 1979).

Sensitivity to Constraint Variations. Let $\mathbf{X}^*$ be a regular point that, together with the multipliers u_j^*, satisfies both the KKT necessary conditions and the sufficient conditions for an isolated local minimum point for the Problem P. If for each $g_j(\mathbf{X}) = 0$ for $j > p$, it is true that $u_j^* > 0$, then the solution $\mathbf{X}^*(\mathbf{e})$ of the modified problem is a continuously differentiable function of $\mathbf{e}$ in some neighborhood of $\mathbf{e} = \mathbf{0}$. Furthermore,

$$\frac{\partial f(\mathbf{X}^*(\mathbf{0}))}{\partial e_j} = -u_j^*; \qquad j = 1, 2, \ldots, m \tag{A-9}$$

Note that the foregoing result is applicable to both equality and inequality constraints; but the inequality constraints must be expressed in the "$\leq$" form and not in the "$\geq$" form. It is useful to note that if conditions stated in the result are not satisfied, existence of implicit derivatives of Eq. (A-9) cannot be ruled out. That is, the derivatives may still exist but their existence cannot be guaranteed.

Using Eq. (A-9), we can estimate a change in the cost function due to a change in the right-hand side of the jth constraint. First-order Taylor expansion for the cost function about the point $e_j = 0$ is given as

$$f(e_j) = f(0) + \frac{\partial f(0)}{\partial e_j} e_j \tag{A-10}$$

where $f(0)$ is the optimum cost function value obtained with $e_j = 0$. Substituting from Eq. (A-9), we obtain the change in the cost function Δf due to the change e_j as

$$\Delta f = f(e_j) - f(0) = -u_j^* e_j \tag{A-11}$$

Equation (A-11) can also be used to show that the Lagrange multiplier corresponding to a "$\leq$ type" constraint must be nonnegative. To see this, let us assume that we want to relax an active inequality constraint $g_j \leq 0$ by selecting $e_j > 0$. Note that when an inequality is relaxed, the feasible set for the problem expands; that is, there are more feasible designs. Thus the minimum value for the cost function should reduce or stay unchanged with the expanded feasible set. However, Eq. (A-11) shows that if $u_j^* < 0$, relaxation of the constraint ($e_j > 0$) results in an increase in cost (i.e., $\Delta f = u_j^* e_j > 0$). This is physically and theoretically impossible. Therefore, the Lagrange multiplier for a "$\leq$ type" constraint cannot be negative.

Equation (A-11) also shows that constraints with larger Lagrange multipliers have greater effect on the cost function if their boundaries are changed, and constraints with smaller Lagrange multipliers have very little effect. This gives a clue to the designer to see if the constraints with larger Lagrange multipliers can be relaxed because, this way, the optimum cost function may be substantially reduced. Thus we see that Lagrange multipliers contain useful information that can have substantial impact in practical applications.

Sometimes changes to a constraint function are dependent on some parameter s; that is, $e_j = e_j(s)$. In this case, we like to calculate change in the cost function Δf if the parameter s is changed by Δs. Using the chain rule of differentiation, the partial derivative of f with respect to s after using Eq. (A-9), is given as

$$\begin{aligned} \frac{\partial f(0)}{\partial s} &= \frac{\partial f(0)}{\partial e_j}\frac{de_j}{ds} \\ &= -u_j^* \frac{de_j}{ds} \end{aligned} \tag{A-12}$$

Now the change in the cost function is given as

$$\Delta f = -u_j^* \Delta s \frac{de_j}{ds} \tag{A-13}$$

Scaling of Cost Function

Many times, the cost function for the problem needs to be scaled by a positive constant. Although this scaling does not affect the optimum point, it does change the cost function value and the Lagrange multipliers for the constraints. Using the KKT conditions of Eq. (A-5), it can be shown that all the Lagrange multipliers also get multiplied by the same scale factor. Let u_j^* be the Lagrange multiplier for the jth constraint with the original cost function. Let the cost function be scaled as $f^{\text{new}} = af$, where $a > 0$ is a given constant. Let u_j^{new} be the new value of the Lagrange multiplier for the jth constraint at optimum. Then the new and old Lagrange multipliers are related as

$$u_j^{\text{new}} = au_j^*; \qquad j = 1, 2, \ldots, m \tag{A-14}$$

Scaling of Constraint

Many times, a constraint is scaled by a positive constant (e.g., to normalize it). The scaling of a constraint does not change its boundary, so it has no effect on the optimum point or the cost function. However, the Lagrange multiplier for the constraint is affected. Using the KKT conditions of Eq. (A-5), it can be shown that the Lagrange multiplier for the scaled constraint gets divided by the same scale factor. Let the jth constraints g_j be scaled by a positive constant b_j as $g_j^{\text{new}} = b_j g_j$, and u_j^* and u_j^{new} be the corresponding Lagrange multipliers for the original and the scaled constraints, respectively. The new Lagrange multipliers are given as

$$u_j^{\text{new}} = \frac{u_j^*}{b_j} \tag{A-15}$$

Note that the scale factor can be any nonzero number, but for the inequality constraints it must be strictly positive.

SUMMARY AND DISCUSSION

Basic concepts related to design optimization of engineering systems are described and discussed. These include optimality conditions for unconstrained and constrained problems, existence of a solution, convexity of a problem, global optimality, and post-optimality analysis. Various results are described and discussed, and their implications for practical applications are delineated. Most of the numerical methods utilize these concepts in their iterative process. Therefore it is useful to understand the concepts for proper application of optimization techniques to practical design problems.

NOTATION

$*$ = a superscript indicating optimum value of a variable
$\mathbf{e}$ = vector of changes in the constraint functions

e_j = small change in the right-hand side of the jth constraint function
f = cost function
g_i = ith constraint function
$\mathbf{H}$ = Hessian matrix (symmetric) consisting of second partial derivatives of a twice continuously differentiable function
KKT = Karush-Kuhn-Tucker
L = Lagrangian function
m = total number of constraints
n = number of design variables
p = number of equality constraints; or total number of active constraints
S = constraint set; set of feasible designs; feasible set
$\mathbf{u}$ = Lagrange multiplier vector
u_i = Lagrange multiplier for the ith constraint
$\mathbf{X}$ = design variable vector of dimension n
Δf = change in the cost function

REFERENCES

Arora, J. S. (1989). *Introduction to optimum design*. McGraw-Hill, New York.

Arora, J. S. (1990). "Computational design optimization: A review and future directions." *Structural Safety*, 7, 131–148.

Arora, J. S. and Thanedar, P. B. (1986). "Computational methods for optimum design of large complex systems." *Computational Mechanics*, 1, 221–242.

Arora, J. S., Huang, M. W., and Hsieh, C. C. (1994). "Methods for optimization of nonlinear problems with discrete variables: A review." *Structural Optimization*, 8, 69–85.

Frangopol, D. M. (1985). "Structural optimization using reliability concepts in design." *J. of Structural Engng.*, ASCE, 111(11), 2288–2301.

Frangopol, D. M. and Moses, F. (1994). "Reliability-based structural optimization." Chapter 13 in *Advances in design optimization*, H. Adeli, ed., Chapman and Hall, New York, 492–570.

Haftka, R. T. and Adelman, H. M. (1989). "Recent developments in structural sensitivity analysis." *Structural Optimization*, 1(3), 137–152.

Haftka, R. T. and Gürdal, Z. (1992). *Elements of structural optimization*. 3rd ed., Kluwer Academic Publishers, Dordrecht.

Haug, E. J. and Arora, J. S. (1979). *Applied optimal design: mechanical and structural systems*. Wiley-Interscience, New York.

Kirsch, U. (1981). *Optimum structural design*. McGraw-Hill, New York.

Osyczka, A. (1984). *Multicriterion optimization in engineering*. Halsted Press, Wiley, New York.

Rao, S. S. (1984). *Optimization: Theory and applications*. 2nd ed., John Wiley, New York.

Rozvany, G. I. N. (1988). *Structural design via optimality criteria*. Kluwer Academic Publishers, Hingham, MA.

Schittkowski, K., ed. (1985). Computational mathematical programming. *Proceedings of NATO ASI*, Bad Windsheim FRG, July 23–August 2 (1984), Springer-Verlag, Berlin.

Vanderplaats, G. N. (1984). *Numerical optimization techniques for engineering design: With applications*. McGraw-Hill, New York.

Vanderplaats, G. N. and Yoshida, N. (1985). "Efficient calculation of optimum design sensitivity." *AIAA J.*, 23(11), 1798–1803.

Appendix B

BASIC CONCEPTS OF COMPUTATIONAL METHODS FOR OPTIMUM DESIGN

This appendix describes some of the basic concepts related to the computational methods for optimum design of systems. Basic philosophy of the iterative methods for continuous variable optimization problems is presented and terminology associated with them is explained. Attributes of a good algorithm are delineated. Linearization of a nonlinear problem is explained as a basic step for many algorithms. Basic ideas of sequential linear programming and sequential quadratic programming algorithms are discussed. It is noted that several good optimization algorithms have become available recently. Therefore designers need not concern themselves with the development of new optimization algorithms. Rather, they should be concerned with the formulation and study of different aspects of their problem to obtain better and more efficient designs.

INTRODUCTION

Computational methods for design optimization of structural and mechanical systems have matured during the 1970s and 1980s and are beginning to be utilized in realistic engineering design applications. Many papers have been published that show applications of the methods. A recent review article (Arora 1990) cites 11 review papers, 19 textbooks, 15 conference proceedings, and 60 other journal articles. Also a recent book chapter on reliability-based optimum structural design cites numerous contributors to this field over the last 20 years (Frangopol 1995). This body of literature contains work

Jasbir S. Arora, M. ASCE, Professor, Civil and Environmental Engineering, The University of Iowa, 1129 EB, Iowa City, IA 52242; email: ⟨Jasbir-Arora@uiowa.edu⟩.

of almost all the researchers in the field. Many applications of optimization are described in these references.

Engineering design problems are usually quite complex because each application has its own design requirements, simulation methods, and constraints to meet. In addition, as more powerful computers become available, the desire to solve more complex and larger problems also grows. Furthermore, since the methods have matured substantially during the last decade, more nonexperts of optimization algorithms are beginning to use this new technology in their design work. All these factors dictate the use of a theoretically sound and numerically reliable algorithm. Use of such an algorithm in a routine design environment can remove uncertainty about the algorithm behavior, allowing the designers to concentrate on the formulation of their design problems. Such theoretically sound algorithms, although computationally more expensive, can be more cost-effective in the long run. In reality they can save human resources and facilitate an investigation of alternate design concepts in a relatively short time, thus resulting in better designs.

In this appendix, some basic concepts related to numerical algorithms for optimization of structural systems are presented and discussed (Arora 1989). The ideas of a descent function, constraint normalization, and potential constraint strategy are introduced. Convergence of an algorithm is discussed and attributes of a good algorithm are delineated. It is important to note that most of the algorithms converge only to a local minimum point for the Problem P defined in Appendix A. Algorithms for finding a global solution require extensive numerical calculations and are not discussed here. Note that this appendix should be read after reading Appendix A.

In many practical applications, the design variables cannot have arbitrary values due to manufacturing and fabrication limitations. For example, the plate thickness and width must be selected from the available ones, the number of bolts used must be an integer, the number of rebars must be an integer and their size must be selected from the available ones, and so on. Design problems with such variables are called discrete variable optimization problems in contrast to the continuous variable optimization problems where design variables can have any value within the specified limits. Several optimization methods have been investigated for such applications; recent articles by Arora et al. (1994), Huang and Arora (1996), and many references cited therein can be consulted for more details on discrete variable optimization methods. Also Appendices D, E, and F describe optimization methods that are suitable for discrete variable structural optimization problems. In this Appendix B, only continuous variable optimization algorithms are covered. It is noted, however, that the continuous variable algorithms are not only important in their own right but are also useful in many discrete variable optimization procedures. Many such procedures have certain phases in their computation: in the first phase a solution is obtained using a continuous variable formulation of the problem, and then in the second phase this solution is manipulated to assign discrete values to the design variables.

A BASIC ALGORITHM

Most of the optimization algorithms are based on the general iterative prescription:

$$\mathbf{X}^{(k+1)} = \mathbf{X}^{(k)} + \alpha_k \mathbf{d}^{(k)}; \qquad k = 0, 1, 2, \ldots \tag{B-1}$$

where the superscript k represents the iteration number, $\mathbf{X}^{(k)}$ is the current estimate of the optimum design, $\mathbf{X}^{(0)}$ is an initial design estimate given by the designer, $\alpha_k \mathbf{d}^{(k)}$ is a change in design, α_k is a step size, and $\mathbf{d}^{(k)}$ is a search direction calculated using function values and their gradients.

Optimization algorithms are broadly classified as direct and transformation methods. In *direct methods* the direction vector $\mathbf{d}^{(k)}$ is calculated using the function values and their gradients at the point $\mathbf{X}^{(k)}$. The step size calculation needs only the function values. Different algorithms can be generated depending on how the direction $\mathbf{d}$ and step size α are calculated. In many algorithms, $\mathbf{d}$ is calculated by solving a linear or quadratic programming subproblem. The *transformation methods* convert the Problem P to a sequence of unconstrained optimization problems whose solutions converge to a solution of the original problem. They include barrier and penalty function methods as well as the augmented Lagrangian or multiplier methods (Arora and Belegundu 1984; Arora et al. 1991; Belegundu and Arora 1984a, 1985). In these methods, a transformed function is constructed by adding a penalty term for the constraint violations to the cost function, as $\Phi(\mathbf{X}, r) = f(\mathbf{X}) + P(g(\mathbf{X}), r)$, where r is a scalar or vector of penalty parameters and P is a real valued function whose action of imposing the penalty is controlled by r.

In direct methods, it is assumed that the optimum solution lies on the boundary of the feasible design set even if there are only inequality constraints in the problem. If the solution is unconstrained, then unconstrained methods can be used to locate it. Several philosophies have been used to develop various algorithms. For example, if an intermediate design or the starting design is infeasible, many methods will iterate through the infeasible region to reach the final solution; many others correct the constraints to reach the feasible region first and then move along the constraint boundary to reach the solution point. Still others make special calculations not to violate constraints during the iterative process. Some algorithms generate and use second-order information about the problem as the iterations progress.

Several algorithms based on the strategies described in the foregoing paragraph have been developed and evaluated. Some algorithms are better for a certain class of problems than others. A few algorithms work well if the problem has only inequality constraints, whereas others can treat both equality and inequality constraints simultaneously. We concentrate only on general algorithms that have no restriction on the form of the functions or the constraints. Such algorithms are usually based on the following basic steps.

1. Linearization of cost and constraint functions about the current design. Some methods require linearization of all the constraints, whereas others only for a subset of them.
2. Definition of a search direction determination subproblem using the linearized cost and constraint functions.
3. Solution of the subproblem which gives a search direction in the design space.
4. Calculation of a step size to minimize a descent function (defined later) in the search direction.

BASIC CONCEPTS RELATED TO COMPUTATIONAL ALGORITHMS

Constraint Normalization

In numerical calculations, it is essential to normalize all the constraint functions. In this way their values can be easily compared to each other. If they are not normalized, it is not easy to determine which constraint is more severely violated compared to others. For example, consider a stress constraint as $\sigma \leq \sigma_a$, or $\sigma - \sigma_a \leq 0$ and a displacement constraint as $\delta \leq \delta_a$, or $\delta - \delta_a \leq 0$, where σ = calculated stress, σ_a = an allowable stress, δ = calculated deflection, and δ_a = an allowable deflection. Since the units for these two constraints are different, their values are of widely differing orders of magnitude. If they are violated during the iterative solution process, it is difficult to judge the severity of their violation. However, if they are normalized as $R - 1.0 \leq 0$, where $R = \sigma/\sigma_a$ for the stress constraint, and $R = \delta/\delta_a$ for the deflection constraint, then it is easy to compare their values (note that σ_a and δ_a are assumed to be positive; otherwise, the sense of the inequality will change).

There are other constraints that must be normalized as $1.0 - R \leq 0$. For example, the fundamental vibration frequency ω of a structure must be above a given threshold value of ω_a; that is, $\omega \geq \omega_a$. When the constraint is normalized and converted to the standard less-than form, it becomes $1.0 - R \leq 0$, where $R = \omega/\omega_a$.

Potential Constraint Strategy

To evaluate the search direction, one needs to know the cost and constraint functions and their gradients at the current design point.

The algorithms can be classified based on whether gradients of all the constraints or only a subset of them are required at a design iteration. The algorithms that use gradients of only a subset of the constraints are said to use a potential (or active) constraint strategy. A potential constraint set can be defined in several different ways. In general, it is comprised of the indices of active, nearly active, and violated constraints, such as the following index set I_k at the kth design point $\mathbf{X}^{(k)}$.

$$I_k = [i = 1 \text{ to } p \quad \text{and all} \quad i > p \text{ with } (g_i + \varepsilon) \geq 0] \tag{B-2}$$

where $\varepsilon > 0$ is a small number used to determine nearly active inequalities. Note that the equality constraints are always included in the index set I_k by definition.

Descent Function

A function used to monitor progress of the iterative optimization process towards the minimum is called the descent function or the merit function. The cost function is the obvious descent function for the unconstrained optimization problems. Use of the cost function, however, as a descent function for constrained optimization can be tedious because it does not account for the constraints of the problem. Therefore many other descent functions have been proposed and used for constrained problems. These functions must include the effect of constraint violations. Basically, the descent function is used in an optimization algorithm to determine an appropriate step size. The basic idea is to compute a step size along the search direction $\mathbf{d}^{(k)}$ such that the descent function is reduced at each iteration of the algorithm (note that the cost function for the problem may actually increase at some iterations). With this requirement proper progress towards the minimum point is maintained. The descent function also has the property that its minimum value is the same as that of the original cost function. A descent function is defined later in this appendix.

Convergence of an Algorithm

An algorithm is said to be convergent if it reaches a local minimum point starting from an arbitrary initial design. An algorithm that has been proven to converge starting from an arbitrary point is called a globally convergent method. Such an algorithm satisfies the following requirements.

1. There is a descent function for the algorithm that is required to decrease at each iteration. This way, progress towards the minimum point can be monitored.

2. The search direction $\mathbf{d}^{(k)}$ is a continuous function of the design variables. This requirement implies that there is no "oscillation," or "zigzagging," in the set of active constraints during the iterative design process (zigzagging implies that a constraint is coming in and out of the active set I_k in alternate iterations).

In addition to the preceding two requirements, existence of a solution for the problem must be guaranteed, which is discussed in Appendix A.

Attributes of a Good Algorithm

Based on the foregoing discussion a good algorithm has the following attributes (Arora 1989, 1990).

1. *Robustness.* The algorithm must be reliable for general design applications and thus must be theoretically guaranteed to converge to a solution point starting from any initial design estimate (even bad designs).
2. *Generality.* The algorithm must be general, implying that it should be able to treat equality as well as inequality constraints and should not impose any restrictions on the form of the cost and constraint functions, such as linear or quadratic functions.
3. *Accuracy.* The ability of an algorithm to converge to the precise mathematical optimum point is important even though a precise optimum may not be needed in practical applications. An accurate optimization algorithm is likely to have a sound mathematical basis and thus higher reliability.
4. *Ease of Use.* The algorithm must be easy to use by experienced as well as inexperienced designers. An algorithm that requires selection of tuning parameters that are problem dependent is difficult to use in practice.
5. *Efficiency.* The algorithm must be efficient for general engineering design applications. To be efficient, the number of repeated analyses of the system (such as finite element analyses) must be kept to a minimum. Thus an efficient algorithm has (i) a faster rate of convergence requiring fewer number of iterations and consequently, fewer system analyses, and (ii) the least number of calculations within one design iteration.

It is important to note that there is a tradeoff between robustness, generality, accuracy, and ease of use of an algorithm versus its efficiency. Some computational penalty must be paid in terms of CPU time to achieve robustness, generality, accuracy, and ease of use of an algorithm (Thanedar et al. 1990).

Optimum design of large scale systems presents special challenges for any numerical algorithm in terms of its efficiency. For such systems, the algorithm should be selected judiciously. For example, if an algorithm does not use the potential constraint strategy in defining the search direction determination subproblem, then it is not suitable for large scale problems. Also an algorithm that requires tuning of certain parameters for its proper performance will not be suitable because it may require several trials before proper values for the parameters are found. This is not only inconvenient but also inefficient. Such algorithms should be avoided for optimization of large scale systems. It is important to note that a good algorithm must be robust (i.e., it must converge to a local minimum) for small as well as large scale problems. The algorithm may be slightly less efficient for small scale problems (small problems require very little CPU time any way), but it must be more efficient for large scale problems.

OVERVIEW OF COMPUTATIONAL ALGORITHMS

Numerous algorithms have been developed and tested for constrained optimization problems. In this section, we describe some of the basic steps common to many algorithms, such as sequential linear programming, gradient projection, feasible directions, generalized reduced gradient, and sequential quadratic programming. Many of the algorithms are described in detail by Arora (1989) and Vanderplaats (1984).

Linearization of the Problem

At each iteration, most numerical methods for constrained optimization problems compute a design change by solving an approximate subproblem which is obtained by writing linear Taylor's expansions for the cost and constraint functions. All search methods start with an initial design estimate and iteratively improve it. Let $\mathbf{X}^{(k)}$ be the design estimate at the kth iteration and $\Delta\mathbf{X}^{(k)}$ be the desired change in design. Instead of using $\Delta\mathbf{X}^{(k)}$ as change in design, usually it is taken as the search direction $\mathbf{d}^{(k)}$ and a step size is calculated along it to determine the new point. We write Taylor's expansion of the cost and constraint functions about the point $\mathbf{X}^{(k)}$ to obtain an approximate linearized subproblem as

Minimize

$$\sum_{i=1}^{N} c_i d_i \tag{B-3}$$

Subject to the linearized equality constraints

$$\sum_{i=1}^{n} \frac{\partial g_j}{\partial X_i} d_i = e_j \qquad j = 1, 2, \ldots, p \tag{B-4}$$

and the linearized inequality constraints

$$\sum_{i=1}^{n} \frac{\partial g_j}{\partial X_i} d_i \leq e_j \qquad j > p \text{ and } j \text{ in the set } I_k \tag{B-5}$$

where $e_j = -g_j(\mathbf{X}^{(k)})$, and $c_i = \partial f(\mathbf{X}^{(k)})/\partial \mathbf{X}_i$. Note that a potential constraint strategy for inequality constraints is used in Eq. (B-5). If it is not to be used, ε can be set to a very large number in defining the index set I_k in Eq. (B-2).

Sequential Linear Programming

Note that all the functions in Eqs. (B-3) to (B-5) are linear in the variables d_i. Therefore linear programming methods can be used to solve for d_i. Such procedures are called Sequential Linear Programming methods or, in short, SLP. Note, however, that the problem defined in Eqs. (B-3) to (B-5) may not have a bounded solution or the changes in design may become too large, thus invalidating the linear approximations. Therefore limits must be imposed on changes in the design variables. Such constraints are usually called *move limits* in the optimization literature and can be expressed as

$$-\Delta_i \leq d_i \leq \Delta_i \qquad i = 1, 2, \ldots, n \tag{B-6}$$

where Δ_i is the maximum allowed decrease or increase in the *i*th design variable, respectively, at the *k*th iteration. The problem is still linear in terms of d_i, so LP methods can be used to solve it. Note that move limits may need to be changed at every iteration. Selection of these move limits is of critical importance because it can mean success or failure of the SLP algorithm.

The sequential linear programming algorithm is a simple and straightforward approach to solve constrained optimization problems. Once a linear approximation for the problem has been developed, existing LP software can be readily used to obtain a new solution at each iteration. However, there are important limitations of the method that should be clearly understood:

1. The method cannot be used as a black box approach for engineering design problems. The selection of move limits is a trial-and-error process and can be best achieved in an interactive mode. The move limits may be too restrictive, resulting in no solution for the LP subproblem. Or they may be too large, resulting in oscillations in the design points.
2. The method is not convergent since no descent function is used. Therefore it is difficult to monitor progress towards the minimum point.

3. The method can cycle between two points if the optimum solution for the nonlinear optimization problem is not at a vertex (intersection of constraints) of the feasible set.

Due to the foregoing limitations, the method can be frustrating to use, so it is not recommended as a technique for the general design optimization environment, although it may be tried to improve specific designs.

Sequential Quadratic Programming

To correct the foregoing drawbacks of SLP, we describe a method where a quadratic programming (QP) subproblem is solved to find a search direction, and a descent function is used to determine an appropriate step size. The QP subproblem is defined as (Arora 1989):

Subproblem QP. Minimize

$$\sum_{i=1}^{n} c_i d_i + \sum_{i=1}^{n} \sum_{j=1}^{n} H_{ij} d_i d_j \tag{B-7}$$

subject to the linearized constraints in Eqs. (B-4) and (B-5) where H is an $n \times n$ matrix that is a positive definite approximation to the Hessian of the Lagrangian (matrix of second partial derivatives).

A descent function for the constrained problems is usually constructed by adding a penalty for constraint violations to the current value of the cost function. One of the properties of the descent function is that its value at the optimum point must be the same as that for the cost function. Also, it must reduce along the search direction at each iteration. We introduce Pshenichny's descent function due to its simplicity and success in solving a large number of engineering design problems (Belegundu and Arora 1984b; Pshenichny and Danilin 1978). Other descent functions have also been developed and used.

Pshenichny's descent function Φ at any point $\mathbf{X}$ is defined as

$$\Phi(\mathbf{X}) = f(\mathbf{X}) + RV(\mathbf{X}) \tag{B-8}$$

where R is a positive number called the penalty parameter and $\mathbf{V}(\mathbf{X}) \geq 0$ is the maximum constraint violation among all the constraints. Note that R is required to be larger than the sum of the magnitude of all the Lagrange multipliers.

In most practical implementations of any algorithm, only an approximate step size can be calculated because determination of an accurate step size can be very time consuming, requiring many analyses of the system. Most algorithms do not require an accurate step size for proof of their convergence. Therefore only an approximate step size needs to be calculated using the so-called inaccurate or inexact line search. The usual procedure is to start with the trial step size as one. If the descent condition is not satisfied, the trial step is taken as half of the previous trial. If the descent condition is still not satisfied, the trial step size is bisected again. The procedure is continued until the descent condition is satisfied.

One of the SQP methods takes **H** as the identity matrix in Eq. (B-7) (Belegundu and Arora 1984b; Arora 1989) whereas others take it as an approximate second derivative matrix that is updated during each iteration (Lim and Arora 1986; Tseng and Arora 1988). Thanedar et al. (1986) have compared performance of several SQP algorithms.

NUMERICAL IMPLEMENTATION ASPECTS

Computational algorithms for design optimization must be properly implemented in a computer program for use in practical applications. Considerable care, judgment, safeguards, and user-friendly features must be designed and incorporated into the software. Numerical calculations must be robustly implemented. Each step of the algorithm must be analyzed and proper numerical procedures developed to implement the intent of the step. The software must be properly evaluated for performance by solving many different problems. Transcribing an algorithm into a computer program is an art to some degree. A theoretically convergent algorithm may be implemented badly such that it is neither convergent nor reliable, and vice versa. Robust implementation of an algorithm requires considerable experience and insight, and it is best to leave the task for the specialist. Proper implementation of algorithms also requires a certain amount of heuristics which can be developed only with experience and by solving many problems of varying difficulty. An excellent discussion on this subject is given in Gill et al. (1981) and Tseng and Arora (1988). Each optimization algorithm has certain parameters that need to be selected for a problem. Specification of these parameters can influence the performance of the algorithm. This has been demonstrated by Tseng and Arora (1988, 1989) and Lim and Arora (1986) in a detailed study of an SQP algorithm. A good optimization algorithm should be relatively insensitive to variations in the parameters. As a minimum requirement, the algorithm must converge with every specification of the parameters within their allowable range. These aspects and attributes, discussed in previous sections, can be used in selecting algorithms and the associated software for practical applications.

SUMMARY AND DISCUSSION

General concepts of computational algorithms for continuous variable optimization are explained and the definition of a good algorithm is given. Some of the concepts that are common to many algorithms are described. Based on experience with several algorithms, it is suggested that designers who are not experts in optimization should use only general and reliable methods. Although such methods require more computational effort compared to other algorithms, they are more cost effective in the longer run. Efficiency, although important, can be sacrificed for reliability. A few more CPU cycles of the machine are worth the wait if the algorithm can reliably give optimum solutions. A detailed discussion on robustness versus efficiency of methods has been presented recently by Thanedar et al. (1990).

Transcription of an algorithm into good software requires considerable experience, time, and resources (Tseng and Arora 1988). This should be a task for optimization experts and software developers. Design engineers need not be involved in algorithm development or its implementation. Their job should be to properly formulate the design optimization problem (Arora 1989), solve it using available software, and interpret the results. They should concentrate on investigating alternate conceptual designs or formulations for their application. Once a concept is properly formulated, well developed and robust software can solve the problem very quickly. In this way optimization methods and programs can be used as tools in the creative process of designing systems.

NOMENCLATURE

$\mathbf{c}$ = gradient of the cost function, ∇f or $\partial f/\partial \mathbf{X}$; $c_i = \partial f/\partial X_i$
d = search direction; d_i is the ith component
e_j = negative of the jth constraint function value
f = cost function
g_i = ith constraint function
$\mathbf{H}$ = Hessian matrix (symmetric) consisting of second partial derivatives of a twice continuously differentiable function
I_k = an index set containing identification numbers for potential constraints
k = superscript or subscript representing the iteration number
LP = linear programming
m = total number of constraints
n = number of design variables
NLP = nonlinear programming
p = number of equality constraints
QP = quadratic programming
R = penalty parameter
SLP = sequential linear programming
SQP = sequential quadratic programming
$\mathbf{X}$ = design variable vector of dimension n

α = step size
$\Delta\mathbf{X}$ = change in the design variable vector, ΔX_i is the ith component

REFERENCES

Arora, J. S. (1989). *Introduction to optimum design*. McGraw-Hill, New York.

Arora, J. S. (1990). "Computational design optimization: A review and future directions," *Structural Safety*, 7, 131–148.

Arora, J. S. and Belegundu, A. D. (1984). "Structural optimization by mathematical programming methods." *AIAA J.*, 22, 854–856.

Arora, J. S., Chahande, A. I., and Paeng, J. K. (1991). "Multiplier methods for engineering optimization." *Int. J. for Numerical Methods in Engng.*, 32(7), 1485–1525.

Arora, J. S., Huang, M. W., and Hsieh, C. C. (1994). "Methods for optimization of nonlinear problems with discrete variables: A review." *Structural Optimization*, 8, 69–85.

Belegundu, A. D. and Arora, J. S. (1984a). "A computational study of transformation methods for optimal design." *AIAA J.*, 22, 535–542.

Belegundu, A. D. and Arora, J. S. (1984b). "A recursive quadratic programming method with active set strategy for optimal design." *Int. J. for Numerical Methods in Engng.*, 20, 803–816.

Belegundu, A. D. and Arora, J. S. (1985). "A study of mathematical programming methods for structural optimization." *Int. J. for Numerical Methods in Engng.*, 21, 1583–1623.

Frangopol, D. M. (1995). "Reliability-based optimum structural design." Chapter 16 in *Probabilistic structural mechanics handbook*, C. Sundararajan, ed., Chapman and Hall, New York, 352–387.

Gill, P. E., Murray, W., and Wright, M. H. (1981). *Practical optimization*. Academic Press, New York.

Huang, M. W. and Arora, J. S. (1996). "Optimal design with discrete variables: Some numerical experiments," *Int. J. Numerical Methods in Engng.* (in press).

Lim, O. K. and Arora, J. S. (1986). "An active set RQP algorithm for engineering design optimization." *Computer Methods Applied Mechanics and Engng.*, 57, 51–65.

Pshenichny, B. N. and Danilin, Y. M. (1978). *Numerical methods in extremal problems*. Mir Publishers, Moscow.

Thanedar, P. B., Arora, J. S., Li, G. Y., and Lin, T. C. (1990). "Robustness, generality and efficiency of optimization algorithms for practical applications." *Structural Optimization*, 2, 203–212.

Thanedar, P. B., Arora, J. S., Tseng, C. H., Lim, O. K., and Park, G. J. (1986). "Performance of some SQP algorithms on structural design problems." *Int. J. for Numerical Methods in Engng.*, 23(12), 2187–2203.

Tseng, C. H. and Arora, J. S. (1988). "On implementation of computational algorithms for optimum design." *Int. J. for Numerical Methods in Engng.*, 26(6), 1365–1402.

Tseng, C. H. and Arora, J. S. (1989). "Optimum design of systems for dynamics and controls using sequential quadratic programming." *AIAA J.*, 27(12), 1793–1800.

Vanderplaats, G. N. (1984). *Numerical optimization techniques for engineering design: With applications*. McGraw-Hill, New York.

Appendix C

OPTIMALITY CRITERION METHOD FOR STRUCTURAL OPTIMIZATION

This appendix briefly describes the details of an Optimality Criterion (OC) method that is applicable for solution of optimization problems involving continuous-valued variables having the following form.

Minimize

$$W(A_i) = \sum_{i=1}^{n} w_i A_i \tag{C-1a}$$

Subject to

$$g_r(A_i) = \sum_{i=1}^{n} \frac{c_{ir}}{A_i} \le \bar{g}_r \qquad (r = 1, 2, \ldots, m) \tag{C-1b}$$

$$A_i^L \le A_i \le A_i^U \qquad (i = 1, 2, \ldots, n) \tag{C-1c}$$

Equations (C-1) are a concise expression of the structural optimization problem presented in Chapter 4. Equation (C-1a) is the objective/weight function expressed in terms of cross-section area variables A_i for the n members of the structure. Equations (C-1b) define the m displacement and stress constraints for the design, where the c_{ir} and g_r quantities are the corresponding gradients and bounds, respectively. Finally, Eqs. (C-1c) impose specified lower and upper bounds A_i^L and A_i^u on the member cross-section areas.

Before developing the OC method for solution of Eqs. (C-1), it is instructive to illustrate the formulation of Eqs. (C-1b) for, say, a dis-

Donald E. Grierson, Civil Engineering, University of Waterloo, Ontario, Canada N2L 3G1; email: ⟨grierson@sail.uwaterloo.ca⟩.

placement constraint

$$\delta_r \leq d_r^u \tag{C-2}$$

where d_r^u is the specified upper bound on displacement δ_r. For simplicity, it is assumed that the displacement δ_r occurs for a truss structure for which all members are prismatic and experience axial force and deformation alone. Therefore, by the principle of virtual work, the displacement can be expressed as

$$\delta_r = \sum_{i=1}^{n} \left(\frac{F_r f_r L}{AE} \right)_i \tag{C-3}$$

where, for each member i, E = Young's material modulus, L = length, F_r = axial force due to actual loading causing displacement δ_r, and f_r = axial force due to virtual loading corresponding to δ_r. From Eq. (C-3), the displacement constraint Eq. (C-2) can be expressed explicitly in terms of the member cross-section areas A_i as

$$\sum_{i=1}^{n} \frac{c_{ir}}{A_i} \leq d_r^u \tag{C-4}$$

where the coefficient

$$c_{ir} = \left(\frac{F_r f_r L}{E} \right)_i \tag{C-5}$$

For simplicity in developing the details of the OC method, the sizing constraints Eqs. (C-1c) are ignored and all the constraints Eqs. (C-1b) are assumed to be active such that $g_r(A_i) = \bar{g}_r$ $(r = 1, 2, \ldots, m)$. The corresponding optimization problem involving Eqs. (C-1a), and Eqs. (C-1b) taken as identities, can be reformulated as the minimization of the Lagrangian function

$$L(A_i, \lambda_r) = \sum_{i=1}^{n} w_i A_i + \sum_{r=1}^{m} \lambda_r \left[\sum_{i=1}^{n} \frac{c_{ir}}{A_i} - \bar{g}_r \right] \tag{C-6}$$

where the Lagrange multipliers are sign-restricted as

$$\lambda_r \geq 0 \qquad (r = 1, 2, \ldots, m) \tag{C-7}$$

The required optimality criteria are obtained by differentiating the Lagrangian function Eq. (C-6) with respect to the design variables A_i to obtain the necessary stationary conditions

$$w_i - \sum_{r=1}^{m} \lambda_r \left(\frac{c_{ir}}{A_i^2} \right) = 0 \qquad (i = 1, 2, \ldots, n) \tag{C-8}$$

Rearranging terms in Eqs. (C-8), the optimality criteria can also be expressed as

$$\sum_{r=1}^{m} \lambda_r \left(\frac{c_{ir}/A_i}{w_i A_i}\right) = \sum_{r=1}^{m} \lambda_r \frac{c_{ir}}{w_i A_i^2} = \sum_{r=1}^{m} \lambda_r \varepsilon_{ir} = 1 \quad i = 1, 2, \ldots, n \qquad \text{(C-9)}$$

Equations (C-9) denote the necessary conditions to be satisfied for an optimal structure. These optimality conditions have a significant physical meaning for design. Each Lagrange multiplier λ_r can be interpreted as a measure of the sensitivity of the optimal design to the rth constraint; that is, λ_r is a weighting factor that measures the importance of the rth constraint to the optimal design. The expression $\varepsilon_{ir} = c_{ir}/w_i A_i^2$ in Eqs. (C-9) represents the so-called virtual strain energy density for member i with respect to the rth constraint. The optimality conditions Eqs. (C-9) specify that the weighted sum of the virtual strain energy densities for each member i must be equal to unity. In other words, for the optimal design of a structure under multiple constraints, each member must contribute the same weighted amount of virtual strain energy density.

The OC method involves the application of a recursive algorithm to find the sizing variables A_i and Lagrange multipliers λ_r satisfying the optimality criteria Eqs. (C-9). To this end, multiply both sides of Eqs. (C-9) by A_i^η to obtain

$$A_i^\eta = A_i^\eta \left(\sum_{r=1}^{m} \lambda_r \varepsilon_{ir}\right) \qquad (i = 1, 2, \ldots, n) \qquad \text{(C-10)}$$

and take the ηth root to get

$$A_i = A_i \left(\sum_{r=1}^{m} \lambda_r \varepsilon_{ir}\right)^{1/\eta} \qquad (i = 1, 2, \ldots, n) \qquad \text{(C-11)}$$

From Eqs. (C-11), an exponential form of the recursive relation for the sizing variables A_i can be obtained as

$$A_i^{v+1} = A_i^{v} \left(\sum_{r=1}^{m} \lambda_r \varepsilon_{ir}\right)^{1/\eta} \qquad (i = 1, 2, \ldots, n) \qquad \text{(C-12)}$$

where $v + 1$ and v indicate successive iterations ($v = 0$ corresponds to the set of cross-section areas A_i^0 prevailing at the beginning of the recursive process), and η is a step-size or relaxation parameter that controls the convergence rate of the recursive process (as η becomes larger, the value of A_i^v is modified to a lesser extent, and vice versa).

A linear form of the recursive relation can be found by expanding Eqs. (C-12) using the Binomial Theorem and retaining only the linear terms to obtain

$$A_i^{\nu+1} = A_i^{\nu}\left[1 + \frac{1}{\eta}\left(\sum_{r=1}^{m} \lambda_r \varepsilon_{ir} - 1\right)_{\nu}\right] \qquad (i = 1, 2, \ldots, n) \qquad \text{(C-13)}$$

Although either recursive relation Eqs. (C-12) or Eqs. (C-13) can be utilized, Eqs. (C-13) are adopted hereafter in this appendix. Note that when the optimality criteria Eqs. (C-9) are satisfied, both Eqs. (C-12) and (C-13) reduce to

$$A_i^{\nu+1} = A_i^{\nu} \qquad (i = 1, 2, \ldots, n) \qquad \text{(C-14)}$$

thereby indicating convergence of the recursive process.

Before the recursive relation Eqs. (C-13) can be utilized to find the sizing variables A_i, the unknown Lagrange multipliers λ_r must first be determined. To establish the current λ_r^{v} values, consider the change Δg_t in the tth constraint due to the changes ΔA_i in the sizing variables; that is,

$$\Delta g_t = \sum_{i=1}^{n} \frac{\partial g_t}{\partial A_i} \Delta A_i \qquad \text{(C-15)}$$

Suppose that constraint t becomes active after the $(v + 1)$th iteration, so that the left-hand side of Eq. (C-15) can be written as

$$\Delta g_t = g_t^{\nu+1} - g_t^{\nu} = \bar{g}_t - g_t^{\nu} \qquad \text{(C-16)}$$

Now, from Eqs. (C-1b),

$$\frac{\partial g_t}{\partial A_i} = -\left(\frac{c_{it}}{A_i^2}\right)_{\nu} \qquad \text{(C-17)}$$

and, from Eqs. (C-9) and (C-13),

$$\Delta A_i = A_i^{\nu+1} - A_i^{\nu} = \frac{A_i^{\nu}}{\eta}\left(\sum_{r=1}^{m} \frac{\lambda_r c_{ir}}{w_i A_i^2} - 1\right)_{\nu} \qquad \text{(C-18)}$$

Therefore, from Eqs. (C-16) through (C-18), Eq. (C-15) for the tth constraint becomes

$$\bar{g}_t - g_t^{\nu} = \sum_{i=1}^{n}\left(\frac{-c_{it}}{A_i^2}\right)_{\nu}\left[\frac{A_i^{\nu}}{\eta}\left(\sum_{r=1}^{m} \frac{\lambda_r c_{ir}}{w_i A_i^2} - 1\right)_{\nu}\right] \qquad \text{(C-19)}$$

which can be rearranged to obtain

$$\sum_{r=1}^{m} \lambda_r^{\nu} \left(\sum_{i=1}^{n} \frac{c_{ir} c_{it}}{w_i A_i^3} \right)_{\nu} = \sum_{i=1}^{n} \frac{c_{it}}{A_i^{\nu}} - \eta(\bar{g}_t - g_t^{\nu}) \qquad \text{(C-20)}$$

From Eq. (C-20), for the effect of changes in the sizing variables A_i accounted for in all ($t = 1, 2, \ldots, m$) constraints, the following set of m simultaneous linear equations in terms of m as yet unknown Lagrange multipliers λ_r is obtained:

$$\sum_{r=1}^{m} \lambda_r^{\nu} \left(\sum_{i=1}^{n} \frac{c_{ir} c_{it}}{w_i A_i^3} \right)_{\nu} = \sum_{i=1}^{n} \frac{c_{it}}{A_i^{\nu}} - \eta(\bar{g}_t - g_t^{\nu}) \qquad (t = 1, 2, \ldots, m) \qquad \text{(C-21)}$$

The set of simultaneous equations Eqs. (C-21) and the resizing relations Eqs. (C-13) together form the basis of the OC iterative method to solve the continuous-valued optimization problem to minimize Eq. (C-1a), subject to Eqs. (C-1b) taken as all being active constraints. Having the current set of sizing variables A_i^v, for example, for the initial "trial" design, the corresponding values for the Lagrange multipliers λ_r^v are readily determined by solving Eqs. (C-21). Having the current values of the λ_r^v, the new set of member section areas A_i^{v+1}, is found through the recursive relations Eqs. (C-13). For each current design stage, it is assumed that the new design does not cause an internal force redistribution for the structure (true for a statically determinate structure). The optimal solution of the minimization problem is obtained through repeated application of Eqs. (C-21) for the Lagrange multipliers λ_r and Eqs. (C-13) for the section areas A_i until their values converge. For a statically indeterminate structure, a further cycle of the OC process is carried out for updated member forces corresponding to the member sizes from the current design stage. This iterative design process is repeated until convergence to the minimum-weight design occurs.

EXAMPLE

The application of the OC method for continuous optimization is illustrated in the following for the simple truss shown in Fig. C-1 (Chan 1993). The truss is loaded at its top node by two lateral point loads taken as two independent loading conditions in the X and Y orthogonal directions. The structure geometry, material properties, and displacement limits at the top node for the two load cases are also shown in Fig. C-1. (Nondimensional units of measurement are used for ease of illustration.)

To facilitate the presentation, a subscript i denotes the member number, subscript x represents information pertaining to the X-direction load case, and subscript y corresponds to the Y-direction

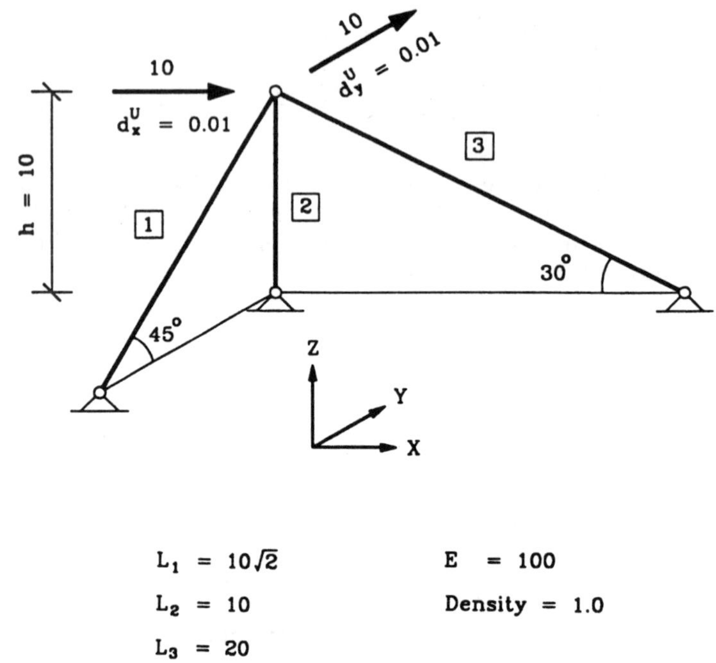

FIG. C-1. Three-Member Truss.

load case. Since the truss structure is statically determinate, member forces can be calculated independently of the member sizes. Therefore no re-analysis is needed since there is no internal force redistribution and the optimal final design is achieved in one design cycle. Analysis results are given in Fig. C-2. Axial member forces F due to the actual loading conditions are printed along the members, and virtual member forces f due to unit virtual loads applied at the free node are provided in parentheses.

It is required to find the cross-section areas A_1, A_2, and A_3 of the three members such that the truss has minimum material weight and the two displacement constraints are satisfied. Continuous-valued member sizes are adopted for the design and are fictitiously assumed to be unbounded. From Eqs. (C-1a,b), the corresponding design optimization problem is

Minimize

$$W(a_i) = \sum_{i=1}^{3} \rho(L_i A_i) \tag{C-22a}$$

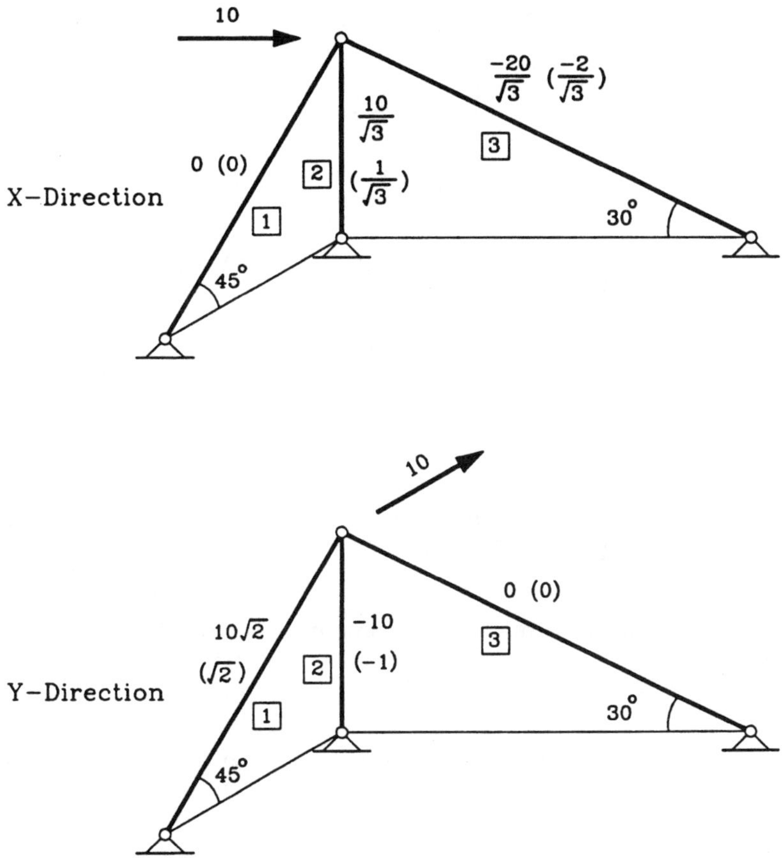

FIG. C-2. Analysis Results for Three-Member Truss [Actual Forces (Virtual Forces)].

Subject to

$$d_x(A_i) = \frac{c_{1x}}{A_1} + \frac{c_{2x}}{A_2} + \frac{c_{3x}}{A_3} \le d_x^u \tag{C-22b}$$

$$d_y(A_i) = \frac{c_{1y}}{A_1} + \frac{c_{2y}}{A_2} + \frac{c_{3y}}{A_3} \le d_y^u \tag{C-22c}$$

where the member lengths L_i are shown in Fig. C-1, ρ is the material density (hereafter, $\rho = 1$ is assumed for simplicity), the drift limits $d_x^u = d_y^u = h/1000$ from Fig. C-1, and, from Eq. (C-5) and Fig. C-2, the coefficients in Eqs. (C-22) are calculated as

$$c_{1x} = \left(\frac{FfL}{E}\right)_{1x} = \frac{0 \cdot 0 \cdot 10\sqrt{2}}{100} = 0$$

$$c_{2x} = \left(\frac{FfL}{E}\right)_{2x} = \frac{(10/\sqrt{3}) \cdot (1/\sqrt{3}) \cdot 10}{100} = \frac{1}{3}$$

$$c_{3x} = \left(\frac{FfL}{E}\right)_{3x} = \frac{(-20/\sqrt{3}) \cdot (-2/\sqrt{3}) \cdot 20}{100} = \frac{4}{1.5}$$

$$c_{1y} = \left(\frac{FfL}{E}\right)_{1y} = \frac{10\sqrt{2} \cdot \sqrt{2} \cdot 10\sqrt{2}}{100} = \frac{\sqrt{2}}{0.5}$$

$$c_{2y} = \left(\frac{FfL}{E}\right)_{2y} = \frac{-10 \cdot -1 \cdot 10}{100} = 1$$

$$c_{3y} = \left(\frac{FfL}{E}\right)_{3y} = \frac{0 \cdot 0 \cdot 20}{100} = 0$$

Therefore the optimization problem Eqs. (C-22) becomes the following.

Minimize

$$W(A_i) = 10\sqrt{2}A_1 + 10A_2 + 20A_3 \tag{C-23a}$$

Subject to

$$d_x(A_i) = \frac{(1/30)}{A_2} + \frac{(4/15)}{A_3} \leq 0.01 \tag{C-23b}$$

$$d_y(A_i) = \frac{(\sqrt{2}/5)}{A_1} + \frac{(1/10)}{A_2} \leq 0.01 \tag{C-23c}$$

To commence the OC optimization process through Eqs. (C-13) and (C-21), initial member sizes are arbitrarily taken to be $A_1^0 = A_2^0 = A_3^0 = 1$. As such, the initial X and Y displacements are obtained from Eqs. (C-23b) and (C-23c) as

$$d_x^0 = \frac{1}{30} + \frac{4}{15} = 0.30$$

$$d_y^0 = \frac{\sqrt{2}}{5} + \frac{1}{10} = 0.38$$

which are noted to violate the displacement constraint limit of 0.01, indicating that the initial design is too flexible and that the members must be resized.

Before the recursive relations Eqs. (C-13) can be utilized to resize the members, a set of two simultaneous linear equations Eqs. (C-21) must first be established in terms of the two Lagrange multipliers λ_x and λ_y associated with the X and Y displacement constraints. To this end, for the initial set of member sizes $A_1^0 = A_2^0 = A_3^0 = 1$, the summation terms on the left-hand side of Eqs. (C-21) are found as

$$\left(\sum_{i=1}^{3} \frac{c_{ix}c_{ix}}{w_i A_i^3}\right)_v = 0 + \frac{(1/30)\cdot(1/30)}{10\cdot 1^3} + \frac{(4/15)\cdot(4/15)}{20\cdot 1^3} = 3.67 \times 10^{-3}$$

$$\left(\sum_{i=1}^{3} \frac{c_{ix}c_{iy}}{w_i A_i^3}\right)_v = 0 + \frac{(1/30)\cdot(1/10)}{10\cdot 1^3} + 0 = 3.33 \times 10^{-4}$$

$$\left(\sum_{i=1}^{3} \frac{c_{iy}c_{ix}}{w_i A_i^3}\right)_v = 0 + \frac{(1/10)\cdot(1/30)}{10\cdot 1^3} + 0 = 3.33 \times 10^{-4}$$

$$\left(\sum_{i=1}^{3} \frac{c_{iy}c_{iy}}{w_i A_i^3}\right)_v = \frac{(2\sqrt{2}/10)\cdot(2\sqrt{2}/10)}{10\sqrt{2}\cdot 1^3} + \frac{(1/15)\cdot(1/10)}{10\cdot 1^3} + 0 = 6.66 \times 10^{-3}$$

Employing a step-size parameter $\eta = 2$, the right-hand sides of Eqs. (C-21) are obtained as

$$\sum_{i=1}^{3} \frac{c_{ix}}{A_i^0} - \eta(d_x^U - d_x^0) = 0.3 - 2(0.01 - 0.3) = 0.88$$

$$\sum_{i=1}^{3} \frac{c_{iy}}{A_i^0} - \eta(d_y^U - d_y^0) = 0.38 - 2(0.01 - 0.38) = 1.12$$

Therefore Eqs. (C-21) in terms of Lagrange multipliers λ_x and λ_y are

$$\begin{bmatrix} 3.67 \times 10^{-3} & 3.33 \times 10^{-4} \\ 3.33 \times 10^{-4} & 6.66 \times 10^{-3} \end{bmatrix} \begin{bmatrix} \lambda_x \\ \lambda_y \end{bmatrix} = \begin{bmatrix} 0.88 \\ 1.12 \end{bmatrix}$$

for which the solution is

$$\lambda_x^0 = 225.62; \qquad \lambda_y^0 = 158.23$$

TABLE C-1 Iteration History for the Three-Member Truss

Iteration ν	A_1^ν	A_2^ν	A_3^ν	d_x^ν	d_y^ν	Simultaneous Eqs. (C-21) L.H.S		R.H.S.	λ_x^ν λ_y^ν	Structure Weight
0	1.000	1.000	1.000	0.300000	0.382843	3.66666*E*-03	3.33333*E*-04	0.88000	225.62	44.14
						3.33333*E*-04	6.65685*E*-03	1.12853	158.23	
1	2.082	1.667	2.004	0.153054	0.195812	4.65699*E*-04	7.19332*E*-05	0.43916	850.18	86.20
						7.19332*E*-05	8.42322*E*-04	0.56744	601.05	
2	3.928	3.486	3.830	0.079184	0.100699	6.59003*E*-05	7.86783*E*-06	0.21755	3037.70	167.01
						7.86783*E*-06	1.16968*E*-04	0.28210	2207.40	
3	7.584	6.361	7.202	0.042265	0.053015	9.94810*E*-06	1.29489*E*-06	0.10679	9758.80	314.92
						1.29489*E*-06	1.68528*E*-05	0.13904	7500.65	
4	13.682	11.633	12.634	0.023972	0.029269	1.83366*E*-06	2.11742*E*-07	0.05192	25781.56	562.51
						2.11742*E*-07	2.84383*E*-06	0.06781	21923.66	
5	22.865	18.933	19.921	0.015147	0.017652	4.66107*E*-07	4.91133*E*-08	0.02544	49395.62	911.11
						4.91133*E*-08	6.20579*E*-07	0.03296	49196.01	
6	32.949	26.807	26.491	0.011310	0.012315	1.97025*E*-07	1.73039*E*-08	0.01393	64078.12	1263.85
						1.73039*E*-08	2.10061*E*-07	0.01694	75385.41	
7	39.354	31.448	29.371	0.010139	0.010367	1.43899*E*-07	1.07174*E*-08	0.01042	66200.18	1458.46
						1.07174*E*-08	1.24965*E*-07	0.01110	83154.49	
8	40.807	32.453	29.712	0.010002	0.010013	1.38809*E*-07	9.75211*E*-09	0.01001	66225.35	1495.86
						9.75211*E*-09	1.12505*E*-07	0.01004	83480.57	
9	40.861	32.489	29.715	0.010000	0.010000	1.38747*E*-07	9.71976*E*-09	0.01000	66225.35	1497.06
						9.71976*E*-09	1.12078*E*-07	0.01000	83481.00	
10	40.861	32.489	29.715	0.010000	0.010000	1.38747*E*-07	9.71972*E*-09	0.01000	66225.35	1497.06
						9.71972*E*-09	1.12077*E*-07	0.01000	83481.00	

Having λ_x^0 and λ_y^0, a new set of member sizes are found from the recursive relations Eqs. (C-13) as

$$A_1^1 = A_1^0 \left[1 + \frac{1}{\eta} \left(\sum_{r=1}^{2} \lambda_r \frac{c_{1r}}{w_1 A_1^2} - 1 \right)_o \right]$$

$$= 1 \left[1 + \frac{1}{2} \left(225.62 \cdot \frac{0}{10\sqrt{2} \cdot 1^2} + 158.23 \cdot \frac{\sqrt{2}/5}{10\sqrt{2} \cdot 1^2} - 1 \right) \right] = 2.082$$

$$A_2^1 = A_2^0 \left[1 + \frac{1}{\eta} \left(\sum_{r=1}^{2} \lambda_r \frac{c_{2r}}{w_2 A_2^2} - 1 \right)_o \right]$$

$$= 1 \left[1 + \frac{1}{2} \left(225.62 \cdot \frac{1/30}{10 \cdot 1^2} + 158.23 \cdot \frac{1/10}{10 \cdot 1^2} - 1 \right) \right] = 1.667$$

$$A_3^1 = A_3^0 \left[1 + \frac{1}{\eta} \left(\sum_{r=1}^{2} \lambda_r \frac{c_{3r}}{w_3 A_3^2} - 1 \right)_o \right]$$

$$= 1 \left[1 + \frac{1}{2} \left(225.62 \cdot \frac{4/15}{20 \cdot 1^2} + 158.23 \cdot \frac{0}{20 \cdot 1^2} - 1 \right) \right] = 2.004$$

After obtaining the new set of member sizes as in the foregoing, the OC process is repeated in an iterative fashion until convergence of both member sizes and Lagrange multipliers occurs. The iteration history of the OC optimization process for the truss design is tabulated in Table C-1. The design process converges to the optimal structure weight of 1497.06 after nine iterations, where both the X and Y displacement constraints are active at their limit of 0.01. By inspection, the 45° diagonal in the Y direction is noted to be structurally less efficient to resist lateral load than the 30° diagonal in the X direction, which means that the truss is more vulnerable under the Y-direction load and the design is influenced more by the Y-direction displacement constraint. Such intuition is confirmed by virtue of the fact that the Y-direction Lagrange multiplier in Table C-1 has a larger value $\lambda_y = 83481.0$ than the value $\lambda_x = 66225.4$ corresponding to the X-direction displacement constraint.

NOTATION

The following notation is used in this appendix.

A = cross-section area
c = sensitivity coefficient; gradient
d = displacement bound
E = Young's modulus
f = axial force
F = axial force

g = constraint function
$\bar{g}$ = constraint bound
L = member length
L = Lagrangian function
m = number of constraints
n = number of variables
OC = optimality criteria
W = weight; weight function
w = weight coefficient
δ = displacement
Δ = increment
ε = virtual strain energy
λ = Lagrange multipliers/variables
ρ = material density
η = step-size parameter
ν = iteration index

REFERENCE

Chan, C. - M. (1993). "An automatic resizing technique for the design of tall steel building frameworks." Ph.D. Thesis, University of Waterloo, Canada.

Appendix D

THE DUAL METHOD FOR DISCRETE OPTIMIZATION

This appendix briefly describes the details of a generalized optimality criterion algorithm, known in the literature as the Dual Method (Schmit and Fleury 1979; Fleury 1979), which is applicable for the solution of optimization problems involving discrete-valued variables having the following form.

Minimize

$$\sum_{i=1}^{n} w_i / x_i \tag{D-1a}$$

Subject to

$$\sum_{i=1}^{n} c_{ir} x_i \leq \bar{g}_r \qquad (r = 1, 2, \ldots, m) \tag{D-1b}$$

$$x_i \in X_i \qquad (i = 1, 2, \ldots, n) \tag{D-1c}$$

Equations (D-1) are a concise expression of the structural optimization problem presented in Chapter 4. Equation (D-1a) is the objective/weight function expressed in terms of reciprocal sizing variables x_i for the n members of the structure. Equations (D-1b) define the m displacement and stress constraints for the design, where the quantities c_{ir} and $\bar{g}_r$ are the corresponding gradients and bounds, respectively. Finally, Eqs. (D-1c) require that each design variable x_i be selected from the discrete set of reciprocal sizes $X_i = \{x_1, x_2, \ldots\}$

Donald E. Grierson, Civil Engineering, University of Waterloo, Ontario, Canada N2L 3G1; email: ⟨grierson@sail.uwaterloo.ca⟩.

prevailing for the section shape specified for member i (e.g., W, T, etc.).

To commence the development of the Dual Method, temporarily ignore the discrete sizing constraints Eqs. (D-1c) and formulate the Lagrangian function

$$L(x, \lambda) = \sum_{i=1}^{n} \frac{w_i}{x_i} + \sum_{r=1}^{m} \lambda_r \left(\sum_{i=1}^{n} c_{ir} x_i - \bar{g}_r \right) \tag{D-2}$$

where the Lagrange variables are sign-restricted as

$$\lambda_r \geq 0 \qquad (r = 1, 2, \ldots, m) \tag{D-3}$$

The Dual Method proceeds by finding a saddle-point (x^*, λ^*) of Eq. (D-2) while accounting for the discrete sizing constraints Eqs. (D-1c), which in turn provides a solution x^* to the original optimization problem Eqs. (D-1) (Zangwill 1969). A saddle-point of Eq. (D-2) is given by a solution of the min-max problem

$$(x^*, \lambda^*) = \max_{\lambda}[\min_{x} L(x, \lambda)] \tag{D-4}$$

The problem posed by Eq. (D-4) is solved by first minimizing the Lagrangian function over the x variables to establish a function in terms of λ variables alone; that is,

$$L_m(\lambda) = \min_{x} L(x, \lambda) = \min_{x} \sum_{i=1}^{n} L_i(x_i, \lambda) - L_o \tag{D-5}$$

where, from Eq. (D-2), $L_o = \Sigma_{r=1}^{m} \lambda_r \bar{g}_r$ and the functions

$$L_i(x_i, \lambda) = \frac{w_i}{x_i} + \sum_{r=1}^{m} \lambda_r c_{ir} x_i \qquad (i = 1, 2, \ldots, n) \tag{D-6}$$

are each expressed in terms of a single sizing variable x_i.

Equation (D-5), which is called the Dual function, is readily established by separately minimizing each function $L_i(x_i, \lambda)$ in Eqs. (D-6) over the corresponding discrete set of values X_i for sizing variable x_i and then summing; that is,

$$L_m(\lambda) = \sum_{i=1}^{n} \min_{x_i \in X_i} L_i(x_i, \lambda) \tag{D-7}$$

As the λ variables vary in value, the value of the sizing variable x_i for which the corresponding function $L_i(x_i, \lambda)$ is minimum may shift from one discrete value x_i^k to the next discrete value x_i^{k+1} in the set

X_i. When this happens, continuity of the Dual function $L_m(\lambda)$ is maintained by requiring that

$$L_i(x_i^k, \lambda) = L_i(x_i^{k+1}, \lambda) \quad \text{(D-8)}$$

which reduces to, from Eqs. (D-6),

$$\sum_{r=1}^{m} \lambda_r c_{ir} = \frac{w_i}{x_i^k x_i^{k+1}} \quad \text{(D-9)}$$

From Eq. (D-9), the specific discrete value of each sizing variable x_i for which the corresponding function $L_i(x_i, \lambda)$ is minimum is given by

$$x_i^o = x_i^k \quad \text{(D-10)}$$

if

$$\frac{w_i}{x_i^{k-1} x_i^k} < \sum_{i=1}^{m} \lambda_r^o c_{ir} < \frac{w_i}{x_i^k x_i^{k+1}} \quad \text{(D-11)}$$

where the values of the Lagrange variables λ_r^o are initially assumed (e.g., if constraint m of Eqs. (D-1b) is the most active for the initial design of the structure, set $\lambda_m^o = 1$ and $\lambda_r^o = 0$ for $r = 1, 2, \ldots, m - 1$); thereafter, the λ_r^o values are known from the previous iterate step of the maximization process conducted to solve Eq. (D-4), as outlined in the following.

Having established the Dual function $L_m(\lambda)$, the Dual Method completes the solution of Eq. (D-4) by solving the maximization problem, from Eqs. (D-3), (D-6), and (D-7).

Maximize

$$L_m(\lambda) = \sum_{i=1}^{n} \frac{w_i}{x_i^o} + \sum_{r=1}^{m} \lambda_r \left(\sum_{i=1}^{n} c_{ir} x_i^o - \bar{g}_r \right) \quad \text{(D-12a)}$$

Subject to

$$\lambda_r \geq 0 \qquad (r = 1, 2, \ldots, m) \quad \text{(D-12b)}$$

where the x_i^o values are known from Eqs. (D-10) and (D-11). The problem posed by Eqs. (D-12) is solved using a gradient projection method (Zangwill 1969; Haftka and Gurdal 1992) to progressively update the Lagrange variables as

$$\lambda = \lambda^o + \alpha \mathbf{s} \quad \text{(D-13)}$$

where **s** = the search direction, α = the step length, and λ^o = the variable values found at the end of the previous search step. Note from Eqs. (D-10) and (D-11) that the x_i^o values in Eq. (D-12a) are subject to change as the λ_r^o values change. The gradient projection search Eq. (D-13) progressively accounts for the bounds on the λ_r variables, Eqs. (D-12b), and the discontinuities in the x_i variables, Eqs. (D-11), to find the solution point (x^*, λ^*) for which Eq. (D-12a) is maximized. The point x^* is the solution of the design optimization problem Eqs. (D-1).

EXAMPLE

The Dual Method for discrete optimization is illustrated in the following for the simple pin-jointed truss loaded as shown in Fig. D-1 (Schmit and Fleury 1979; Haftka and Gurdal 1992; Grierson 1993). The truss has two members of cross-section area a_1 and a_2, and two independent nodal displacements u_1 and u_2. It is required to find the least-weight truss by selecting each of the section areas a_1 and a_2 from the discrete set of areas

$$A = \{1, 3/2, 2\} \tag{D-14}$$

while satisfying the displacement constraints

$$u_1 \leq 0.75\ (PL/E); \qquad u_2 \leq 0.25\ (PL/E) \tag{D-15}$$

where, from Fig. D-1, P = applied load, L = member length, and E = Young's modulus. As the truss is statically determinate, the displacements are readily found to be

$$u_1 = \frac{PL}{2E}\left(\frac{1}{a_1} + \frac{1}{a_2}\right); \qquad u_2 = \frac{PL}{2E}\left(\frac{1}{a_1} - \frac{1}{a_2}\right) \tag{D-16}$$

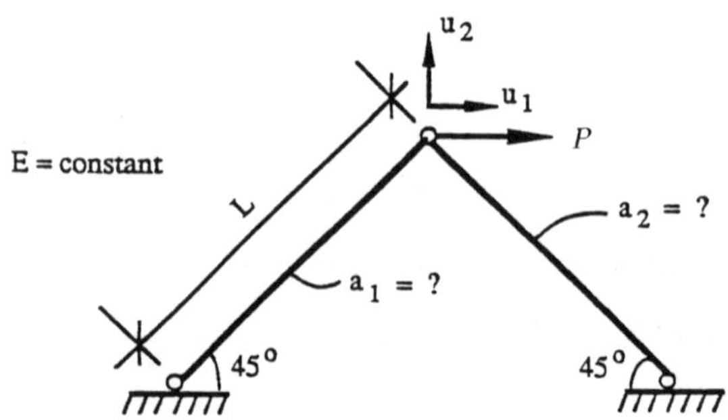

FIG. D-1. Two-Bar Pin-Jointed Truss.

The weight of the truss is given by

$$W = \rho L(a_1 + a_2) \tag{D-17}$$

where ρ is the material density.

From Eqs. (D-14) through (D-17), the discrete optimization problem Eqs. (D-1) expressed in terms of reciprocal sizing variables $x_1 = 1/a_1$ and $x_2 = 1/a_2$ is as follows.

Minimize

$$\frac{1}{x_1} + \frac{1}{x_2} \tag{D-18a}$$

Subject to

$$x_1 + x_2 \leq 1.5 \tag{D-18b}$$

$$x_1 - x_2 \leq 0.5 \tag{D-18c}$$

$$x_1, x_2 \in \left\{\frac{1}{2}, \frac{2}{3}, 1\right\} \tag{D-18d}$$

The Dual Method is applied to solve the problem posed by Eqs. (D-18). To that end, the Lagrangian function Eq. (D-2) is formed as

$$L(x, \lambda) = \frac{1}{x_1} + \frac{1}{x_2} + \lambda_1(x_1 + x_2 - 1.5) + \lambda_2(x_1 - x_2 - 0.5) \tag{D-19}$$

where, from Eqs. (D-3), the Lagrange variables are sign-restricted as

$$\lambda_1 \geq 0; \qquad \lambda_2 \geq 0 \tag{D-20}$$

From Eqs. (D-4) and (D-5), a saddle-point (x^*, λ^*) of Eq. (D-19) is given by a solution of the min-max problem

$$(x^*, \lambda^*) = \max_{\lambda}[\min_{x} L(x, \lambda)] = \max_{\lambda} [L_m(\lambda)] \tag{D-21}$$

where, from Eq. (D-17), the Dual function is

$$L_m(\lambda) = \sum_{i=1}^{2} \min_{x_i} L_i(x_i, \lambda) \tag{D-22}$$

in which the individual functions of the variables x_i are, from Eqs. (D-6) and (D-19),

$$L_1(x_1, \lambda) = \frac{1}{x_1} + (\lambda_1 + \lambda_2)x_1 \tag{D-23}$$

$$L_2(x_2, \lambda) = \frac{1}{x_2} + (\lambda_1 - \lambda_2)x_2 \tag{D-24}$$

When the Dual function is maximized over the λ variables in Eq. (D-21), its continuity is maintained by requiring that, from Eq. (D-8),

$$L_i(x_i^k, \lambda) = L_i(x_i^{k+1}, \lambda) \qquad (i = 1, 2) \tag{D-25}$$

where x_i^k and x_i^{k+1} are successive discrete values of each variable x_i from Eq. (D-18d). Therefore from Eqs. (D-18d), (D-23), and (D-25), the boundaries for changes in the values of x_1 are

$$\frac{1}{1/2} + \frac{1}{2}(\lambda_1 + \lambda_2) = \frac{1}{2/3} + \frac{2}{3}(\lambda_1 + \lambda_2)$$

$$\frac{1}{2/3} + \frac{2}{3}(\lambda_1 + \lambda_2) = \frac{1}{1} + (\lambda_1 + \lambda_2)$$

which reduce to

$$\lambda_1 + \lambda_2 = 3; \qquad \lambda_1 + \lambda_2 = 1.5 \tag{D-26}$$

Similarly, from Eqs. (D-18d), (D-24), and (D-25), the boundaries for changes in x_2 are

$$\lambda_1 - \lambda_2 = 3; \qquad \lambda_1 - \lambda_2 = 1.5 \tag{D-27}$$

As Eqs. (D-20) require variables λ_1 and λ_2 to be nonnegative, Eqs. (D-26) and (D-27) divide the positive quadrant of the (λ_1, λ_2) plane into the six regions shown in Fig. D-2, where given in parentheses for each region are the fixed values of the variables x_1, x_2 prevailing for the Dual function $L_m(\lambda)$. For example, suppose that the search to maximize $L_m(\lambda)$ in Eq. (D-21) commences at the origin $\lambda = (0,0)$. From Eq. (D-19), the Lagrangian function $L(x, \lambda) = 1/x_1 + 1/x_2$, which is minimized for the discrete values $x_1 = x_2 = 1$, as indicated in Fig. D-2, such that the Dual function is, from Eqs. (D-22) through (D-24),

$$L_m(\lambda) = 2 + 0.5\lambda_1 - 0.5\lambda_2 \tag{D-28}$$

Since λ_2 cannot be reduced below zero, the right-hand side of Eq. (D-28) is maximized by increasing λ_1 to the boundary of the region

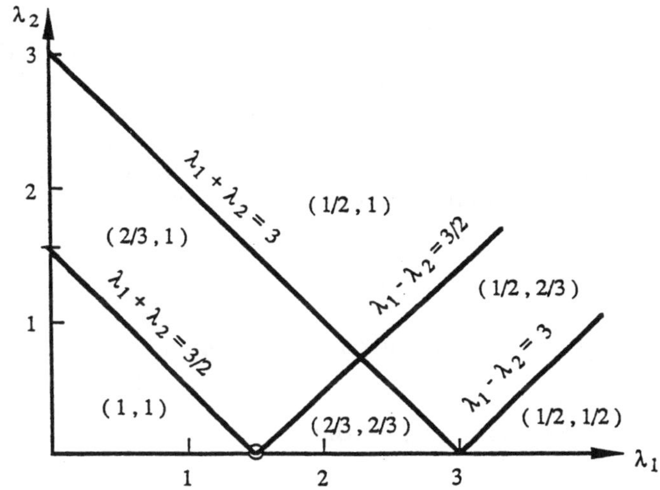

FIG. D-2. Constant (x_1, x_2) Regions in λ-Space.

at $\lambda = (1.5, 0)$, at which point $L_m(\lambda) = 2.75$. From Fig. D-2, it is now possible to move into one of the two regions for which $x = (2/3, 1)$ or $x = (2/3, 2/3)$ in a further attempt to maximize the Dual function. In the former region, from Eqs. (D-22) through (D-24), the Dual function is

$$L_m(\lambda) = 2.5 + \lambda_1/6 - 5\lambda_2/6 \tag{D-29}$$

whereas in the latter region it is

$$L_m(X) = 3 - \lambda_1/6 - \lambda_2/2 \tag{D-30}$$

It is not possible to further maximize either of the Dual function Eqs. (D-29) or (D-30) for nonnegative λ values different than $\lambda^* = (1.5, 0)$, which means that the solution of the min-max problem Eq. (D-21) occurs for any one of the three pairs of x_i values (1, 1), (2/3, 1), or (2/3, 2/3). However, only $x^* = (2/3, 2/3)$ satisfies the constraints Eqs. (D-18b) and (D-18c), for which the optimal cross-section areas are $a_1 = a_2 = 3/2$ and the value of the objective function Eq. (D-18a) is 3.

A final comment is in order concerning the fact that the minimum value of the objective function Eq. (D-18a) is 3, whereas the maximum value of the corresponding Dual function Eq. (D-22) is 2.75. This difference serves as a reminder that the discrete optimization problem is not convex, and therefore that there is no guarantee that the solution found is optimal and/or unique [as it happens, the solution point $x^* = (2/3, 2/3)$ found for the present example is one of two optima, the other being $x^* = (1/2, 1)$]. The consequence of this for the design of structural steel frameworks is that the set of discrete standard sections found by the Dual Method may correspond to but one of several near-optimal designs for the structure. Moreover, although all such designs are likely to be quite similar to each other from the viewpoint of both structure weight and standard section selection, it is theoretically possible that there may exist designs of similar weight but with radically different standard sections for the members.

NOTATION

The following symbols are used in this appendix.

a = cross-section area
c = sensitivity coefficient; gradient
E = Young's modulus
$\bar{g}$ = constraint bound
$\bar{L}$ = member length
L = Lagrangian function
m = number of constraints
n = number of variables
P = load
s = search direction
$\mathbf{u}$ = displacement
W = weight
w = weight coefficient
X = set of reciprocal sizing variables
x = reciprocal sizing variable
α = step length
λ = dual variable; Lagrange variable
ρ = material density

REFERENCES

Fleury, C. (1979). "Structural weight optimization by dual methods of convex programming." *Int. J. of Numerical Methods in Engng.*, 14, 1761–1783.

Grierson, D. E., (1993) "Practical optimization of structural steel frameworks." In Chapter 12, *Advances in design optimization,* H. Adeli, ed., Chapman and Hall, London.

Haftka, R. T. and Gurdal, Z. (1992). *Elements of structural optimization.* 3rd ed., Kluwer Academic Publishers, Netherlands.

Schmit, L. A., Jr. and Fleury, C. (1979). "An improved analysis/synthesis capability based on dual methods—ACCESS 3." *Proceedings of AIAA/ASME/AHS 20th Structures, Structural Dynamics and Materials Conference*, St. Louis, MO.

Zangwill, W. I. (1969). *Nonlinear programming: A unified approach.* Prentice-Hall, Englewood Cliffs, NJ.

Appendix E

COMBINATORIAL SEARCH

Combinatorial search algorithms are used to solve discrete optimization problems. These are problems where the values of the design variables must be selected from a finite set of discrete values. For example, the number of reinforcing bars in a reinforced concrete member must be an integer, and the diameter of the reinforcing bars must correspond to one of the standard bar sizes. Only two combinatorial search algorithms are presented in this appendix—the exhaustive search algorithm and the branch and bound algorithm.

EXHAUSTIVE SEARCH

The simplest combinatorial search strategy is exhaustive search. Consider the design of a steel frame consisting of four groups of members. All members in each member group must be the same size. Suppose the size for each member group must be selected from among three possible sizes: Size *A*, Size *B*, and Size *C*. Suppose cost is minimized and there are several constraints on displacement, stress, and stability at various locations throughout the frame. In this problem we have four discrete design variables each of which must be selected from among three discrete values. In the exhaustive search algorithm, we form all possible designs, analyze each of them, and take the best design as the optimum. In this example problem, there are $3^4 = 81$ possible designs, and therefore 81 required analyses to find the optimum design. Obviously, the exhaustive search algorithm will guarantee finding the global optimum to the discrete optimization problem.

Richard J. Balling, Department of Civil and Environmental Engineering, Brigham Young University, Provo, Utah 84602; email: ⟨rick@et.byu.edu⟩.

BRANCH AND BOUND

The operation of combinatorial search algorithms may be displayed with search trees. For example, the branch and bound search tree for the frame design problem involving four member groups and three sizes is shown in Fig. E-1. The search tree has four levels corresponding to the four design variables. The top level corresponding to member group 1 shows boxes or "nodes" for each of the three possible sizes, *A*, *B*, and *C*. In the search tree for exhaustive search, one would have three branches emanating from each of these three nodes leading to a total of 9 nodes at the level corresponding to member group 2. The exhaustive search tree would have 27 nodes at the level corresponding to member group 3, and 81 nodes at the level corresponding to member group 4.

Note that the branch and bound search tree has significantly fewer nodes than the exhaustive search tree would have. This is because many of the branches have been cut off or "fathomed" as indicated by double lines beneath the nodes. The optimum design consists of size *C* for member group 1, size *A* for member group 2, size *B* for member group 3, and size *C* for member group 4. The number to the left of each node is the node number. Nodes are numbered in the order in which they are generated. The number to the right of each node is its cost bound which is now explained.

Branch and bound search begins by fixing member group 1 to size *A*. The sizes of member groups 2, 3, 4 are treated as continuous variables, and a continuous optimization is performed to find their op-

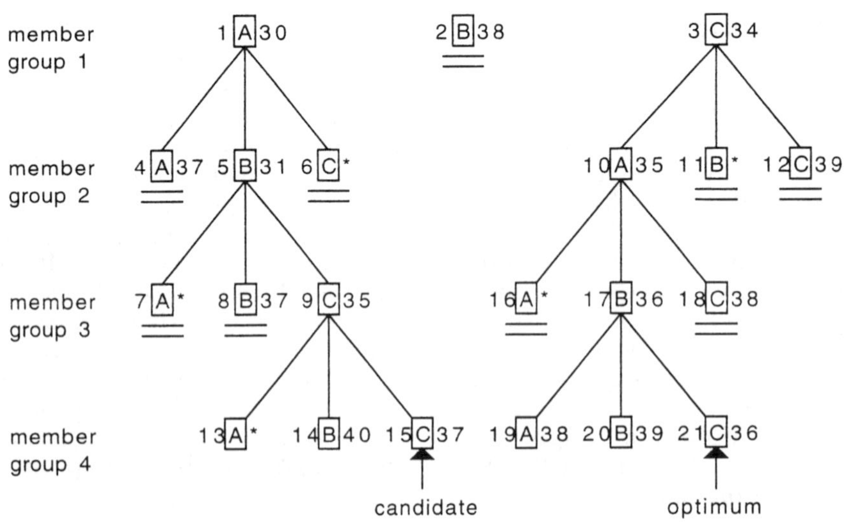

FIG. E-1. Branch and Bound Search Tree.

timal values. Suppose that the optimal cost from this continuous optimization is 30. This represents a lower bound on the cost since it is surely lower than the cost one would get from discrete optimization of the sizes of member groups 2, 3, 4. The cost bound of 30 is tabulated to the right of node 1. Similarly, continuous optimizations are performed at nodes 2 and 3 to get their cost bounds of 38 and 34, respectively.

Since node 1 has the lowest cost bound, it is expanded to nodes 4, 5, 6. Cost bounds are generated for these three nodes by holding the sizes of member groups 1 and 2 fixed and continuously optimizing the sizes of member groups 3 and 4. Note the asterisk for the cost bound of node 6. This indicates that no feasible design could be found in the continuous optimization. Then surely no feasible design would be found in discrete optimization, and therefore this branch may be fathomed.

Node 5 now has the lowest cost bound to this point, and it is expanded to nodes 7, 8, 9 whose cost bounds are then determined. Node 3 now has the lowest cost bound to this point, and it is expanded to nodes 10, 11, 12 whose cost bounds are then determined. Node 9 now has the lowest cost bound to this point, and it is expanded to nodes 13, 14, 15. These three nodes represent discrete designs. The discrete design for node 13 is infeasible. The discrete designs for nodes 14 and 15 have costs of 40 and 37, respectively. Node 15 then becomes the candidate design since it is the best discrete design encountered thus far.

Since we now have a candidate design in hand with a cost of 37, all nodes with cost bounds greater than or equal to 37 may be fathomed. Thus we can fathom nodes 2, 4, 8, 12. The only unfathomed-unexpanded node remaining is node 10. It is expanded to nodes 16, 17, 18 whose cost bounds are computed. Node 16 is fathomed since it is infeasible, and node 18 is fathomed since its cost bound is greater than the cost of the candidate design at node 15. Node 17 is expanded to nodes 19, 20, 21, whose costs are computed. Node 21 now becomes the candidate design since it has a lower cost than the previous candidate at node 15. At this point any nodes with cost bounds greater than or equal to the new candidate cost of 36 could be fathomed. However, there are no such nodes. In fact, the branch and bound search is now complete since there are no remaining unfathomed-unexpanded nodes. Therefore the candidate design (node 21) is the optimum design.

SUMMARY AND DISCUSSION

The combinatorial search algorithms guarantee finding the optimum to the discrete optimization problem. As the number of design variables and the number of discrete values increase, the number of analyses required by exhaustive search increases exponentially. Consider the discrete optimization of a steel frame involving 12 member groups with 40 possible sizes for each member group. Suppose the structural analysis for such a frame required 15 minutes on the computer. The amount of time required to execute the exhaustive search would be $(40)^{12}(15 \text{ min}) = 2.52 \times 10^{20}$ minutes $= 4.78 \times 10^{14}$ years.

The branch and bound potentially eliminates many of the designs that must be analyzed by exhaustive search. To do so requires performing continuous optimizations to establish cost bounds. Nonlinear programming (NLP) algorithms must be executed to perform these continuous optimizations, and the NLP algorithms will require several analyses. Thus it is still computationally unrealistic to use branch and bound to solve a problem with 12 member groups and 40 possible sizes for each. Some modifications to the branch and bound algorithm are necessary so that such problems can be treated.

First, one may perform an optimization treating all design variables as continuous using an NLP algorithm before beginning the branch and bound algorithm. For each member group, one could identify the 4 discrete sizes nearest to the continuous optimum size. Branch and bound could then be limited to searching among 4 discrete sizes for each member group rather than searching among 40 discrete sizes.

Second, one could use the gradients at the continuous optimum design to make linear approximations of the cost and constraint functions. These linear approximations would convert the continuous optimizations in the branch and bound algorithm from NLP problems to LP problems. LP problems can be solved very efficiently without analysis via the simplex method. Nevertheless, every time a discrete candidate design is found in the linearized branch and bound algorithm, it would be wise to perform an exact analysis to check its true, rather than approximate, feasibility.

These two modifications were made to the branch and bound algorithm, and a three-dimensional, six-story frame with 12 member groups was optimized in about 1.5 hours on a DECSTATION 3100 workstation (May and Balling 1992).

REFERENCE

May, S. A. and Balling, R. J. (1992). "A filtered simulated annealing strategy for discrete optimization of 3D steel frameworks." *Structural Optimization*, 4, 142.

Appendix F

STOCHASTIC SEARCH

A class of optimization algorithms known as "stochastic search" algorithms is described in this appendix. These algorithms employ the generation of random numbers as they search for the optimum. Although they do not require the evaluation of gradients of the objective and constraint functions, they typically require many more function evaluations than do the gradient-based NP algorithms. Unlike NP algorithms, stochastic search algorithms may be applied to optimization problems involving discrete variables. They may also be used on difficult problems that are large, noisy, fraught with local minima, or discontinuous. Two simple random search algorithms are described from which a "simulated annealing" algorithm and a "genetic algorithm" are developed. These last two algorithms imitate optimization processes found in nature. The simulated annealing algorithm imitates the process of annealing in metals as they are cooled wherein the internal energy is minimized. The genetic algorithm imitates the process of evolution of biological species wherein fitness is maximized.

EXAMPLE PROBLEM

In this appendix, we refer to the following discrete optimization problem as the "example" problem.

Find

$$\mathbf{x} = \begin{Bmatrix} x_1 \\ x_2 \end{Bmatrix}$$

Richard J. Balling, Department of Civil and Environmental Engineering, Brigham Young University, Provo, Utah 84602; email: ⟨rick@et.byu.edu⟩.

Restrict

x_1 and x_2 to the set {0.5, 1.0, 1.5, 2.0, 2.5, 3.0, . . . , 10.0}

Minimize

$$f(\mathbf{x}) = 5x_1^2 - 9x_1x_2 + 5x_2^2$$

Satisfy

$$g_1(\mathbf{x}) = 25 - 16x_1x_2 \leq 0$$

The continuous optimum to the example problem is $\begin{Bmatrix} 1.25 \\ 1.25 \end{Bmatrix}$ which could be verified from the Kuhn-Tucker conditions. However, the discrete optimum to the example problem is $\begin{Bmatrix} 1.5 \\ 1.5 \end{Bmatrix}$, which could be verified via exhaustive search over the $20 \times 20 = 400$ possible discrete designs.

SERIAL RANDOM SEARCH

One of the simplest stochastic search algorithms for the example problem is the following "serial random search" algorithm.

Step 1.
Obtain a starting feasible design $\mathbf{x}^*$ and analyze it.
Step 2.
Randomly generate a new candidate design $\mathbf{x}$ close to $\mathbf{x}^*$.
Step 3.
Analyze the candidate. If $g_1(\mathbf{x}) \leq 0$ and $f(\mathbf{x}) < f(\mathbf{x}^*)$, let $\mathbf{x}^* = \mathbf{x}$.
Step 4.
Stop if $\mathbf{x}^*$ has not changed for several iterations; otherwise return to Step 2.

How does one randomly generate a candidate design $\mathbf{x}$ from the current design $\mathbf{x}^*$ in Step 2? The approach may vary from problem to problem. For the example problem, one could use the following approach.

1. Generate a random number r between zero and one. If $r < 0.5$, we will perturb variable x_1; otherwise we will perturb variable x_2.

2. Generate another random number r between zero and one. If $r < 0.5$, perturb the variable to the next smallest value; otherwise perturb it to the next greatest value.

We now illustrate a few iterations of the serial random search algorithm on the example problem. Suppose we begin with the starting design $\mathbf{x}^* = \begin{Bmatrix} 2.5 \\ 1.5 \end{Bmatrix}$ and analyze it to get $f = 8.75$ and $g_1 = -35$.

Step 2.

$r = 0.810 \Rightarrow$ perturb x_2. $r = 0.781 \Rightarrow x_2 = 2.0 \Rightarrow \mathbf{x} = \begin{Bmatrix} 2.5 \\ 2.0 \end{Bmatrix}$

Step 3.

$f = 6.25$, $g_1 = -55 \Rightarrow$ feasible and lower $f \Rightarrow$ accept $\mathbf{x} \Rightarrow \mathbf{x}^* = \begin{Bmatrix} 2.5 \\ 2.0 \end{Bmatrix}$

Step 2.

$r = 0.156 \Rightarrow$ perturb x_1. $r = 0.659 \Rightarrow x_1 = 3.0 \Rightarrow \mathbf{x} = \begin{Bmatrix} 3.0 \\ 2.0 \end{Bmatrix}$

Step 3.

$f = 11.00$, $g_1 = -71 \Rightarrow$ feasible but higher $f \Rightarrow$ reject $\mathbf{x} \Rightarrow \mathbf{x}^* = \begin{Bmatrix} 2.5 \\ 2.0 \end{Bmatrix}$

Step 2.

$r = 0.210 \Rightarrow$ perturb x_1. $r = 0.402 \Rightarrow x_1 = 2.0 \Rightarrow \mathbf{x} = \begin{Bmatrix} 2.0 \\ 2.0 \end{Bmatrix}$

Step 3.

$f = 4.00$, $g_1 = -39 \Rightarrow$ feasible and lower $f \Rightarrow$ accept $\mathbf{x} \Rightarrow \mathbf{x}^* = \begin{Bmatrix} 2.0 \\ 2.0 \end{Bmatrix}$

The outlined serial random search algorithm will get stuck at the nonoptimal design $\mathbf{x}^* = \begin{Bmatrix} 2.0 \\ 2.0 \end{Bmatrix}$. This is because all possible candidates that could be generated in Step 2 $\left(\begin{Bmatrix} 2.5 \\ 2.0 \end{Bmatrix}, \begin{Bmatrix} 1.5 \\ 2.0 \end{Bmatrix}, \begin{Bmatrix} 2.0 \\ 2.5 \end{Bmatrix}, \text{ or } \begin{Bmatrix} 2.0 \\ 1.5 \end{Bmatrix} \right)$ have higher objective values. To get to the true optimum design $\mathbf{x}^* = \begin{Bmatrix} 1.5 \\ 1.5 \end{Bmatrix}$, one would have to either: (1) modify the random perturbation in Step 2 so that candidates can be generated with both variables perturbed simultaneously, or (2) occasionally accept candidates with higher f.

SIMULATED ANNEALING ALGORITHM

The simulated annealing algorithm modifies the serial random search algorithm so that designs with higher f are occasionally accepted. The steps for the algorithm are as follows.

Step 1.
Obtain a starting temperature T and a starting feasible design $\mathbf{x}^*$ and analyze it.
Step 2.
Randomly generate a new candidate design $\mathbf{x}$ close to $\mathbf{x}^*$.
Step 3.
Analyze the candidate. If it is feasible and $f(\mathbf{x}) < f(\mathbf{x}^*)$, let $\mathbf{x}^* = \mathbf{x}$. If the candidate is feasible but $f(\mathbf{x}) > f(\mathbf{x}^*)$, then generate a random number r. If $r < P$ where

$$P = \exp\left(\frac{f(\mathbf{x}^*) - f(\mathbf{x})}{T}\right)$$

let $\mathbf{x}^* = \mathbf{x}$.
Step 4.
Stop if $\mathbf{x}^*$ has not changed for several iterations; otherwise update $T = \alpha T$ where α is a number less than one, and return to Step 2.

We now execute a few iterations of the simulated annealing algorithm starting with $\mathbf{x}^* = \begin{Bmatrix} 2.0 \\ 2.0 \end{Bmatrix}$ and $T = 1.000$. We take $\alpha = 0.9$.

Step 2.

$$r = 0.437 \Rightarrow \text{perturb } x_1.\ r = 0.144 \Rightarrow x_1 = 1.5 \Rightarrow \mathbf{x} = \begin{Bmatrix} 1.5 \\ 2.0 \end{Bmatrix}$$

Step 3.

$$f = 4.25\ g_1 = -23 \Rightarrow P = 0.779,\ r = 0.387 \Rightarrow \text{accept } \mathbf{x} \Rightarrow \mathbf{x}^* = \begin{Bmatrix} 1.5 \\ 2.0 \end{Bmatrix}$$

Step 4.
$T = 0.9(1.00) = 0.900$
Step 2.

$$r = 0.876 \Rightarrow \text{perturb } x_2.\ r = 0.651 \Rightarrow x_2 = 2.5 \Rightarrow \mathbf{x} = \begin{Bmatrix} 1.5 \\ 2.5 \end{Bmatrix}$$

Step 3.
$f = 8.75\ g_1 = -35 \Rightarrow P = 0.00674,\ r = 0.365 \Rightarrow \text{reject } \mathbf{x} \Rightarrow \mathbf{x}^* = \begin{Bmatrix} 1.5 \\ 2.0 \end{Bmatrix}$

Step 4.
$T = 0.9(0.900) = 0.810$
Step 2.

$$r = 0.537 \Rightarrow \text{perturb } x_2.\ r = 0.174 \Rightarrow x_2 = 1.5 \Rightarrow \mathbf{x} = \begin{Bmatrix} 1.5 \\ 1.5 \end{Bmatrix}$$

Step 3.

$$f = 2.25\ g_1 = -11 \Rightarrow \text{accept } \mathbf{x} \Rightarrow \mathbf{x}^* = \begin{Bmatrix} 1.5 \\ 1.5 \end{Bmatrix}$$

Step 4.
$T = 0.9(0.810) = 0.729$
Step 2.

$$r = 0.964 \Rightarrow \text{perturb } x_2.\ r = 0.326 \Rightarrow x_2 = 1.0 \Rightarrow \mathbf{x} = \begin{Bmatrix} 1.5 \\ 1.0 \end{Bmatrix}$$

Step 3.

$$f = 2.75\ g_1 = 1 \Rightarrow \text{reject } \mathbf{x} \Rightarrow \mathbf{x}^* = \begin{Bmatrix} 1.5 \\ 1.5 \end{Bmatrix}$$

Step 4.
$T = 0.9(0.729) = 0.6561$

Note that as the temperature T decreases from iteration to iteration it decreases the probability P of accepting designs with higher f. This means that the algorithm is likely to accept designs with higher f in the initial iterations, but it is less likely to accept worse designs in the final iterations as it converges to the global minimum. This simulates the annealing of metals as they cool from liquid to solid states. If the cooling is performed very slowly, the metal will solidify in a crystalline state which is the global minimum of the internal energy function. On the other hand, if cooling is performed very rapidly, the metal will solidify in a glass state which is a local minimum of the energy function. If the temperature drops rapidly, the acceptance probability for designs with higher f goes to zero which is essentially the serial random search algorithm.

PARALLEL RANDOM SEARCH

In the following table we list in increasing order the possible discrete values for either of the two variables in the example problem. In the second column we number them with a five-digit binary number.

0.5	00000	5.5	01010
1.0	00001	6.0	01011
1.5	00010	6.5	01100
2.0	00011	7.0	01101
2.5	00100	7.5	01110
3.0	00101	8.0	01111
3.5	00110	8.5	10000
4.0	00111	9.0	10001
4.5	01000	9.5	10010
5.0	01001	10.0	10011 through 11111

Thus each five-digit binary number is associated with one of the possible discrete values. Note that since there are 20 discrete values, and since $2^5 = 32$ possible five-digit binary numbers, we must assign several binary numbers (10011 through 11111) to the discrete value of 10.0.

Since there are two design variables, we may have one binary number corresponding to the discrete value of the first design variable, and another binary number corresponding to the discrete value of the second design variable. We may place these two binary numbers end to end to form a 10-digit "chromosome" representing a particular design. For example, the chromosome "00110 01101" corresponds to the design $\begin{Bmatrix} 3.5 \\ 7.0 \end{Bmatrix}$ since the first five digits correspond to 3.5 and the last five digits correspond to 7.0. Thus discrete designs may be encoded as chromosomes.

A random search algorithm can be constructed that randomly generates a population of chromosomes, evaluates f and g_1 for each design, and selects the best design from the population as the optimum. For example, suppose we randomly generate a population of six chromosomes. This could be accomplished by generating a random number for each digit in each chromosome. If the random number is greater than 0.5, the digit is a "1"; otherwise it is a "0." Thus for a population of six chromosomes, we would have to generate 60 random numbers since each chromosome has 10 digits. After generating the six chromosomes in this fashion, they are decoded into designs, and the f and g_1 functions are evaluated as shown in the following.

chromosome	$\mathbf{x}$	$f(\mathbf{x})$	$g_1(\mathbf{x})$
00010 11111	{1.5, 10.0}	376.25	−215.00
00000 00110	{0.5, 3.5}	46.75	−3.00
10111 01110	{10.0, 7.5}	106.25	−1175.00
00000 01110	{0.5, 7.5}	248.75	−35.00
10111 01100	{10.0, 6.5}	126.25	−1015.00
10001 10101	{9.0, 10.0}	95.00	−1415.00

Since the second chromosome is the feasible design with lowest f

in the population, it is returned as the optimum. For this algorithm to find the true optimum to the problem, the population size would have to be much greater than six. In this algorithm, the chromosomes in the population could be generated and analyzed in parallel. Thus the algorithm is called a parallel random search algorithm. Contrast it to the serial random search algorithm given earlier which generates one candidate design at a time.

GENETIC ALGORITHM

In genetic algorithms, one maximizes the "fitness" of chromosomes in accordance with the idea of survival of the fittest. We need a positive scalar-valued function that can be evaluated for each design. In optimization problems where one minimizes a positive-valued objective, one may define the fitness function as the reciprocal of the objective. If constraints are present in the optimization problem, they must be factored into the fitness function since designs that violate constraints are not very "fit" designs. For the example problem, we define the fitness function as follows.

$$d(\mathbf{x}) = \frac{1}{f(\mathbf{x}) + \max(0, g_1(\mathbf{x}))} = \frac{1}{5x_1^2 - 9x_1x_2 + 5x_2^2 + \max(0, 25 - 16x_1x_2)}$$

The genetic algorithm begins by executing the parallel random search algorithm to create a starting population called the "first generation." For the example problem, we take the population of six chromosomes given earlier as the first generation. We now "select" members from this generation to become mating pairs for the second generation. Selection is based on fitness. It is necessary to partition the interval from zero to one into six subintervals corresponding to the six chromosomes in the first generation. The length of each subinterval is proportional to the fitness of the chromosome to which it corresponds, giving the most fit chromosomes the biggest subintervals. Thus the length of the subinterval corresponding to a chromosome is equal to its fitness divided by the sum of the fitnesses in the population:

chromosome	$d(\mathbf{x})$	$d(\mathbf{x})/\Sigma\, d(\mathbf{x})$	subinterval
00010 11111	0.0026578	0.047522	0.0 to .047522
00000 00110	0.0213904	0.382469	.047522 to .429991
10111 01110	0.0094118	0.168287	.429991 to .598278
00000 01110	0.0040201	0.071881	.598287 to .670159
10111 01100	0.0079208	0.141627	.670159 to .811786
10001 10101	0.0105263	0.188214	.811786 to 1.0
$\Sigma\, d(\mathbf{x}) =$	0.0559272		

We now generate six random numbers and select chromosomes from the first generation according to the intervals in which the random numbers fall:

random number	corresponding chromosome
0.770	10111 01100
0.392	00000 00110
0.130	00000 00110
0.233	00000 00110
0.556	10111 01100
0.414	00000 00110

Note that the second chromosome was selected four times, and the third and fifth chromosomes were each selected once. We have grouped the preceding chromosomes into three "mating pairs."

The "crossover" operation is the mating operation. We begin with the first mating pair. We must randomly select a "crossover point." Since there are 10 digits in the chromosome, we divide the interval from zero to one into 10 equal subintervals each with length 0.1. For the first mating pair we generate the random number 0.828. Since this number falls in the ninth subinterval, the crossover point is taken between the eighth and ninth digits. We cut both chromosomes between their eighth and ninth digits. The mating operation consists of switching the portions after the cut. In other words, two new chromosomes are formed by attaching the succeeding part of the second chromosome to the preceding part of the first chromosome, and attaching the succeeding part of the first chromosome to the preceding part of the second chromosome. For the first mating pair:

before crossover	after crossover
10111 011|00	10111 011|10
00000 001|10	00000 001|00

For the second mating pair, the random number 0.693 is generated corresponding to a crossover point between the sixth and seventh digits. However, since both chromosomes in this mating pair are the same, the crossover operation leaves this pair unchanged. For the third mating pair, the random number 0.269 is generated corresponding to a crossover point between the second and third digits. Following the crossover operation for all three mating pairs, we are left with the following six chromosomes.

10111 01110
00000 00100
00000 00110
00000 00110
10000 00110
00111 01100

The "mutation" operation makes it possible for new genetic material to occasionally enter into the population. The probability for mutation must be low. For the example problem we assume a mutation probability of 5%. Since there are 60 digits in our population, we generate 60 random numbers. If a random number is less than 0.05, we reverse its corresponding digit (if the corresponding digit is a "0," change it to a "1," and if the corresponding digit is a "1," change it to a "0"). For our example, we generated 60 random numbers. Only the 38th and 59th were less than 0.5, and the corresponding digits are underlined in the following population. These corresponding digits were reversed. The resulting population is labeled the "second generation."

chromosome	x	$f(\mathbf{x})$	$g_1(\mathbf{x})$	$d(\mathbf{x})$
10111 01110	{10.0, 7.5}	106.25	−1175.00	0.0094118
00000 00100	{0.5, 2.5}	21.25	5.00	0.0380952
00000 00110	{0.5, 3.5}	46.75	−3.00	0.0213904
00000 00010	{0.5, 1.5}	5.75	13.00	0.0533333
10000 00110	{8.5, 3.5}	154.75	−451.00	0.0064620
00111 01110	{4.0, 7.5}	91.25	−455.00	0.0109589

We compare the second generation to the first generation, and note the improvement in fitness:

generation	maximum d	average d	minimum d
first	0.0213904	0.0093212	0.0026578
second	0.0533333	0.0232753	0.0064620

At this point, the operations of selection, crossover, and mutation could be repeated again to form the third generation. This process is repeated several times. The best design in the final generation could be returned as the optimal design. Note, however, that the genetic algorithm not only returns one final design, but a whole population of good designs which may be significantly different from each other.

SUMMARY AND DISCUSSION

Stochastic search algorithms can be used to solve discrete and other difficult optimization problems. Since the stopping criteria for such algorithms is somewhat informal, they should be viewed as strategies for producing good designs that are not necessarily the optimal design.

The simulated annealing algorithm has been applied to the discrete optimization of a three-dimensional six-story steel frame (May and Balling 1992). Linearization of analysis and the use of a "filter" were employed to speed up execution of the simulated annealing algorithm. The filter used gradient information to filter out poor candidates before analysis was performed. The same optimum designs reached via combinatorial search methods were achieved by the filtered simulated annealing algorithm in about an hour on a DECSTATION 3100 workstation.

In a classic paper (Kirkpatrick et al. 1983), the simulated annealing algorithm was applied to a traveling salesman problem (TSP) to find the minimum-length path connecting over 6,000 cities! The previous record for the largest TSP solved by combinatorial search methods involved 318 cities.

An article in the May 8, 1995 issue of *Newsweek Magazine* reported that genetic algorithms have been used by: (1) John Deere to optimize production scheduling; (2) Texas Instruments to minimize computer chip size; and (3) USWest to design fiber-optic cable networks (design time reduced from two months to two days). Genetic algorithms have also been used to optimize the design of plane frames (Jenkins 1992).

REFERENCES

Jenkins, W. M. (1992). "Plane frame optimum design environment based on genetic algorithm." *J. of Structural Engng*, ASCE, 118, 3103.

Kirkpatrick, S., Gelatt, C. D., and Vecchi, M. P. (1983). "Optimization by simulated annealing." *Science*, 220, 671.

May, S. A. and Balling, R. J. (1992). "A filtered simulated annealing strategy for discrete optimization of 3D steel frameworks." *Structural Optimization*, 4, 142.

INDEX